United States
Department of
Agriculture

Forest Service

Engineering Staff

Washington Office

Forest Service Specifications for Construction of Roads and Bridges

EM–7720–100

Revised
August 1996

Preface

The Forest Service, U.S. Department of Agriculture, developed this book for use in the preparation and administration of contracts for the construction of roads and bridges on the National Forest Development Road System under Public Works Contracts, or as a condition of Timber Sale Contracts.

This book contains the Standard Specifications that commonly apply in Public Works and Timber Sale Contracts. When designated in a Forest Service contract, the specifications in this book are binding on the parties signing the contract and become a part of the contract. Provisions and clauses pertinent to Public Works Contracts are located in the Federal Acquisition Regulation (FAR). Standard Provisions and Special Provisions for timber sale projects are located in the Timber Sale Contract, Divisions B (BT) and C (CT) respectively. Conditions and requirements peculiar to individual projects, including additions and revisions to Standard Specifications, are located in the SPECIAL PROJECT SPECIFICATIONS section of the Contract.

In the development of this book, some section numbers were not utilized. The section numbers that are not included in this book may be used when it is necessary to develop a specification that is not adequately covered in one of the sections contained in this book, as long as it utilizes the same format and is included within the appropriate division.

TABLE OF CONTENTS

Page

Page

Page

DIVISION 100
General Specifications

Section 101—Abbreviations, Acronyms, & Terms

101.01 Terms, Organizations, & Standards

These specifications are generally written in the imperative mood. In sentences using the imperative mood, the subject, "the Contractor," is implied. Also implied in this language is "shall," "shall be," or similar words and phrases. In material specifications, the subject may also be the supplier, fabricator, or manufacturer supplying material, products, or equipment for use on the project.

Wherever "directed," "required," "prescribed," or similar words are used, the "direction," "requirement," or "order" of the Contracting Officer (CO) is intended. Similarly, wherever "approved," "acceptable," "suitable," "satisfactory," or similar words are used, they mean "approved by," "acceptable to," or "satisfactory to" the CO.

The word "will" generally pertains to decisions or actions of the CO.

Whenever in these specifications, or in other contract documents, the following terms (or pronouns in place of them) are used, the intent and meaning shall be interpreted as follows: reference to a specific standard, test, testing method, or specification shall mean the latest published edition or amendment that is in effect at the solicitation issue date for Public Works Contracts or the sale advertisement date for Timber Sale Contracts.

These specifications are divided into the following divisions:

- Division 100 consists of general specifications for which no direct payment is made. These requirements are applicable to all contracts.

- Division 150 consists of engineering requirements that are applicable to some contracts. Work under this division is paid for directly when there is a PAY ITEM IN THE SCHEDULE OF ITEMS. When there is no PAY ITEM IN THE SCHEDULE OF ITEMS, no direct payment is made.

- Divisions 200–600 consist of construction contract requirements for specific items of work. Work under these divisions is paid for directly or indirectly according to Section 106 and the section for ordering the work when there is a PAY ITEM IN THE SCHEDULE OF ITEMS.

- Division 700 contains the material requirements for Divisions 200–600. No direct payment is made under Division 700. Payment for material is included as part of the work required in Divisions 200–600.

(a) Acronyms. The following acronyms are used in these specifications:

AA	Aluminum Association
AASHTO	American Association of State Highway and Transportation Officials
ACI	American Concrete Institute
ADA	Americans With Disabilities Act
AGC	Associated General Contractors of America
AI	Asphalt Institute
AISC	American Institute of Steel Construction
AITC	American Institute of Timber Construction
ALSC	American Lumber Standards Committee
ANSI	American National Standards Institute
APA	American Plywood Association
ARTBA	American Road and Transportation Builders Association
ASTM	American Society for Testing and Material
AWPA	American Wood Preservers Association
AWS	American Welding Society
AWWA	American Water Works Association
CFR	Code of Federal Regulations
CRSI	Concrete Reinforcing Steel Institute
DEMA	Diesel Engine Manufacturers Association
DOT	U.S. Department of Transportation
FAR	Federal Acquisition Regulation
FHWA	Federal Highway Administration
FLH	Federal Lands Highway (Federal Highway Administration)
FSS	Federal Specifications and Standards
ISSA	International Slurry Surfacing Association
MIL	Military Specification(s)
MSHA	Mine Safety and Health Administration
MUTCD	Manual on Uniform Traffic Control Devices
NBS	National Bureau of Standards
OSHA	Occupational Safety and Health Administration
PCI	Prestressed Concrete Institute
PTI	Post-Tensioning Institute
SI	International System of Units
SSPC	Steel Structures Painting Council
WWPI	Western Wood Preservation Institute

(b) **System of International Units (SI) Symbols.** The following SI symbols are used in these specifications:

A	-	ampere	electric current
cd	-	candela	luminous intensity
°C	-	degree Celsius (K – 273.15)	temperature
d	-	day	time
g	-	gram	mass
h	-	hour	time
ha	-	hectare	area
Hz	-	hertz (s^{-1})	frequency
J	-	joule (N•m)	energy
K	-	kelvin	temperature
L	-	liter	volume
lx	-	lux	illuminance
m	-	meter	length
m^2	-	square meter	area
m^3	-	cubic meter	volume
min	-	minute	time
N	-	newton ($kg•m/s^2$)	force
Pa	-	pascal (N/m^2)	pressure
s	-	second	time
t	-	ton	mass
V	-	volt (W/A)	electric potential
W	-	watt (J/s)	power
Ω	-	ohm (V/A)	electric resistance
°	-	degree	plane angle
′	-	minute	plane angle
″	-	second	plane angle

(c) **SI Prefix Symbols.** The following SI prefix symbols are used in these specifications:

E	-	exa	10^{18}
P	-	peta	10^{15}
T	-	tera	10^{12}
G	-	giga	10^{9}
M	-	mega	10^{6}
k	-	kilo	10^{3}
c	-	centi	10^{-2}
m	-	milli	10^{-3}
µ	-	micro	10^{-6}

n - nano 10^{-9}

p - pico 10^{-12}

f - femto 10^{-15}

a - atto 10^{-18}

(d) SI Slope Notation (vertical : horizontal). For slopes flatter than 1:1, express the slope as the ratio of one unit vertical to a number of units horizontal. For slopes steeper than 1:1, express the slope as the ratio of a number of units vertical to one unit horizontal.

101.02 Abbreviations

ABS	Acrylonitrile-butadiene-styrene
ACA	Ammoniacal copper arsenate
ACZA	Ammoniacal copper zinc arsenate
Agg	Aggregate
Al	Aluminum
AOS	Apparent opening size
AQ	Actual quantities
AQL	Acceptable Quality Level
BMP	Best Management Practice
CAPWAP	Case pile wave analysis program
CCA	Chromated copper arsenate
CMP	Corrugated metal pipe
CMPA	Corrugated metal pipe arch
CO	Contracting Officer
CPF	Composite pay factor
CSP	Corrugated steel pipe
CSPA	Corrugated steel pipe arch
CTB	Cement-treated base
DAR	Durability Absorption Ratio
Dia	Diameter
DQ	Designed quantities
DTI	Direct tension indicator
Dwgs	Drawings
FM	Fineness modulus
GFM	Government-furnished materials
Gr	Grade
h	Hour
H	Height
ha	Hectare

HDO	High-density overlay
HDPE	High-density polyethylene
Hor	Horizontal
HSLA	High-strength low-alloy
kg	Kilogram
kL	Kiloliter
kL km	Kiloliter kilometer
km	Kilometer
L	Length
L	Liter
LSL	Lower specification limit
m	Meter
m^2	Square meter
m^3	Cubic meter
m^3 km	Cubic meter kilometer
Matl	Material
max.	Maximum
Mbf	Thousand board feet
min.	Minimum
Misc	Miscellaneous
mm	Millimeter
N/C	Numerically controlled
PG	Performance-graded
PI	Plasticity index
ppm	Parts per million
PS	Product Standard (issued by the U.S. Department of Commerce)
PVC	Polyvinylchloride
SQ	Staked quantities
t	Ton (1,000 kg)
t km	Ton kilometer
T	Temperature
T&L	Tops and limbs
TFE	Tetraflouroethylene
Th	Thickness
TV	Target value
USL	Upper specification limit
Vert	Vertical
VMA	Voids in Mineral Aggregate
VOC	Volatile organic compound
W	Width
W/	With
W/O	Without
WW	Woven wire
WWF	Welded wire fabric

Section 102—Definitions

Wherever the following terms, or pronouns in place of them, are used in these specifications or in other contract documents, the intent and meaning are as follows:

Adjustment in Contract Price. "Equitable adjustment," as used in the Federal Acquisition Regulations, or "construction cost adjustment," as used in the Timber Sale Contract, as applicable.

Arch. A culvert section, usually formed of bolted structural plates, that is an arc of a circle (usually one-half or less); that is, a bottomless culvert.

Base Course. The layer or layers of specified or selected material of designed thickness placed on a subbase or subgrade to support a surface course. (See figure 102-1.)

Bearings. The portion of a beam, girder, or truss that transmits the bridge superstructure load to the substructure.

Berm. Curb or dike constructed to control roadway runoff water. (See figure 102-1.)

Best Management Practice. A series of water quality protection practices and procedures approved or certified by the State water quality agency under the provisions of sections 319 and 402 of the Clean Water Act, as amended.

Bridge. A structure, including supports, erected over a depression or an obstruction, such as water, a road, a trail, or a railway, and having a floor for carrying traffic or other moving loads.

Bridge Length. The overall length measured along the centerline of road to the back of abutment backwalls, if present; otherwise, end to end of the bridge floor, but in no case less than the total clear opening of the structure.

Bridge Traveled Way Width. The clear width measured at right angles to the longitudinal centerline of the bridge between the bottom of curbs or, if curbs are not used, between the inner faces of parapet or railing.

Certificate of Compliance. A signed statement by a person with legal authority to bind a company or supplier to its product. The certificate states that the material or assemblies furnished fully comply with the requirements of the contract.

Change. "Change" means "change order" as used in the Federal Acquisition Regulations, or "design change" as used in the Timber Sale Contract.

Clearing Limits. The limits of clearing as designated on the ground or on the drawings. (See figure 102-1.)

Cofferdam. A cofferdam is an enclosed single or double wall braced structure with walls sheeted with timber, concrete, or steel, and extending well below the bottom of excavation, when practical. Earthen or rockfill dikes, dams, or embankments are not considered cribs or cofferdams for this purpose.

Conduit. A natural or artificial channel for carrying fluids, such as water pipe, canal, or aqueduct.

Construction Slash. All vegetative material not meeting Utilization Standards, such as tops and limbs, timber, brush, and grubbed stumps associated with construction or reconstruction of a facility.

Contracting Officer (CO). The person with the authority to enter into, administer, and/or terminate contracts and make related determinations and findings. The term includes certain authorized representatives of the CO acting within the limits of their authority as delegated by the CO. Authorized representatives include the Forest Service Representative, Engineering Representative, Contracting Officer's Representative, and Inspector.

Contractor. The individual, partnership, joint venture, or corporation undertaking the execution of the work under the terms of the contract and acting directly or through agents, employees, or subcontractors. As used in specifications and drawings for specified roads (Timber Sale Contracts), "Contractor" is "purchaser."

Controlled Felling. Directing the placement of trees in felling by using wedges, jacks, cable tension, or distribution of holding wood, or any combinations of these, to ensure that trees are dropped into previously cleared areas, or clear of any objects that are to remain.

Culvert. A conduit or passageway under a road, trail, or other obstruction. A culvert differs from a bridge in that it is usually constructed entirely below the elevation of the traveled way.

Curve Widening. Additional width added to curves to allow for vehicle offtracking.

Cushion Material. Native or imported material generally placed over rocky sections of unsurfaced roads to provide a usable and maintainable traveled way.

Defect. A failure to meet a requirement with respect to a single quality characteristic.

Drawings. The documents, including plan and profile sheets, plans, cross sections, diagrams, layouts, schematics, descriptive literature, illustrations, schedules, performance and test data, and similar materials showing details for construction of a facility.

Embankment. A structure of soil, aggregate, or rock material placed on the prepared ground surface and constructed to subgrade.

Equipment. All machinery and equipment, together with the necessary supplies for upkeep and maintenance, as well as tools and apparatus necessary for the proper construction and acceptable completion of the work.

Excess Excavation. Material from the roadway in excess of that needed for construction of designed roadways.

Falsework. Any temporary construction work used to support the permanent structure until it becomes self-supporting. Falsework includes steel or timber beams, girders, columns, piles, foundations, and any proprietary equipment including modular shoring frames, post shores, and adjustable horizontal shoring.

Forest Service. The United States of America, acting through the Forest Service, U.S. Department of Agriculture.

Government Land. National Forest System lands, and other lands controlled or administered by the Forest Service or other Federal agencies.

Inspector. The Government-authorized representative designated in writing by the Contracting Officer, Contracting Officer's Representative, or Engineering Representative responsible for detailed inspection.

Invert. The lowest point of the internal cross section of culvert or pipe arch.

Job-Mix Formula. The percentage of each material in a mixture intended for a particular use.

Laboratory. A testing laboratory of the Government, or any other testing laboratory approved by the Contracting Officer.

Live Stream. A defined streambed with flowing water.

Lot. An isolated quantity of material from a single source; a measured amount of construction assumed to be produced by the same process.

Materials. Any substance specified for use in the construction of the project and its appurtenances.

Maximum Density. The highest density that can be obtained for a specific material using the stated test procedure.

Measurement. Determining and expressing the quantities of work or materials.

Multibeam Girder. A precast, prestressed concrete member where the concrete deck is precast as an integral part of the member.

Neat Line. A line defining the proposed or specified limits of an excavation or structure.

Nominal Dimensions or Weights. The numerical values shown on the drawings or in the specifications as measurements of material to be used in the construction.

Nominal Maximum Particle Size. The largest sieve size listed in the applicable specification upon which any material is permitted to be retained.

Overbreak. Material beyond the neat line of an excavation that is removed in the process of excavation, usually by blasting.

Pass. A pass shall consist of one complete coverage of the surface.

Pavement Structure. Subbase, base, or surface course, or combination thereof, placed on a subgrade to support the traffic load and distribute it to the roadbed.

Pioneer Road. Temporary construction access built along the route of the project.

Pipe. A culvert that is circular (round) in cross section.

Pipe Arch. A pipe that has been factory-deformed from a circular shape such that the width (or span) is larger than the vertical dimension (or rise).

Profile Grade. The trace of a vertical plane, as shown on the drawings, intersecting the top surface at the centerline of the proposed facility construction.

Purchaser. The individual, partnership, joint venture, or corporation contracting with the Government under the terms of a Timber Sale Contract and acting independently or through agents, employees, or subcontractors.

Random Sampling. A sample of material chosen such that each increment of a population of material has an equal probability of being selected.

Reasonably Close Conformity. Compliance with reasonable and customary manufacturing and construction tolerances, performing all work and furnishing all materials in "reasonably close conformity" with lines, grades, cross sections, dimensions, and material requirements shown on the drawings, indicated in the specifications, or designated on the ground.

Right-of-Way. A general term denoting (1) the privilege to pass over land in some particular line (including easement, lease, permit or license to occupy, use, or traverse public or private lands), or (2) land, appurtenances thereto, or interest therein, usually in a strip, acquired for public or private passageway. (See figure 102-1.)

Road Order. An order affecting and controlling traffic on roads under Forest Service jurisdiction. Road Orders are issued by a designated Forest Officer under the authorities of 36 CFR, part 260.

Road Template. The shape and cross-sectional dimensions of the roadway to be constructed, as defined by the construction staking notes and the characteristics of the typical sections.

Roadbed. The graded portion of a road between the intersection of subgrade and side slopes, excluding that portion of the ditch below subgrade. (See figure 102-1.)

Roadside. A general term denoting the area adjoining the outer edge of the roadway. (See figure 102-1.)

Roadway. The portion of the road within the limits of excavation and embankment, including slope rounding. (See figure 102-1.)

Schedule of Items. A schedule in the contract that contains a listing and description of construction items, quantities, units of measure, methods of measurement, unit price, and amount.

Second Samples. A sample taken when the initial sample indicates that the material is defective.

Shoulder. The portion of the roadway contiguous to the traveled way for accommodation of stopped vehicles, emergency use, and lateral support of pavement structure. (See figure 102-1.)

Sidewalk. The portion of the roadway constructed primarily for pedestrian use.

Special Project Specifications. The specifications that detail the conditions and requirements specific to the individual project, including additions and revisions to Standard Specifications.

Specifications. A description of the technical requirements for a material, product, or service that includes criteria for determining whether these requirements are met.

Spring Line. The point of contact between arch and footing.

Standard Specifications. Specifications approved for general application and repetitive use.

Station. (1) A measure of distance used for highways and railroads equal to 1 kilometer. (2) A precise location along a survey line.

Subbase. The layers of specified or selected material of designed thickness placed on a subgrade to support a base course.

Subgrade Treatment. Modification of roadbed material by stabilization.

Subgrade. The prepared surface, including widening for curves, turnouts, and other areas upon which a subbase, base, or surface course is constructed. For roads without base course or surface course, that portion of roadbed prepared as the finished wearing surface. (See figure 102-1.)

Substructure (Bridge). All of that part of the structure below the bearings of simple and continuous spans, skewbacks of arches, and tops of footings of rigid frames, together with the backwalls, wingwalls, and wing protection railings.

Superstructure (Bridge). The entire structure, except the substructure.

Surface Course. The top layer of a pavement structure, sometimes called the wearing course, usually designed to resist skidding, traffic abrasion, and the disintegrating effects of climate. (See figure 102-1.)

Tackifier. Binder for vegetative mulch.

Target Value. Values that are established according to contract, and from which allowable variations are measured.

Timber Sale Contract. A written contract for the removal of national forest timber.

Tops and Limbs. All woody material including bushes, vines, and portions of trees smaller than the dimensions for timber shown in Subsection 201.03.

Traveled Way. The portion of the roadway for the movement of vehicles, exclusive of shoulders and auxiliary lanes. (See figure 102-1.)

Turnout. A short auxiliary lane on a one-lane road provided for the passage of meeting vehicles.

Unit of Measurement. The unit and fractions of units DESIGNATED IN THE SCHEDULE OF ITEMS.

Unsuitable Material. The material excavated during roadway construction that is not usable in embankment and must be disposed of, or that can be used only in certain locations or for limited purposes.

Utilization Standards. The minimum size and percent soundness of trees described in the specifications to determine merchantable timber.

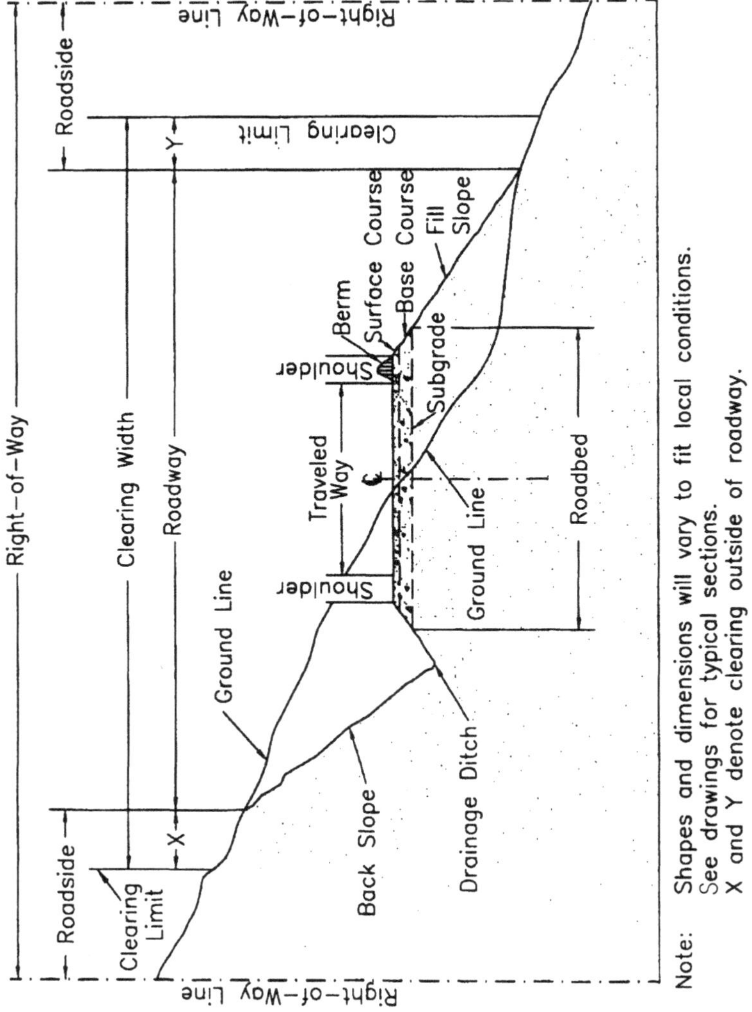

Figure 102-1.—Illustration of road structure terms.

Section 103—Intent of Contract

The intent of the contract is to provide for the complete construction of the project described in the contract. Unless otherwise provided, furnish all labor, materials, equipment, tools, transportation, and supplies, and perform all work required to complete the project in reasonably close conformity with drawings and specifications, and in accordance with provisions of the contract.

Section 104—Maintenance for Traffic

104.01 Roads To Be Constructed

Unless otherwise SHOWN ON THE DRAWINGS or described in the SPECIAL PROJECT SPECIFICATIONS, keep existing roads open to all traffic during road improvement work, and maintain them in a condition that will adequately accommodate traffic. Perform no work that interferes or conflicts with traffic or existing access to the roadway surface until a plan for the satisfactory handling of traffic has been approved. Specific requirements for temporary closures, detours, part-width construction, and access to adjacent or intersecting facilities will be SHOWN ON THE DRAWINGS or described in the SPECIAL PROJECT SPECIFICATIONS. Post construction signs and traffic control devices in conformance with the "Manual on Uniform Traffic Control Devices" (MUTCD). Do not proceed with work on the project until all required signs are in place and approved.

Before shutting down any operations, take all necessary precautions to prevent damage to the project, such as temporary detours, approaches, crossings, or intersections; and provide for normal drainage and minimization of erosion. Leave all travelways in a condition suitable for traffic.

The Government may permit use of portions of the project during periods when operations have shut down All maintenance attributable to permitted use during periods of work suspension will be provided by the Government, except for maintenance needed through the fault or negligence of the Contractor. The Contractor shall be responsible for any maintenance not attributable to use, or that is necessary during suspensions through the fault or negligence of the Contractor.

When SHOWN ON THE DRAWINGS or described in the SPECIAL PROJECT SPECIFICATIONS, road segments may be closed to all traffic during the period(s) when construction is in progress. If any of the listed roads are to be closed during construction operations, give at least 14 days advance notice.

Unless otherwise provided, when construction activity is in progress and total closure has not been provided for herein, delays may not exceed *30 minutes*, in order to reasonably accommodate traffic.

104.02 Use of Roads by Contractor

The Contractor is authorized to use roads under the jurisdiction of the Forest Service for all activities necessary to complete this contract, subject to the limitations and authorizations SHOWN ON THE DRAWINGS, designated in the Road Order, or described in the SPECIAL PROJECT SPECIFICATIONS, when such use will not damage the roads or national forest resources, and when traffic can be accommodated safely.

Section 105—Control of Materials

105.01 Handling Materials

Transport and handle all materials to preserve their quality and fitness for the work. Stockpile, load, and transport aggregates in a manner that will preserve specified gradation and avoid contamination. Do not intermingle stockpiles of aggregate with different gradations. Stockpile crushed or screened aggregate in accordance with Section 305.

105.02 Weighing Devices

When the measurement is by weight, provide weigh scales and transport the materials so they can be weighed. Perform all weighing.

(a) Platform Scales. Provide platform scales of sufficient length and capacity to permit simultaneous weighing of all axle loads of each hauling unit.

Provide scales accurate to within 1 percent of the correct weight throughout the range of use. Before using the scales and as frequently thereafter as necessary to ensure accuracy, have the scales checked, adjusted, and certified by a representative of the State agency responsible for weights and measures or by a qualified manufacturer's representative.

Provide copies of weight tickets from a certified scale.

Material may be weighed on other certified scales without additional compensation. In this case, furnish certified weight tickets for all material delivered to the project, and guarantee permission to periodically check the weighing procedure and records.

(b) Belt Conveyor Scales. Belt conveyor weighing will be accepted in lieu of platform scales, provided this method or device meets the requirements specified below and is compatible with the provisions of measurement and payment in the applicable specifications.

Use a belt conveyor scale that meets the design, marking, installation, and tolerance requirements of the National Bureau of Standards (NBS) Handbook No. 44. Provide a copy of a NBS Prototype Examination Report of Test to certify the scale.

Use a weighing mechanism that contains a weight totalizer and a self-printing device that legibly imprints the load-out weight on appropriate serially numbered and time-dated tickets. Time date manually, or use an automatic printing device. Deliver each ticket at the job site or point of use. Furnish the totalizer calibration adjustment and ticket imprinting device with a security lock and key.

Under observation of the CO, run a daily zero-load test in accordance with NBS Handbook No. 44.

105.03 Sampling & Testing of Aggregate

When Designated Sources or Contractor-Furnished Sources are specified (see Subsections 105.06(a) and (b)), submit test results and a Certificate of Compliance that states that the aggregate meets the contract requirements. Equip crushing, screening, and mixing plants with sampling devices. Submit test results within 16 working hours of obtaining sample(s). Take additional samples, if required by the CO, to validate the certification.

Before incorporating material into the work, ensure that sampling and testing of material conform to the American Association of State Highway and Transportation Officials (AASHTO) requirements, and occur as follows:

(a) For onsite-produced materials at crushing or screening plants: after additions of any necessary blending material.

(b) For commercially produced aggregates: at the producer's plant or stockpile.

(c) For gradation of combined aggregate in bituminous plant mixtures: either before or after introduction of bituminous material.

These test results shall not preclude later sampling and testing for final acceptance after final processing of the material.

105.04 Certification & Sampling of Asphalt Materials

(a) Certification With Shipments. When each load of asphalt material is delivered, furnish one copy of the Bill of Lading; a fully executed Certificate of Compliance in the format shown in figure 105-1; and a copy of the refinery test reports. The refinery test reports should include the following information:

(1) Consignee.
(2) Contract number.
(3) Date of shipment.
(4) Type and grade of material.
(5) Test results as follows:
  (a) For performance-based asphalt and performance-graded asphalt cements:
    (1) Flash point.
    (2) Absolute viscosity of the original asphalt at 60 °C.

Consignee ____________ Designation ____________
Contract Number ____________ Date ____________
Identification (truck no., car no., etc.)____________
Type and Grade ____________ With Additive (percent and brand)____________
Loading Temp. ____________ Net Weight ____________
Net Liters ____________ Specific Gravity____________

This shipment of bituminous material identified above and covered by this Bill of Lading complies with Forest Service STANDARD SPECIFICATIONS as modified by SPECIAL PROJECT SPECIFICATIONS applicable to this project.

Producer ____________
Signed ____________
Producer's Representative

Figure 105-1.—Sample Bill of Lading and Certificate of Compliance.

(3) Absolute viscosity of the residue from the Rolling Thin Film Oven test at 60 °C.
(4) Penetration of the residue from the Rolling Thin Film Oven test at 4 °C.

(b) For aged-residue graded asphalt cements:
(1) Flash point.
(2) Absolute viscosity of the original asphalt at 60 °C.
(3) Absolute viscosity of the residue from the Rolling Thin Film Oven test at 60 °C.
(4) Penetration of the residue from the Rolling Thin Film Oven test at 25 °C.

(c) For asphalt concrete graded asphalt cements:
(1) Flash point.
(2) Absolute viscosity of original and residue asphalt at 60 °C.
(3) Penetration of the original asphalt at 25 °C.

(d) For pen graded asphalt cements:
(1) Flash point.
(2) Penetration of the original and residue asphalt at 25 °C.

(e) For emulsified asphalt:
(1) Percent residue from distillation.
(2) Saybolt furol viscosity of the emulsion at the specified temperature.
(3) Penetration of the residue from distillation at 25 °C.
(4) Oil distillate, by volume of emulsion.
(5) If applicable, torsional recovery or toughness tenacity.

(f) For cutback asphalt:
(1) Flash point.
(2) Percent residue from distillation.

(3) Original kinematic viscosity at 60 °C.
(4) Absolute viscosity of the residue from distillation at 60 °C.

A separate Certificate of Compliance will not be required if the standard Bill of Lading contains the essential information required by the certificate.

(b) Sampling. Unless otherwise directed, take at least two samples of bituminous material from each hauling unit of the transporting vehicle, or samples representing each hauling unit taken from the distribution truck. Obtain samples in the presence of the CO; samples will become property of the Government. Obtain polymer-modified emulsified asphalt samples from the distribution truck just prior to application.

Construct all delivery and plant equipment to permit sampling in conformance with AASHTO T 40 test procedure.

105.05 Rights in & Use of Materials Found or Produced on the Work

(a) With the written approval of the CO, suitable stone, gravel, sand, or other material found in the excavation can be used on the project. Payment will be made both for the excavation of such materials at the corresponding contract unit price and for the pay items for which the excavated material is used. Replace, without additional compensation, sufficient suitable materials to complete the portion of the work that was originally contemplated to be constructed with such material.

(b) Materials produced or processed from Government lands in excess of the quantities required for performance of this contract are the property of the Government. The Government is not obligated to make reimbursement for the cost of producing these materials.

105.06 Material Sources

(a) Designated Sources. Sources of local materials designated in the SPECIAL PROJECT SPECIFICATIONS or SHOWN ON THE DRAWINGS are guaranteed by the Government for the quality and quantity of material in the source. Determine the equipment and work required to produce the specified product. Submit test results and a Certificate of Compliance that states that the gradation of the aggregate meets the contract requirements.

Utilize all suitable material in the source. The designation of a source includes the Contractor's right to use areas SHOWN ON THE DRAWINGS for the purposes designated (that is, plant sites, stockpiles, and haul roads). Unless otherwise indicated or approved, no additional operating area shall be allowed. In this case, operate only in the confines of the area(s) designated.

The weight/volume relationship used for determining designed quantities (DQ) of material in designated sources subject to weight measurement is SHOWN ON THE DRAWINGS.

Should the designated source contain insufficient suitable material due to causes beyond the Contractor's control, the Government will provide another source, with an adjustment in contract price, in accordance with applicable contract provisions.

Designated sources will be available for the Contractor's use during the periods SHOWN ON THE DRAWINGS. Use at any other time will require an agreement with the party scheduled for that period, with the CO's approval.

(b) **Contractor-Furnished Sources.** When the material sources are not designated as provided above, or when designated sources are not used, furnish material that produces an end product equivalent in performance to that originally designated. An adjustment in contract price shall be made where the weight/volume relationship differences between designated source material and Contractor-furnished source material result in a financial disadvantage to the Government. When SHOWN ON THE DRAWINGS, complete any pit development specified for a designated source, even when material is not obtained from the source.

Test for quality in conformance with applicable requirements, to establish the equivalency of the end product. Furnish test results and a Certificate of Compliance.

Section 106—Measurement & Payment

106.01 Measurement & Payment

Compensation provided for in the contract is full payment for performing all contract work in a complete and acceptable manner. All risk, loss, damage, or expense arising out of the nature or prosecution of the work is included in the compensation provided by the contract.

Work required by the contract will not be paid for directly unless a PAY ITEM for the work is DESIGNATED IN THE SCHEDULE OF ITEMS.

Work referenced for measurement under another section will not be paid for directly unless a PAY ITEM for the work is DESIGNATED IN THE SCHEDULE OF ITEMS for the referenced section.

Work not paid for directly is considered to be included under the other contract PAY ITEMS.

Unless otherwise shown, work measured and paid for under one PAY ITEM will not be paid for under any other PAY ITEM.

The quantity to be paid for is the quantity DESIGNATED IN THE SCHEDULE OF ITEMS. No payment will be made for work performed in excess of that staked, ordered, or otherwise authorized.

When more than one class, size, or thickness is specified in the SCHEDULE OF ITEMS for any PAY ITEM, suffixes will be added to the item number to differentiate between the items.

106.02 Determination of Quantities

The following measurements and calculations are used to determine contract quantities.

For individual construction items, longitudinal and lateral measurements for area computations shall be made horizontally or corrected to horizontal measurement unless otherwise specified. Measurements for seeding, mulching, geotextiles, netting, erosion control blankets, and sodding shall be along slope lines.

The average end area method shall be used to compute volumes of excavation or embankment. However, if in the judgment of the CO the average end area method is impractical, measurement shall be made by volume in hauling vehicles, or by other three-dimensional methods.

Structures shall be measured according to neat lines SHOWN ON THE DRAWINGS, or as altered by the CO in writing to fit field conditions.

For items that have linear measurements, such as pipe culverts, fencing, guardrails, and underdrains, measurements shall be made parallel to the base or foundation upon which the structures are placed. Pipe and pipe arch culverts shall be measured along center of invert, and arches shall be measured at spring line.

For aggregates weighed for payment, the tonnage weight shall not be adjusted for moisture content, unless otherwise provided in SPECIAL PROJECT SPECIFICATIONS.

For asphalt material, volumes shall be measured at 15.6 °C, or shall be corrected to the volume at 15.6 °C by using ASTM D 1250 for asphalt. Emulsified asphalt shall be measured at 15.6 °C, or by converting the volume at another temperature to volume at 15.6 °C by means of the following formula:

$$\text{L at } 15.6\ ^\circ\text{C} = \frac{\text{L at } T\ ^\circ\text{C}}{1 + 0.00045\,(T\ ^\circ\text{C} - 15.6\ ^\circ\text{C})}$$

where

T °C = temperature of the emulsified asphalt at the time the volume is measured

For vehicular shipments, net certified scale weights or weights based on certified volumes shall be used as a basis of measurement. Measurements shall be adjusted when asphalt material has been lost from the vehicle or from the distributor, has been wasted, or has otherwise not been incorporated into the work. True weights of hauling vehicles shall be determined by weighing the empty vehicles at least once a day at the times the CO specifies. Each vehicle shall bear a plainly legible identification mark.

When asphalt materials are shipped, net certified weights, or volume corrected for loss of foaming, can be used for computing quantities.

For standard manufactured items—such as fence, wire, plates, rolled shapes, and pipe conduits—identified by gauge, weight, section dimensions, and so forth, such identifications shall be considered the nominal weights or dimensions. Unless controlled by tolerances in cited specifications, manufacturer's tolerances shall be accepted.

106.03 Units of Measurement

Payment will be by units defined and determined according to measure. Unless otherwise specified, the meanings of the following terms are as follows:

(a) Cubic Meter in Place (m^3). Measure solid volumes by the average end area method as follows:

(1) Measure cross sections of the original ground and use with design or staked templates, or take other comparable measurements to determine the end areas. Do not measure work outside of the established lines or slopes.

(2) If any portion of the work is acceptable, but is not completed to the established lines and slopes, remeasure cross sections or comparable measurements of that portion of the work. Deduct any quantity outside the designated or staked limits. Use these measurements to calculate new end areas.

(3) Compute the quantity using the average end areas multiplied by the horizontal distance along a centerline or reference line between the end areas. Deduct any quantity outside the designed or staked limits.

Where it is impractical to measure material by the average end area method, other methods involving three-dimensional measurements may be used.

Measure liquid volumes in accordance with Subsection 106.03(h).

(b) Cubic Meter in the Hauling Vehicle. Measure the cubic meter volume in the hauling vehicle using three-dimensional measurements at the point of delivery. Use vehicles bearing a legible identification mark with the body shaped so the actual contents may be readily and accurately determined. Before use, mutually agree in writing upon the volume of material to be hauled by each vehicle. Vehicles carrying less than the agreed volume may be rejected or accepted at the reduced volume.

Level selected loads. If leveling reveals that the vehicle was hauling less than the approved volume, reduce the quantity of all material received since the last leveled load by the same ratio as the ratio of the current leveled load volume to the agreed volume. Payment will not be made for material in excess of the agreed volume.

Material measured in the hauling vehicle may be weighed and converted to cubic meters for payment purposes if the conversion factors are mutually agreed to in writing.

Compute measurement using measurements of material in the hauling vehicles at the point of delivery. Load vehicles to at least their water-level capacity. Leveling of the loads may be required when vehicles arrive at the delivery point.

(c) Each. One entire unit, which may consist of one or more parts. The quantity is the actual number of units completed and accepted.

(d) Hectare (ha). 10,000 m^2. Make longitudinal and transverse measurements for area computations horizontally. Do not make deductions from the area computation for individual exclusions having an area of 50 m^2 or less.

(e) Hour (h). Measurement will be for the actual number of hours ordered and performed by the Contractor.

(f) Kilogram (kg). 1,000 g. If sacked or packaged material is furnished, the net weight as packed by the manufacturer may be used.

(g) Kilometer (km). 1,000 m. Measure horizontal along the centerline of each roadway, approach road, or ramp.

(h) Liter (L). The quantity may be measured by any of the following methods:

(1) Measured volume container.

(2) Metered volume. Use an approved metering system.

(3) Commercially packaged volumes.

(i) Lump Sum. Do not measure directly. The bid amount is complete payment for all work described in the contract and necessary to complete the work for that item.

(j) Meter (m). Measure from end to end, parallel to the base or foundation being measured, or horizontal.

(k) Station. 1,000 m measured horizontally.

(l) Square Meter (m^2). Measure on a plane parallel to the surface being measured or horizontal.

Where measurement is horizontal, make no deductions from the area computation for individual exclusions having an area of 1 m^2 or less.

For pavement structure courses, measure the width horizontally to include the top design width and allowable curve widening. Do not include side slopes. Measure the length horizontally along the centerline of each roadway, approach road, or ramp.

(m) Thousand Board Feet (Mbf). 1,000 board feet based on nominal widths, thickness, and extreme usable length of each piece of lumber or timber actually incorporated in the job. For glued laminated timber, 1,000 board feet based on actual width, thickness, and length of each piece actually incorporated in the job.

(n) Ton (t). 1,000 kg.

No adjustment in contract unit price will be made for variations in quantity due to differences in the specific gravity or moisture content.

Use net certified scale weights, or weights based on certified volumes.

106.04 Methods of Measurement

One of the following methods of measurement for determining final payment is DESIGNATED IN THE SCHEDULE OF ITEMS for each PAY ITEM:

(a) Designed Quantities (DQ). These quantities denote the final number of units to be paid for under the terms of the contract. They are based upon the original design data available prior to advertising the project. Original design data include the preliminary survey information, design assumptions, calculations, drawings, and the presentation in the contract. Changes in the number of units DESIGNATED IN THE SCHEDULE OF ITEMS may be authorized under any of the following conditions:

(1) Changes in the work authorized by the CO.

(2) A determination by the CO that errors exist in the original design that cause a PAY ITEM quantity to change by 15 percent or more.

(3) A written request submitted to the CO showing evidence of errors in the original design that cause the quantity of a PAY ITEM to change by 15 percent or more. The evidence must be verifiable and consist of calculations, drawings, or other data that show how the designed quantity is in error.

(b) Staked Quantities (SQ). These quantities are determined from staked measurements prior to construction.

(c) Actual Quantities (AQ). These quantities are determined from measurements of completed work.

(d) Vehicle Quantities (VQ). These quantities are measured or weighed in hauling vehicles.

(e) Lump Sum Quantities (LSQ). These quantities denote one complete unit of work as required by or described in the contract, including necessary materials, equipment, and labor to complete the job.

106.05 Price Adjustment for Asphalt Materials

Asphalt materials are defined as all types and grades of asphalt cement, cutback, and emulsified asphalt.

The refinery test reports and the Certificate of Compliance required in Subsection 105.04 will be reviewed and used to support acceptance of the asphalt material incorporated into the project.

If materials are found not to be in conformance with the specified tolerances or within the specification limits, the CO will determine an equitable adjustment in payment.

If the CO elects to test field samples, the test results may be used for acceptance.

106.06 Earthwork Tolerances

Where tolerances are shown in the contract, they are intended to define "reasonably close conformity." Make adjustments of horizontal or vertical alignment within the tolerances specified in this contract, or shifts of balance points up to 30 m, as necessary to produce the designed roadway section and to balance earthwork. Such adjustments will not be considered "changes."

DIVISION 150
Engineering

Section 160—Quality Control & Quantity Measurement

Description

160.01 Work. Provide quality control in conformance with the Inspection of Construction provisions of this contract to ensure compliance with the drawings, specifications, and provisions of the contract. Measure the quantities of completed work in conformance with the provisions of the applicable specification. Provide all personnel, equipment, tests, and reports necessary to meet the requirements of this specification.

Construction

160.02 Quality Control & Quantity Measurement System. Provide and maintain a quality control system that will ensure that all services, supplies, and construction required under this contract conform to the contract requirements. Perform, or cause to be performed, the sampling, inspection, and testing required to substantiate that all supplies, services, and construction conform to the contract requirements.

In addition, perform, or cause to be performed, all measurement of quantities of materials incorporated into the work or work processes that are to be measured under the provisions of the contract.

(a) Quality Control Plan. Submit in writing the following:

(1) Authorities and responsibilities of inspection and testing personnel.

(2) Experience and qualifications of inspection and testing personnel to be assigned and name and location of any (for hire) testing facility to be used.

(b) Approval of Quality Control Plan. Before beginning work, submit proposed quality control plan for all items requiring quality control to the CO for review. Within 5 days of receipt of the plan, the CO will determine whether the plan adequately covers quality control requirements. Do not perform construction work before receiving written approval of the proposed plan. Submit to the CO in writing any proposed changes in the approved quality control plan. Do not put proposed changes into effect until approved in writing by the CO.

160.03 Sampling, Testing, Inspection, & Measurement of Quantities. Provide and maintain appropriate measuring and testing devices, equipment, and supplies to accomplish the required measurement, testing, and inspection in a timely manner. Make all tests, measurements, and certifications as required by the drawings and specifications. Take samples and perform inspections and tests as necessary to achieve the quality of construction required by the contract, and make required measurements of work performed onsite or offsite under this contract. Sampling and testing frequency for specific items will be SHOWN ON THE DRAWINGS, in the Standard Specifications, or in a SPECIAL PROJECT SPECIFICATION.

Where random sample or random measurement is specified, provide a stratified statistically random sample. Determine random numbers in accordance with ASTM D 3665, sections 5.1 through 5.7, or use a computer-generated random number program approved by the CO. Ensure that the sampling is stratified to eliminate the possibility that sample points are "clustered." Perform stratification by dividing the total quantity for the applicable bid item by the sample frequency. This process divides the total project quantity of one lot into sublots. Use the random number to obtain a random sampling point within each sublot. A lot may be terminated and a new lot started when approved by the CO. After a lot is terminated, do not combine it with any other lot. If material within a sublot fails to meet specification requirement, the CO may allow the sublot materials to be reworked and resampling to be performed at new randomly selected locations.

The CO may reject any quantity of material that appears to be defective based on visual inspection or test results. Do not use such rejected material in the work. Results of tests run on this rejected material will not be included in results of lot acceptance tests.

160.04 Records of Inspection, Tests, & Measurement. Meet the following requirements for inspection and tests, and as-built drawings:

(a) Inspection and Tests. Maintain current records of all inspections and tests performed. The following format, or one with the following information, will be acceptable to the Government:

Road No. ______________ Contract No. ____________________

Pay Item No.	Test	Date	Station	Standard	Results	Test By (Initials)

Certify in writing that all inspections and tests were performed in accordance with specifications.

(b) As-Built Drawings. Maintain a set of the contract drawings depicting as-built conditions. Maintain these drawings in current condition, and make them available for review. Indicate all variations from contract drawings in red on the drawings. Upon completion of the contract work, submit as-built drawings to the CO.

160.05 Certifications & Measurements. Meet the following requirements for offsite-produced materials and quantity measurements:

(a) Offsite-Produced Materials. Furnish certificates executed by the manufacturer, supplier, or vendor, stipulating that all materials produced offsite that are incorporated into the work meet the applicable requirements SHOWN ON THE DRAWINGS or stated in the specifications. Certify all incidental purchases needed to remedy minor shortages of material.

(b) Quantity Measurements. Make all measurements for computation of quantities for all work items, except those specified for payment by designed quantity or lump sum. Compute the quantities for periodic progress payments; the CO will compute the quantities for the final payment based on measurements taken. All Contractor measurements are subject to verification. Submit all field notes, calculation sheets, and other data used to determine quantities, and certify in writing as to the accuracy of the measurements and computations submitted.

The following format, or one containing the following information, will be acceptable to the Government:

Road No. ______________ Contract No. ____________________

Pay Item No.	Date	Station	Quantity or Measurement	Measured By (Initials)

Measurement

160.06 Method. Do not make separate measurements for this section.

Payment

160.07 Basis. The accepted quantities will be paid for at the contract unit price for each PAY ITEM DESIGNATED IN THE SCHEDULE OF ITEMS. Otherwise, quality control and quantity measurement will be incidental to other specified work.

Payment will be prorated based on the percentage of work accomplished on the related PAY ITEM that meets specifications.

Section 161—Certification for Quality & Quantity

Description

161.01 Work. Provide certification that the quality and quantity of construction conforms to the drawings, specifications, and requirements of the contract.

Construction

161.02 Certifications & Measurements. Meet the following requirements for offsite-produced materials and quantity measurements:

(a) Offsite-Produced Materials. Furnish certificates executed by the manufacturer, supplier, or vendor, stipulating that all materials produced offsite that are incorporated into the work meet the applicable requirements SHOWN ON THE DRAWINGS or stated in the specifications. Make each certificate apply to a single commodity or invoice. Certify all incidental purchases needed to remedy minor shortages of material.

(b) Quantity Measurements. Make all measurements for computation of quantities for all work items, except those specified for payment by designed quantity or lump sum. Compute the quantities for periodic progress payments; the CO will compute the quantities for the final payment based on measurements taken. All Contractor measurements are subject to verification. Submit all field notes, calculation sheets, and other data used to determine quantities, and certify in writing as to the accuracy of the measurements and computations submitted.

The following format, or one containing the following information, will be acceptable to the Government:

Road No. ____________		Contract No. ________________		
Pay Item No.	Date	Station	Quantity or Measurement	Measured By (Initials)

161.03 Records. Meet the following requirements for as-built drawings:

As-Built Drawings. Maintain a set of the contract drawings depicting as-built conditions. Maintain these drawings in current condition, and make them available for review. Indicate all variations from contract drawings in red on the drawings. Upon completion of the contract work, submit as-built drawings to the CO.

Measurement

161.04 Method. Make no separate measurements for this item.

Payment

161.05 Basis. Payment will be considered incidental to other pay items in this contract.

Section 170—Construction Staking, L-Line

Description

170.01 Work. Complete the construction staking of a road by the L-line method in accordance with the drawings and specifications. Furnish all labor, equipment, instruments, materials, transportation, and other incidentals necessary to complete the construction staking in accordance with these specifications and acceptable engineering practice. In addition, set grade-finishing stakes, and stake major structures.

Conduct construction staking under the direction of a licensed professional engineer or land surveyor who is closely associated and familiar with the construction staking. Periodic visits to the project site are required.

Materials

170.02 Stakes. Provide stakes that all have the nominal dimensions SHOWN ON THE DRAWINGS or stated in the SPECIAL PROJECT SPECIFICATIONS. Ensure that identification stakes and hubs are of sufficient length to provide a solid set in the ground and space for marking above ground when applicable. Other dimensions and materials may be used, such as steel reinforcing bars, wire flagging and markers, and metal pins, if approved in writing by the CO. Paint the top 50 mm of all slope, guard, reference, clearing, and structure stakes, or mark them with plastic flagging. Use colors on stakes or for flagging as SHOWN ON THE DRAWINGS or stated in the SPECIAL PROJECT SPECIFICATIONS.

170.03 Survey Notes. Furnish field notebooks or note papers. Use moisture-resistant paper for survey notes. Keep notes in books with covers that will protect the contents and retain the pages in numerical sequence during field use. When using electronic data collectors, provide electronic and hardcopy notes as listed in the SPECIAL PROJECT SPECIFICATIONS.

170.04 Government-Furnished Documents. Drawings, P-line survey notes, P-line to L-line offset data, construction staking notes, and the projected locations of catch points will be furnished by the Government. Return one set of "as-staked" drawings and all documents to the CO.

Survey Requirements

170.05 Precision. Accuracy and precision requirements are contained in tables 170-1 and 170-2. Perform all work under this specification to meet the

Table 170-1.—Accuracy requirements for reestablishing P-line, traverse, and elevations.

Precision Class	Minimum Position Closure	Angular Accuracy (±)	L-Line Tangent Control Points[a] (±)	Vertical Closure[b] (±)
A (Bridges)	1/10,000	2 sets, direct/reverse; 10" rejection limit	N/A	5 mm or 20 mm/km[c]
B	1/5,000	2 sets, direct/reverse; 20" rejection limit	30 mm	50 mm or 200 mm/km[c]
C	1/1,000	1 set, direct/reverse; 1' rejection limit	60 mm	500 mm/km

a. Accuracy of offset measurement.
b. Determine vertical closures at intervals not to exceed 1 km, as measured along centerline.
c. Use greatest value.

precision requirements DESIGNATED IN THE SCHEDULE OF ITEMS or stated in the SPECIAL PROJECT SPECIFICATIONS.

170.06 Survey Notes Format. All notes will become the property of the Forest Service. Use the slope stake note format shown in figure 170-1. The sketch in the lower half of figure 170-1 is for information only, and is not required as part of actual slope stake note format. Other formats may be used if approved by the CO.

Print all manually recorded survey notes in characters at least 4 mm high, and make them legible at a distance of 750 mm. Delete errors by lining out. At the beginning of each day's work, record date, crew names and positions, instrumentation, and weather in the notes. Ensure that the party chief signs or initials each page of the notes immediately after the last entry for each day's work.

Consecutively number electronically recorded survey notes, and use headings to identify the contents. Support and accompany the notes with a bound Day Book that records the project name and, for each day, the date; crew names and positions; instrumentation; weather; type of survey; stationing of sections between which the survey was performed; and survey data or sketches that cannot be electronically recorded. Ensure that the party chief signs or initials the electronically recorded notes and Day Book immediately after the last entry for each day's work.

Table 170-2.—Cross section and slope stake precision.

Item	Precision A (±)	Precision B (±)	Precision C (±)
Allowable deviation of cross-section line projection from a true perpendicular to tangents, a true bisector of angle points, or a true radius of curves	2°	3°	3°
Take cross-section topography measurements so that variations in ground from a straight line connecting the cross-section points will not exceed	150 mm	300 mm	600 mm
Horizontal and vertical accuracy for cross sections, in millimeters or percentage of horizontal distance measured from traverse line, whichever is greater	30 mm or 0.4%	45 mm or 0.6%	60 mm or 1.0%
Horizontal and vertical accuracy for slope stake, slope stake references, and clearing limits, in millimeters or percentage of horizontal distance measured from centerline or reference stake, whichever is greater:			
Slope reference stakes and slope stakes	30 mm or 0.4%	45 mm or 0.6%	60 mm or 1.0%
Clearing limits	300 mm	300 mm	300 mm

170.07 Reestablishing Preliminary Survey Line. A preliminary survey line has been established on the ground for this project, with initial and specific succeeding survey points referenced. Reestablish missing P-line points necessary to control subsequent construction staking operations to the precision DESIGNATED IN THE SCHEDULE OF ITEMS.

170.08 Establishing Centerline. Determine the direction of centerline (L-line) tangents using coordinates furnished by the CO. Locate at least two points on each tangent to establish its direction. Do not change the location of tangent lines established on the ground.

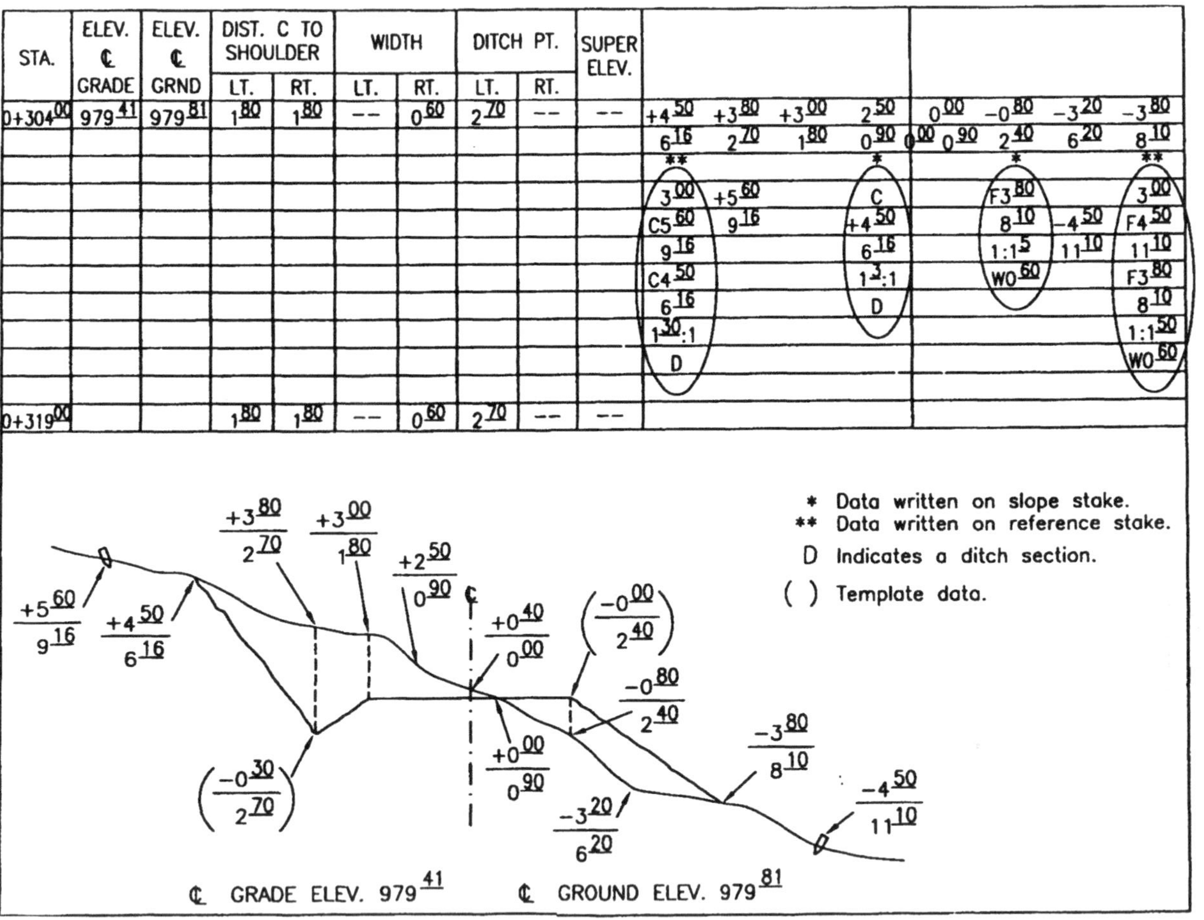

STA.	ELEV. ℄ GRADE	ELEV. ℄ GRND	DIST. ℄ TO SHOULDER LT.	DIST. ℄ TO SHOULDER RT.	WIDTH LT.	WIDTH RT.	DITCH PT. LT.	DITCH PT. RT.	SUPER ELEV.								
0+304.00	979.41	979.81	1.80	1.80	--	0.60	2.70	--	--	+4.50	+3.80	+3.00	2.50	0.00	-0.80	-3.20	-3.80
										6.16	2.70	1.80	0.90	0.00 0.90	2.40	6.20	8.10
										**			*		*		**
										3.00	+5.60		C		F3.80		3.00
										C5.60	9.16		+4.50		8.10	-4.50	F4.50
										9.16			6.16		1:1.5	11.10	11.10
										C4.50			1.3:1		WO.60		F3.80
										6.16			D				8.10
										1.30:1							1:1.50
										D							WO.60
0+319.00			1.80	1.80	--	0.60	2.70	--	--								

Figure 170-1.—Slope stake note entries related to actual ground elevations.

Measure the deflection angle from one tangent to another. When the measured deflection angle differs from the one SHOWN ON THE DRAWINGS, use the measured angle and the curve external (E) SHOWN ON THE DRAWINGS to compute new curve data. Compute the new curve data and note them in the field books, and on the "as-staked" drawings. Establish the new control points (P.I., or P.C.'s & P.T.'s) on the ground using hubs and tacks.

Throughout the project, continuously establish centerline points using horizontal distance measurements, and stake them to the nearest 3 mm for control points, and 30 mm for other points. Mark centerline stakes as shown in figure 170-2. Introduce equations at the P.T. of curves to adjust field stationing to that SHOWN ON THE DRAWINGS or in the staking notes when the difference between designed and located centerline stationing exceeds 1.5 m. Set centerline stakes at even 15-m intervals when practicable, at significant breaks in the ground, at culvert locations, at equation points, or at other locations indicated in the staking notes. Set stakes no more than 15 m apart. Stake all curves of 20 degrees or more every 10 m. Stake all other curves every 15 m.

Where centerline stations fall in an existing trail, obstruction, or roadway, offset the stakes left or right from centerline (perpendicular to tangents and on the radial lines of curves) clear of the trail, obstruction, or roadway, and mark the offset distance on the side facing the centerline. As centerline point, use a 20-penny or larger nail that is flagged and driven at least 25 mm below the road surface.

Clear the survey line to facilitate travel and surveying. Remove clearing slash from the travel or work area. Cut all brush and trees as near to the ground as possible.

170.09 Referencing Centerline. Reference centerline control points, and make them intervisible after clearing is completed to facilitate reestablishment of the centerline. Measure references to the precision of the centerline survey. Establish references consisting of two intersecting lines with an included angle of at least 30°. Place the forward reference a minimum of 8 m outside the clearing limits as computed from the preliminary slope stake printout notes, and place the rear hub or point on each line not less than 10 m beyond the forward hub or point. Mark reference points with hubs and tacks.

170.10 Vertical Control & "L" Profile Levels. Relocate bench marks that were established during the P-line survey that are within the clearing limits to points 6 m or more outside the clearing limits. Determine elevation of relocated bench marks by the precision class specified, as listed in table 170-1. Construct bench marks to be permanent and to allow a level rod to stand vertically and squarely on the mark. Bench marks may be established by driving a 40-penny or larger nail into a notch cut in the base of a tree, by marking a point on a stable rock, or by other approved means. Drive spikes into trees less than 300 mm above the ground. Record location and descriptions of relocated bench marks in the level notes. Set at least two bench marks at each bridge and structural-plate culvert site.

Use appropriate survey equipment between bench marks to determine centerline ground elevations on L-line stations, to the nearest 30 mm, and to verify bench marks.

170.11 Discrepancies. Compare the staked centerline horizontal and vertical alignment with the design data. Refer to the CO any differences found between previously recorded and observed elevations of bench marks, and any differences exceeding 1 degree in angle found between the horizontal alignment data SHOWN ON THE DRAWINGS and the alignment observed on the ground. Report differences in centerline profile elevations exceeding 300 mm at any two or more consecutive points to the CO for evaluation and possible revision. Defer the staking of these areas until these differences are resolved by the CO.

170.12 "L" Topography Cross Sections. Take cross sections at right angles to tangents and normal to curves at every staked point on the "L" profile line. Determine the elevations of significant breaks in topography, breaks in the designed roadway template, and cross-section reference points. Record ground shots for these cross sections in terms of meters plus or minus from ground at centerline, and horizontal distances from centerline. Measure cross sections and record them to the nearest 10 mm in elevation and to the nearest 100 mm in horizontal distance. Ensure that cross sections extend approximately 6 m beyond the designed clearing-and-grubbing limit on cut sections, and approximately 6 m beyond the toe of fill on fill sections.

Identify cross sections at each end with lath marked to show centerline station and the horizontal and vertical distance to the centerline.

Return cross-section data to the CO for recomputation of earthwork quantities and slope stake catchpoint printouts. If the PAY ITEM for earthwork under Section 203 is Staked Quantities, submit the cross-section data to the CO for recomputation of earthwork quantities and slope stake "catchpoints."

Slope stakes established during the "L" topography cross section phase of the work may be subject to relocation to adjust earthwork quantities.

170.13 Slope Stakes, Clearing Limits, & Reference Stakes. Establish slope catchpoints, clearing limits, and slope reference stakes on both sides of the centerline at each "L" station established. Determine the position of these stakes by methods that will produce on the ground the designed template shown in the slope stake survey notes to the precision shown in table 170-2 and DESIGNATED IN THE SCHEDULE OF ITEMS.

Record the cut or fill and horizontal distance to centerline, to bottom of ditch, or to shoulder as designated by the CO on the slope stakes and in the slope stake notes, as shown in figure 170-1.

Set clearing limits on both sides of the centerline at each established "L" station within the tolerance shown in table 170-2. Locate the clearing limit on the ground to the dimensions SHOWN ON THE DRAWINGS and mark with lath, flagging, or other methods approved by the CO. Record the total horizontal distance from the centerline to the clearing limit at each section to the nearest 300 mm in the field book.

Establish slope reference stakes at a minimum horizontal distance of 3 m outside the clearing limits, and record on the stakes the horizontal distance to centerline and the vertical distance to the construction grade at centerline. In addition, record on the reference stake, and in the slope stake book, the offset from the slope stake catchpoint, and slope stake catchpoint information, as shown in figure 170-1.

Ensure that the elevation and location of slope reference stakes comply with the precision class specified in table 170-2.

Reset the slope stake where the difference in reference stake elevation between that established by slope staking and that observed by an elevation survey exceeds the allowed tolerance.

170.14 Monuments of Property Boundaries or Surveys. If property boundary or survey monuments, or survey markers, are found within or adjacent to the construction limits, immediately notify the CO.

170.15 Staking Culverts. Set slope stakes and slope reference stakes at all culvert locations. Set a culvert reference stake and hub on the centerline of the culvert 3 m from each end, or beyond the clearing limit, whichever is greater. Record the following on these stakes:

(a) Diameter, actual field measured length, and type of culvert.

(b) The vertical and horizontal distance from hubs to the invert at the ends of the culvert.

(c) Stationing of centerline point.

When SHOWN ON THE DRAWINGS, stake headwalls for culverts by setting a hub with a guard stake on each side of the culvert on line with the face of the headwall. (Perform this work after clearing is completed.)

170.16 Staking Drain Dips. Establish slope stakes and slope reference stakes on the projected centerline of the bottom of the dip at all drain dip locations, as SHOWN ON THE DRAWINGS.

170.17 Staking Major Structures. Meet the following requirements for staking bridges, cattleguards, and other structures:

(a) Bridges. Designate bridge locations on the ground by establishing reference points for the bridge centerline and the transverse centerline of each pier or abutment. Use hubs and tacks that are set on line beyond the construction limits as reference points, and mark them to identify the point and distance to the point referenced. Set at least one bench mark on each side of the stream beyond construction limits, but close enough to the bridge site to allow direct leveling between the bench marks and the bridge without an intermediate setup. Record all of the above information in a separate book that includes a sketch showing the stream, bridge, and location of all construction stakes set. Perform staking to the accuracy standards shown in table 170-1.

(b) Cattleguards. Stake cattleguards as SHOWN ON THE DRAWINGS.

(c) Other Structures. When required, stake other structures as described in SPECIAL PROJECT SPECIFICATIONS and/or as SHOWN ON THE DRAWINGS.

170.18 Grade Finishing Stakes. Set finishing stakes when DESIGNATED IN THE SCHEDULE OF ITEMS. Use blue tops for subgrade finishing stakes, and red tops for base course finishing stakes.

Ensure that stakes are nominal 25 × 25-mm hubs of sufficient length to provide a solid set.

Place finishing stakes on the staked cross section and road template line. Set a stake at each shoulder and at centerline. Set additional stakes when SHOWN ON THE DRAWINGS.

Set finishing stakes when subgrade is within 150 mm, or when base course is within 60 mm of final grade. Set stakes to the nearest 6 mm of the measured grade line.

170.19 Marking Stakes. Legibly mark all stakes in the format that is shown in figure 170-2, or as SHOWN ON THE DRAWINGS, with a stake pencil that leaves an imprint or with waterproof ink. Mark in conformance with the following nomenclature:

PI	Point of intersection of tangents
PC	Point of curvature
POC	Point on curve

PT	Point of tangency
POT	Point on tangent
RP	Reference point
P	P-line (preliminary location line)
L	L-line (final location line)
BM	Bench mark
TBM	Temporary bench mark
BT	Begin taper (any)
ET	End taper (any)
BFTO	Begin full turnout
EFTO	End full turnout
BFEW	Begin full extra widening
EFEW	End full extra widening
DD	Drain dip
C	Cut
F	Fill
℄	Centerline
D	Ditch
W	Width
CW	Curve widening
FW	Fill widening
H	Horizontal
M	Metric
SE	Superelevation
V	Vertical

170.20 Stake Approval & Maintenance. Do not begin construction work within a roadway segment until the stakes, marks, and controls established have been approved in writing by the CO. The minimum segment for approval shall be 1 km or the length of the project, whichever is less.

Approval of construction staking shall not relieve the Contractor of responsibility for maintaining the survey work and for correcting errors, whether the errors are discovered during the actual survey work or in subsequent phases of the project. Stakes within the roadway need not be maintained after clearing operations have started.

Measurement

170.21 Method. Use the method of measurement that is DESIGNATED IN THE SCHEDULE OF ITEMS.

Reestablishing P-line includes all work needed to replace missing portions of the P-line that are necessary for the determination of L-line tangents. When DESIGNATED IN THE SCHEDULE OF ITEMS, the quantity shall be the number of

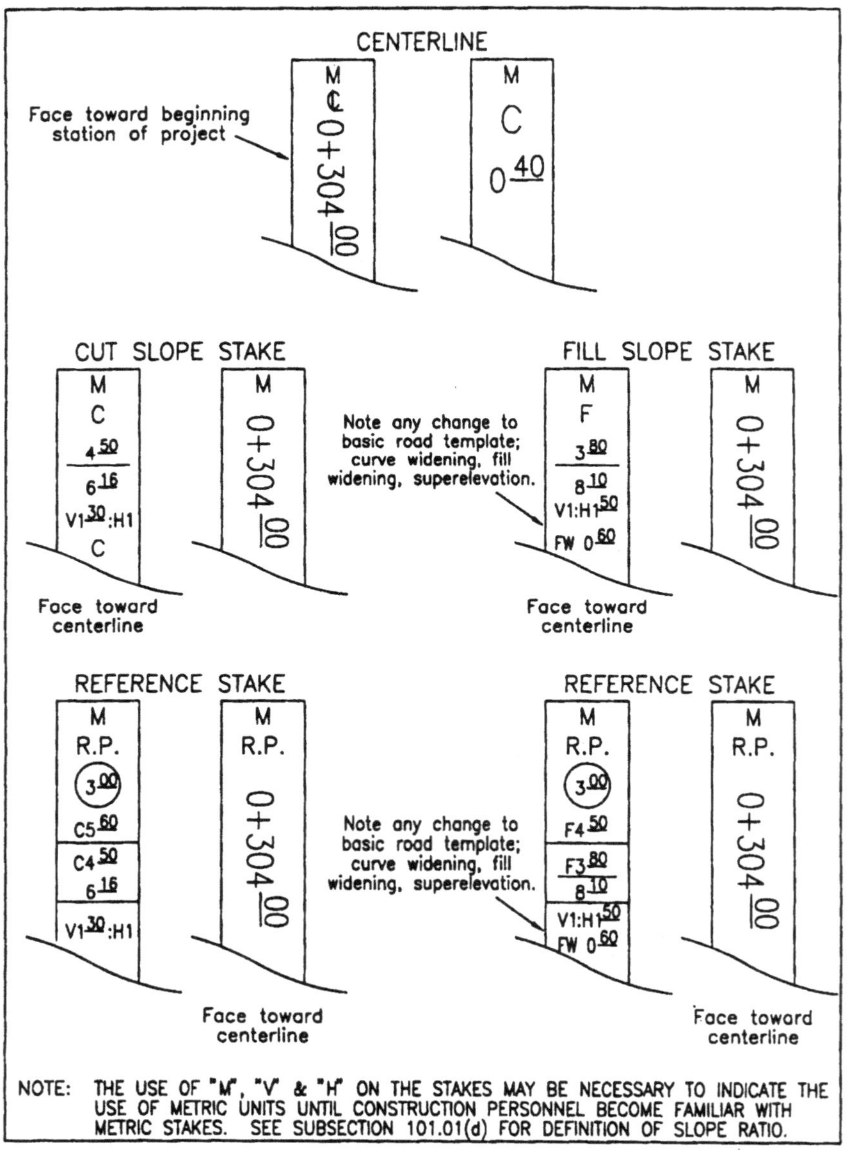

Figure 170-2.—Construction stakes.

kilometers of P-line reestablished, as measured along centerline to the nearest 5 m. When the length of P-line to be replaced does not exceed 10 percent of the measured length of the L-line, reestablishing P-line shall be considered incidental to establishing centerline, and no separate payment will be made.

Establishing centerline includes all work necessary to establish and reference the centerline, establish vertical controls, determine the centerline profile elevations, and cross-section the original ground from the centerline datum established by this survey. The quantity shall be the number of kilometers of centerline completed and accepted, as measured along centerline to the nearest 5 m.

Slope staking includes all work necessary to establish slope stakes, clearing limits, and reference stakes from a previously established centerline. The quantity shall be the number of kilometers of previously established centerline completed and accepted, as measured along centerline to the nearest 5 m.

Finish staking includes all work necessary to reestablish the centerline to control placement of finish stakes and set the finish stakes. The quantity shall be the number of kilometers of previously established centerline that were completed and accepted, as measured along centerline to the nearest 5 m.

Staking major structures includes all work necessary to establish lines and grades for the construction of the structure(s). The quantity shall be the actual number of structures completed and accepted of the type DESIGNATED IN THE SCHEDULE OF ITEMS.

Payment

170.22 Basis. The accepted quantities will be paid for at the contract unit price for each PAY ITEM DESIGNATED IN THE SCHEDULE OF ITEMS.

Payment will be made under:

Pay Item	Pay Unit
170 (01) Reestablish P-line, precision __________	Kilometer
170 (02) Establish centerline, precision __________	Kilometer
170 (03) Slope staking, precision __________	Kilometer
170 (04) Finish staking, subgrade, precision _________	Kilometer
170 (05) Finish staking, base course, precision _______	Kilometer
170 (06) Staking major structure(s), type _____ precision _____	Each

Section 171—Construction Staking, Offset L-Line

Description

171.01 Work. Complete the construction staking of a road project in accordance with the drawings and specifications. Furnish all labor, equipment, instruments, materials and transportation, and other incidentals necessary to complete the construction staking in accordance with these specifications and acceptable engineering practice.

Conduct construction staking under the direction of a licensed professional engineer or land surveyor who is closely associated and familiar with the construction staking. Periodic visits to the project site are required.

Materials

171.02 Stakes. Provide stakes and hubs that have the nominal dimensions SHOWN ON THE DRAWINGS or stated in the SPECIAL PROJECT SPECIFICATIONS. Ensure that identification stakes and hubs are of sufficient length to provide a solid set in the ground and space for marking above ground when applicable. Other dimensions and materials may be used if approved in writing by the CO. Paint the top 50 mm of all stakes and lath, or mark them with plastic flagging. Use colors for paint or flagging as SHOWN ON THE DRAWINGS or as stated in the SPECIAL PROJECT SPECIFICATIONS.

171.03 Survey Note Paper & Books. Furnish field notebooks or note papers. Use moisture-resistant paper for survey notes. Keep notes in books with covers that will protect the contents and retain the pages in numerical sequence during field use.

171.04 Government-Furnished Documents. Drawings, P-line survey notes, P-line to L-line offset data, construction staking notes, and the projected location of catch points will be furnished by the Government. Return one set of "as-staked" drawings and all documents to the CO.

Survey Requirements

171.05 Precision. Accuracy and precision requirements are contained in tables 171-1 and 171-2. Ensure that all work performed under this specification meets the requirements of the survey precision DESIGNATED IN THE SCHEDULE OF ITEMS or stated in the SPECIAL PROJECT SPECIFICATIONS.

171.06 Survey Notes. All notes will become the property of the Forest Service. Format slope stake notes in conformance with the format shown in figure 171-1.

Table 171-1.—Accuracy requirements for reestablishing P-line, traverse, and elevations.

Precision Class	Minimum Position Closure	Angular Accuracy (±)	L-Line Tangent Control Points[a] (±)	Vertical Closure[b] (±)
C	1/1,000	1 set, direct/reverse; 1′ rejection limit	60 mm	500 mm/km
D	1/300	Foresight and backsight; 15′ rejection limit[c]	120 mm	1,000 mm/km
E	1/100	Foresight and backsight; 30′ rejection limit[c]	240 mm	1,000 mm/km

a. Accuracy of offset measurement.
b. Determine vertical closures at intervals not to exceed 1 km, as measured along centerline.
c. Magnetic attraction will require a deflection angle traverse.

The sketch in the lower half of figure 171-1 is for information only and is not required as part of actual slope stake note format. Other formats may be used if approved by the CO.

Print manually recorded survey notes in characters at least 4 mm high, and make them legible at a distance of 750 mm. Delete errors by lining out. At the beginning of each day's work, record date, crew names and positions, instrumentation, and weather in the notes.

Consecutively number electronically recorded survey notes, and use headings to identify the contents. Support and accompany the notes with a bound Day Book that records the project name and, for each day, the date; crew names and positions; instrumentation; weather; type of survey; stationing of sections between which the survey was performed; and survey data or sketches that cannot be electronically recorded. Ensure that the party chief signs or initials the Day Book immediately after the last entry for each day's work. When using electronic data collectors, have them approved by the CO, or provide suitable hardcopy notes.

Ensure that the party chief signs or initials each page of electronically recorded notes and the final page of bound notes immediately after the last entry for each day's work.

Table 171-2.—Cross section and slope stake precision.

Item	Precision C (±)	D (±)	E (±)
Allowable deviation of cross-section line projection from a true perpendicular to tangents, a true bisector of angle points, or a true radius of curves	3°	5°	5°
Take cross-section topography measurements so that variations in ground from a straight line connecting the cross section points will not exceed	300 mm	450 mm	750 mm
Staking by computed method:			
Horizontal and vertical accuracy for cross sections, in millimeters or percentage of horizontal distance measured from traverse line, whichever is greater	45 mm or 0.6%	60 mm or 0.8%	90 mm or 1.0%
Horizontal and vertical accuracy for slope stake, slope stake references, and clearing limits, in millimeters or percentage of horizontal distance measured from centerline or reference stake, whichever is greater—			
Slope reference stakes and slope stakes	45 mm or 0.6%	60 mm or 0.8%	75 mm or 1.5%
Clearing limits	300 mm	450 mm	600 mm
Staking by catchpoint measurement method:			
Accuracy for setting slope catchpoints, reference points, and clearing limits, in millimeters or percentage of slope distance, measured from centerline, whichever is greater—			
Slope catchpoint stakes and reference points	45 mm or 0.5%	60 mm or 0.7%	90 mm or 2.0%
Clearing limits	300 mm	450 mm	600 mm

171.07 Reestablishing Preliminary Survey Line. A preliminary survey line has been established on the ground for this project, with initial and specific succeeding survey points referenced. Reestablish missing P-line points as necessary to control subsequent construction staking operations to the precision DESIGNATED IN THE SCHEDULE OF ITEMS, SHOWN ON THE DRAWINGS, or stated in the SPECIAL PROJECT SPECIFICATIONS.

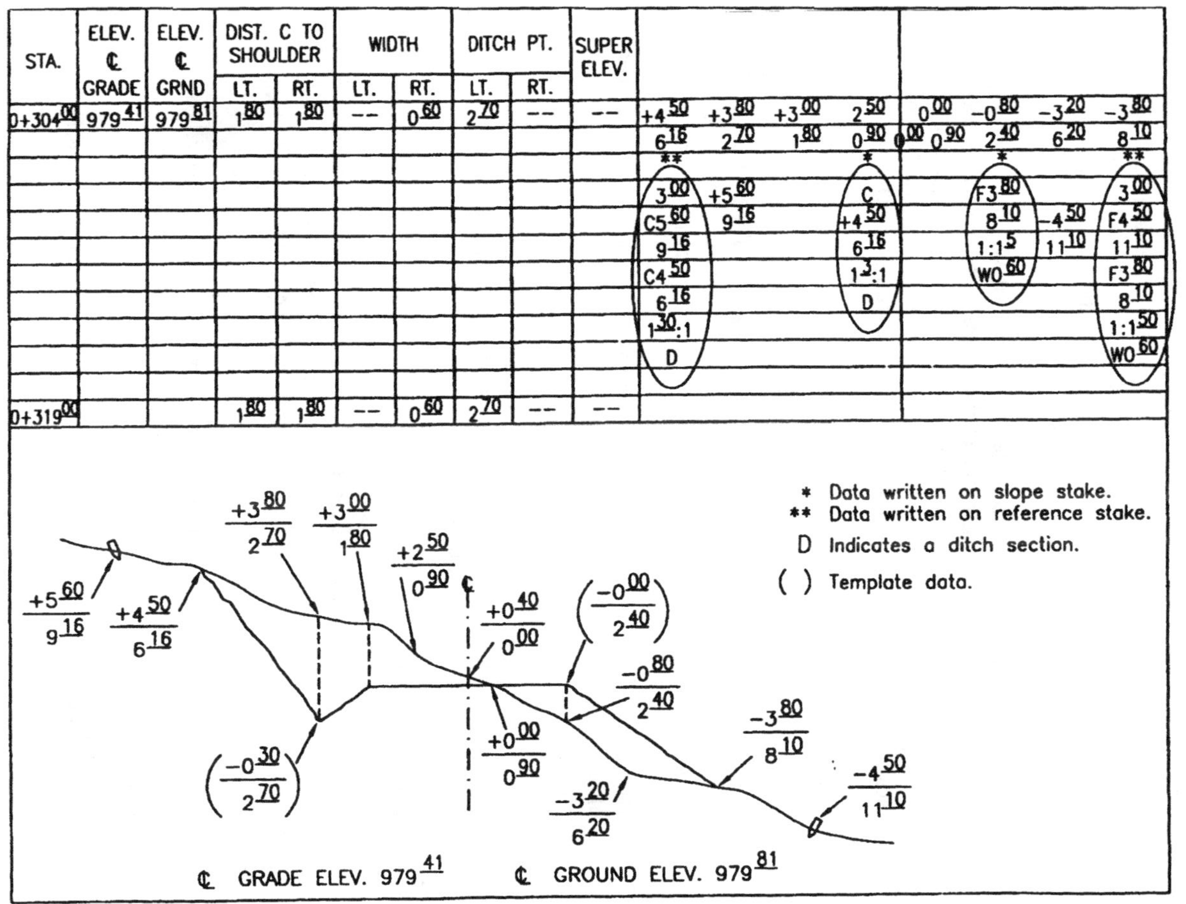

Figure 171-1.—Slope stake note entries related to actual ground elevations.

171.08 Establishing Centerline. Establish the position of the centerline (L-line) by measuring right or left from the preliminary survey line (P-line) the horizontal distance shown in the "offset listing" furnished by the Forest Service. Adjust the centerline established in alignment only to correct misalignment created by measured offsets along skewed sections. Ensure that the station of the centerline point is that listed in the P-line to L-line offset data.

Set additional intermediate centerline stakes at locations SHOWN ON THE DRAWINGS and listed in the construction staking notes, as needed to establish control for beginning and ending of extra widening and turnout tapers; for the beginning and end of full-width extra widening and turnouts; for crest and sag of drainage dips; for culvert catch basins; and for turnarounds. Establish the position and ground elevation of these additional stakes by measuring from the nearest established centerline stake.

Where centerline stations fall in an existing trail, roadway, or obstruction, offset the stakes right or left from centerline (perpendicular to tangents and on the bisector of angle points) and the distance marked on the side of the stake facing centerline. Drive suitable markers on the centerline to denote the actual centerline point.

Clear the survey line to facilitate surveying. Remove clearing slash from the travel or work area. Cut brush and trees as near to the ground as possible.

171.09 Vertical Control & "L" Profile Levels. Relocate bench marks that were established during the P-line survey within the clearing limits to points 6 m or more outside the clearing limits. Determine elevation of relocated bench marks by the precision class specified, as listed in table 171-1.

Construct bench marks to be permanent and to allow a level rod to stand vertically and squarely on the mark. Bench marks may be established by driving a 40-penny or larger nail into a notch cut in the base of a tree, by marking a point on a stable rock, or by other approved means. Drive spikes into trees less than 300 mm above the ground. Record location and descriptions of relocated bench marks in the level notes.

Use appropriate survey equipment between bench marks to determine centerline ground elevations on L-line stations, to the nearest 30 mm, and to verify bench marks.

For precision C and D, determine elevation of centerline stations by leveling from the listed elevation of the P-line station from which they were offset.

171.10 Discrepancies. Refer to the CO any differences that exceed 5° of angle in horizontal alignment of curves with less than 30 m radius found between the data SHOWN ON THE DRAWINGS and those observed on the ground. Compare the found centerline cut and fill depth with design data. Report differences in centerline profile elevations exceeding 300 mm at any two or more consecutive points to the CO, who will determine whether revision is needed. Defer staking of these areas until these differences are resolved by the CO.

Check horizontal distances on the centerline by measuring between stakes on L-line. If discrepancies in actual distances measured are greater than 1.5 m, report them to the CO for possible corrective action.

171.11 Slope Stakes, Clearing Limits, & Reference Points. Establish slope stakes, clearing limits, and slope stake references at each side centerline station, as SHOWN ON THE DRAWINGS; at each centerline station; on a line at right angles to tangents; and on the radial lines of curves. Use a method to establish the slope stake catchpoint that conforms to the METHOD described below, as DESIGNATED IN THE SCHEDULE OF ITEMS.

(a) Method I—Computed Method. Locate slope stake catchpoints by using the template information shown in the slope stake notes to calculate the actual location of the catchpoint. The slope stake "catchpoint distance" shown in the stake notes may be used as a trial location to initiate slope staking.

Where SHOWN ON THE DRAWINGS, measure topography of the cross section at each centerline stake. Record the horizontal and vertical distance to the centerline ground for each break in ground slope between the centerline and the reference point(s).

(b) Method II—Slope Distance Measurement Method. Locate slope stake catchpoints by measuring the slope distance shown in the slope stake notes.

(c) Method III—Catchpoint Measurement Method. Locate slope stake catchpoints and clearing limits by measuring the "catchpoint distance" shown in the slope stake notes.

(d) Method IV—Reestablishing Slope Stakes, Clearing Limits, & References. Slope stakes and marks previously established for this project have either been destroyed or have become unreadable. Reestablish the missing stakes and marks from the original slope stake notes, as described in the SPECIAL PROJECT SPECIFICATIONS.

Mark clearing limits with colored plastic ribbon or tags on trees to be left standing, or on lath.

Place a reference stake or tag for each slope stake 3 m outside the clearing limit as SHOWN ON THE DRAWINGS. Remove slope stakes to the reference stake prior to clearing, and replace them after clearing is completed.

Set slope stakes temporarily at the slope stake catchpoint location established under method I, II, or III above for use in determining clearing limits and slope stake references.

Locate clearing limits at the distance SHOWN ON THE DRAWINGS from either the slope stake catchpoint or road shoulder, whichever is greater. Mark clearing limits with plastic flagging or tags on trees to be left standing, or on lath.

After clearing limits and references are established, move the slope stakes to the reference stake, and replace them at the catchpoint after clearing is completed.

171.12 Resetting Slope Stakes. Reestablish slope stakes after clearing and grubbing is completed and before excavation is started. Recheck the original catchpoint location from the reference stake to determine whether revisions are needed because of ground disturbance; recheck slope stakes and reset them to the original precision requirements.

171.13 Monuments of Property Boundaries or Surveys of Other Agencies. If property boundary or survey monuments or survey markers of other agencies are found within or adjacent to the construction limits, immediately notify the CO.

171.14 Staking Culverts. Set slope stakes and slope stake references at all culvert locations. Set a culvert reference stake on the centerline of the culvert 3 m from each end or beyond the clearing limit, whichever is greater. Record the following on these stakes:

(a) Diameter, actual field measured length, and type of culvert.

(b) The vertical and horizontal distance from the reference stake to the invert at the ends of the culvert.

(c) Stationing of centerline point.

171.15 Staking Drain Dips. Establish slope stakes and slope stake references on the projected centerline of the bottom of the dip at all drain dip locations, as SHOWN ON THE DRAWINGS.

171.16 Staking Structures. Stake cattleguards and other structures as described in the SPECIAL PROJECT SPECIFICATIONS or as SHOWN ON THE DRAWINGS.

171.17 Marking Stakes. Legibly mark all stakes in the format shown in figure 171-2, or as SHOWN ON THE DRAWINGS, with a stake pencil that leaves an imprint or with waterproof ink. Mark in conformance with the following nomenclature:

RP	Reference point
P	P-line (preliminary location line)
L	L-line (final location line)
BM	Bench mark
TBM	Temporary bench mark
BT	Begin taper (any)
ET	End taper (any)
BFTO	Begin full turnout
EFTO	End full turnout
BFEW	Begin full extra widening
EFEW	End full extra widening
DD	Drain dip
C	Cut
F	Fill
C	Centerline
D	Ditch
W	Width
CW	Curve widening
FW	Fill widening
H	Horizontal
M	Metric
SE	Superelevation
V	Vertical

171.18 Stake Approval & Maintenance. Do not begin construction work within a roadway segment until the stakes, marks, and controls established have been approved in writing by the CO. The minimum segment for approval shall be 1 km or the length of the project, whichever is less.

Approval of construction staking shall not relieve the Contractor of responsibility for maintaining the survey work until construction has been completed, and for correcting errors, whether the errors are discovered during the performance of survey or in subsequent phases of the project. Centerline stakes need not be maintained after clearing operations have started.

Measurement

171.19 Method. Use the method of measurement that is DESIGNATED IN THE SCHEDULE OF ITEMS.

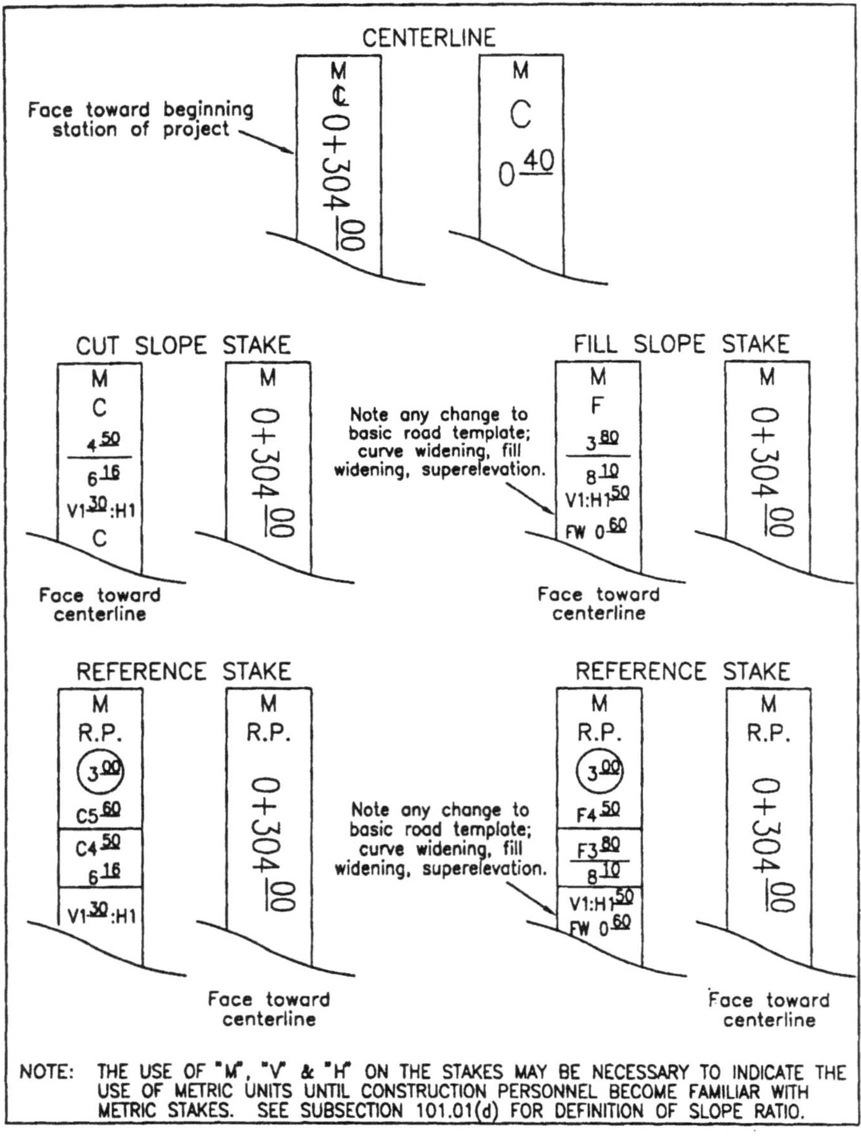

Figure 171-2.—Construction stakes.

Reestablishing P-line includes all work needed to replace missing portions of the P-line that are necessary for the determination of L-line tangents. When listed in the SCHEDULE OF ITEMS, the quantity shall be the number of kilometers of P-line reestablished, as measured along centerline to the nearest 5 m. When the length of P-line to be replaced does not exceed 10 percent of the length of the P-line, reestablishing P-line shall be considered incidental to construction staking, and no separate payment will be made.

Construction staking includes all work necessary to establish the project centerline and to establish slope stakes, clearing limits, and reference stakes in accordance with the METHOD DESIGNATED IN THE SCHEDULE OF ITEMS. The quantity shall be the number of kilometers of construction staking completed and accepted, to be measured along the centerline to the nearest 10 m.

Payment

171.20 Basis. The accepted quantities will be paid for at the contract unit price for each PAY ITEM DESIGNATED IN THE SCHEDULE OF ITEMS.

Payment will be made under:

Pay Item	Pay Unit
171 (01) Reestablish P-line, precision ________________	Kilometer
171 (02) Construction staking, precision ________________, method ________________	Kilometer
171 (03) Staking structures, precision _______, method ______	Each

Section 173—Construction Staking, Location Line

Description

173.01 Work. Complete the construction staking of a road project that will be constructed predominantly by sidecast or end-dump construction methods. Establish clearing limits and staking of drainage structures. Establish slope stakes for cuts and fills when SHOWN ON THE DRAWINGS. The survey required by this specification may be used in conjunction with other specifications in specific areas SHOWN ON THE DRAWINGS in order to produce staking adequate for construction needs.

Furnish all labor, equipment, instruments, materials, transportation, and other incidentals necessary to complete the construction staking in accordance with these specifications and acceptable engineering practice.

Perform construction staking under the direction of a licensed professional engineer or land surveyor who is closely associated and familiar with the construction staking. Periodic visits to the project site are required.

Materials

173.02 Stakes. Provide stakes that have the nominal dimensions SHOWN ON THE DRAWINGS or stated in the SPECIAL PROJECT SPECIFICATIONS. Ensure that identification stakes are of sufficient length to provide a solid set in the ground and space for marking above ground when applicable. Other dimensions and materials may be used if approved in writing by the CO. Paint the top 50 mm of all stakes and lath, or mark them with plastic flagging. Use colors for paint or flagging as SHOWN ON THE DRAWINGS or as stated in the SPECIAL PROJECT SPECIFICATIONS.

173.03 Survey Note Paper & Books. Use moisture-resistant paper for survey notes. Keep notes in books with covers that will protect the contents and retain the pages in numerical sequence during field use.

173.04 Government-Furnished Documents. Drawings, P-line survey notes, and, where applicable, construction staking notes will be furnished by the Government. Return all documents to the CO.

Survey Requirements

173.05 Precision. Use a woven or fiberglass/plastic tape in good condition and a hand level or abney, or other instruments capable of attaining the same accuracy of measurement.

173.06 Survey Notes. Neatly record survey notes in a standard format approved by the CO. Make lettering at least 4 mm high and legible at a distance of 750 mm from the eye. Delete errors by lining out. Certify all field notes as to originality. All field notes will become the property of the Forest Service.

173.07 Location Survey Line. A location line for this project has been established on the ground.

173.08 Clearing Limits. Establish clearing limits on each side of the location line by measuring the horizontal or slope distances as shown in the stake notes. Mark the clearing limits with flagging or tags on trees to be left standing, or on lath. Make markings intervisible, and in no case more than 30 m apart. Use colors for flagging or tags as SHOWN ON THE DRAWINGS or as stated in the SPECIAL PROJECT SPECIFICATIONS.

After establishing clearing limits, move the location line stake outside the clearing limits for station identification purposes, and mark it with horizontal distance to location line.

173.09 Slope Stakes & References. When SHOWN ON THE DRAWINGS, locate slope stakes on designated portions of the road. Locate the slope stake catchpoints and use them to establish clearing limits and slope stake references. Use a method to establish the slope stake catchpoint that conforms to the METHOD described below, as DESIGNATED IN THE SCHEDULE OF ITEMS.

(a) Method I—Computed Method. Establish slope stake catchpoints by using the template information shown in the slope stake notes to calculate the actual location of the catchpoint. The slope stake "catchpoint distance" shown in the stake notes may be used as a trial location to initiate slope staking.

(b) Method II—Catchpoint Measurement Method. Determine the location of slope stake catchpoints by measuring the catchpoint distances shown in the stake notes.

Place slope stakes as SHOWN ON THE DRAWINGS. Ensure that slope stakes indicate the station, the amount of cut or fill in meters, the horizontal distance to centerline in meters, and the cutslope or fillslope ratios.

Place slope reference stakes a minimum of 3 m outside the clearing line, and marked with the offset distance to the slope stake.

Prior to clearing and grubbing operations, move the slope stake outside the clearing limit to the slope reference stake. After clearing and grubbing and before excavation, reset the slope stakes in their original position.

173.10 Monuments of Property Boundaries or Surveys. If property boundary or survey monuments, or survey markers, are found within or adjacent to the construction limits, immediately notify the CO.

173.11 Staking Culverts. Set culvert reference stakes at all culvert locations. Set a culvert reference stake on the centerline of the culvert 3 m from each end or beyond the clearing limit, whichever is greater. Record the following on these stakes:

(a) Diameter, design length, and type of culvert.

(b) The horizontal distance from the reference stake to the invert at the ends of the culvert.

(c) Stationing of centerline point.

173.12 Staking Drain Dips. Establish reference stakes outside the clearing limits on the projected centerline of the bottom of the drain dip at all drain dip locations, as SHOWN ON THE DRAWINGS.

173.13 Staking Cattleguards. Stake cattleguards as SHOWN ON THE DRAWINGS.

173.14 Marking Stakes. Mark all stakes with a stake pencil that leaves an imprint, or with waterproof ink, in a format approved by the CO.

173.15 Approval & Maintenance. Do not begin construction work within a roadway segment until the stakes and marks established are approved in writing by the CO. The minimum segment for approval shall be 1 km or the length of the project, whichever is less.

Approval of construction marking or staking shall not relieve the Contractor of responsibility for maintaining the survey work until construction has been completed and accepted, and for correcting errors, whether the errors are discovered during the performance of the survey or during subsequent phases of the project. Location line stakes need not be maintained after clearing operations have started.

Measurement

173.16 Method. Use the method of measurement that is DESIGNATED IN THE SCHEDULE OF ITEMS.

Construction staking includes all work necessary to establish slope stakes, clearing limits, and slope stake references. The quantity shall be the number of kilometers of construction staking completed and accepted, as measured along centerline to the nearest 10 m.

Payment

173.17 Basis. The accepted quantities will be paid for at the contract unit price for each PAY ITEM DESIGNATED IN THE SCHEDULE OF ITEMS.

Payment will be made under:

Pay Item	Pay Unit
173 (01) Establish clearing limits ___________	Kilometer
173 (02) Establish slope stakes, method _________	Kilometer

DIVISION 200
Earthwork

Section 201—Clearing & Grubbing

Description

201.01 Work. Clear and grub; treat timber, construction slash, and debris; and preserve vegetation and objects designated to remain free from injury or defacement.

Construction

201.02 Clearing & Grubbing. Clear and grub in accordance with the following:

(a) Exceptions. Within the clearing limits, clear and treat trees, debris, stumps, roots, and other protruding vegetative material not designated to remain, except the following:

(1) Undisturbed stumps outside the roadway or in embankment areas, provided they do not extend more than 300 mm above the original ground (measured from the uphill side); they are no closer than 600 mm to the finished subgrade or 300 mm to any slope surface, or as otherwise SHOWN ON THE DRAWINGS; and they do not interfere with the placement or compaction of embankments.

(2) Material in channel changes, rock sections, and ditches that is below the depth of the proposed excavation.

(3) Uncut vegetation less than 1 m in height and less than 75 mm in diameter, that is within the clearing limits but beyond the roadway and not in a decking area, and that does not interfere with sight distance along the road.

(b) Performance. Grub all roots over 75 mm in diameter within the roadbed area to a minimum depth of 150 mm below subgrade. Cut flush with the excavated road surface all roots over 75 mm in diameter that protrude from the excavated slope.

Clear slash treatment areas and treat debris in accordance with Subsections 201.02(a) and 201.05.

Clear decking areas and treat debris in accordance with Subsection 201.05.

Unless shown otherwise in the SPECIAL PROJECT SPECIFICATIONS, fell trees into the area being cleared when ground conditions, tree lean, and shape of clearing permit.

Use controlled felling to ensure the direction of fall to prevent damage to property, structures, trees designated to remain, and traffic.

Dead trees over 150 mm in diameter measured 300 mm above the ground that lean toward the road and are sufficiently tall to reach the roadbed are designated for cutting. Fell hazard trees or unstable live trees that are tall enough to reach the roadbed, when marked, before felling timber in the immediate clearing vicinity. Maximum stump height is 300 mm or one-third of the stump diameter, whichever is higher, measured on the side adjacent to the highest ground. Leave trees felled outside the clearing limits in place, and treat them no further unless otherwise SHOWN ON THE DRAWINGS.

Trim branches on remaining trees or shrubs to give a clear height of 5 m above the roadbed, unless otherwise SHOWN ON THE DRAWINGS. Trim tree limbs as near flush with the trunk as practicable.

201.03 Utilization of Timber. Merchantable timber is timber that meets Utilization Standards in the SPECIAL PROJECT SPECIFICATIONS. Conform logging methods and utilization to the following:

(a) Felling & Bucking. Fell trees to minimize damage to merchantable timber and to remaining trees located outside of clearing limits. Fell trees with saws or shears unless shown otherwise in the SPECIAL PROJECT SPECIFICATIONS.
Buck logs to permit removal of all minimum pieces set forth in the SPECIAL PROJECT SPECIFICATIONS.

(b) Utilization & Removal of Timber. Remove or treat trees that equal or exceed the diameters and minimum lengths listed in the SPECIAL PROJECT SPECIFICATIONS, and that contain one minimum piece, using one of the following methods, as DESIGNATED IN THE SCHEDULE OF ITEMS:

(1) Dispose of merchantable timber designated for removal in accordance with the B(BT) provisions of the Timber Sale Contract.

(2) Limb and deck logs that meet Utilization Standards at locations approved by the CO or SHOWN ON THE DRAWINGS. Deck logs such that logs are piled parallel one to the other; can reasonably be removed by standard log-loading equipment; will not damage standing trees; and will not roll. Log decks are to be free of brush and soil.

(3) Remove from Government land merchantable timber designated for removal, without charge to the Government. This timber becomes the property of the Contractor, but may not be exported from the United States or used as substitution (as defined in 23 CFR 223.10) for timber from private lands exported by the Contractor or an affiliate, directly or indirectly.

(4) Dispose of unmerchantable timber in accordance with Subsection 201.05 by the treatment methods SHOWN ON THE DRAWINGS and DESIGNATED IN THE SCHEDULE OF ITEMS.

201.04 Pioneer Roads. During pioneering operations, prevent undercutting of the final excavation slope. Avoid any restriction of drainages while pioneering the road. Keep all materials within the roadway limits unless otherwise SHOWN ON THE DRAWINGS.

201.05 Slash Treatment. Use or treat construction slash larger than 75 mm in diameter and 1 m in length by one or more of the following methods, as DESIGNATED IN THE SCHEDULE OF ITEMS:

(1) Windrowing construction slash.

(2) Windrowing large material.

(3) Windrowing and covering.

(4) Scattering.

(5) Burying.

(6) Chipping or grinding.

(7) Piling and burning.

(8) Decking unmerchantable material.

(9) Placement in cutting units.

(10) Removal.

(11) Piling.

(12) Placing slash on embankment slopes.

(13) Debris mat.

Pieces of wood less than 75 mm in diameter and 1 m in length may be scattered within the clearing limits.

(a) All Methods. Construction slash placement will not be allowed in lakes, meadows, streams, or streambeds. Immediately remove construction slash that interferes with drainage structures.

Fell and dispose of trees that are scorched or damaged beyond recovery, and adjacent to the clearing limits, in accordance with Subsection 201.03; or treat these trees as construction slash.

(b) Specific Methods. When using one or more of the following slash treatment methods, meet requirements specified below:

(1) Windrowing Construction Slash. Windrow according to the following requirements unless otherwise specified in the SPECIAL PROJECT SPECIFICATIONS. Clear areas to accommodate the windrow slash. Place construction slash outside the roadway in neat, compacted windrows laid approximately parallel to and along the toeline of embankment slopes. Do not permit the top of the windrows to extend above the top of subgrade. Use construction equipment to matt down all material in a windrow to form a compact and uniform pile. Construct breaks of at least 5 m at least every 60 m in a windrow. Do not place windrows against trees. A pioneer road may be constructed to provide an area for placement of windrows, provided the excavated material is kept within the clearing limits and does not adversely affect the road construction.

(2) Windrowing Large Material. Windrow construction slash that is 250 mm or more in diameter at the small end, and 2 m or more in length, as specified in Subsection 201.05(b)(1). Treat smaller material by one or more of the other included methods for slash treatment.

(3) Windrowing & Covering. Place and compact construction slash as specified in Subsection 201.05(b)(1), and cover with at least 150 mm of rock and soil to form a smooth and uniform windrow.

(4) Scattering. Scatter according to the following requirements unless otherwise specified in the SPECIAL PROJECT SPECIFICATIONS. Scatter construction slash outside the clearing limits without damaging trees. Limb all logs. Place logs and stumps away from trees, positioned so they will not roll, and are not on top of one another. Limb and scatter other construction slash to reduce slash concentrations.

(5) Burying. Bury construction slash at the locations SHOWN ON THE DRAWINGS and designated on the ground. Mat construction slash down in layers, and cover it with at least 600 mm of rock and soil. Smooth and slope the final surface to drain.

(6) Chipping or Grinding. Process construction slash that is up to at least 100 mm in diameter and longer than 1 m through a chipping machine or machine designed and operated to grind slash and stumps into pieces, such as a tub grinder. Deposit chips or ground woody material on embankment slopes or outside the roadway to a loose depth not exceeding 150 mm. Minor amounts of chips or ground woody material may be permitted within the roadway if they are thoroughly mixed with soil and do not form a layer.

(7) Piling & Burning. Deposit construction slash in areas SHOWN ON THE DRAWINGS and designated on the ground. Construct piles so that burning does not damage standing trees. If burning is incomplete, repile and burn the slash remaining until pieces are reduced to less than 75 mm in diameter and 1 m in length. Scatter the remaining pieces.

(8) Decking Unmerchantable Material. Deck logs that do not meet Utilization Standards specified in Subsection 201.03, and other material that exceeds the diameter and length shown in the SPECIAL PROJECT SPECIFICATIONS in areas SHOWN ON THE DRAWINGS. Other locations may be approved by the CO.

Cut material into lengths not exceeding 9.7 m, and remove all limbs. Decks are to be stable and free of brush and soil. Treat other material according to slash treatment methods SHOWN ON THE DRAWINGS and in the SCHEDULE OF ITEMS.

(9) Placement in Cutting Units. Place construction slash from within cutting units and the adjacent 60 m with cutting unit logging slash. Place construction slash at least 15 m inside the cutting unit boundary such that it will not inhibit logging of the unit and may be treated by the prescribed logging slash treatment method.

(10) Removal. Remove or haul construction slash to locations SHOWN ON THE DRAWINGS or designated on the ground.

(11) Piling. Pile construction slash in areas SHOWN ON THE DRAWINGS or designated on the ground. Place and construct piles so future burning will not damage remaining trees. Keep piles reasonably free of dirt from stumps. Cut unmerchantable logs into lengths of less than 6 m prior to placement in the pile.

(12) Placing Slash on Embankment Slopes. Place construction slash on completed embankment slopes to reduce soil erosion where SHOWN ON THE DRAWINGS. Place construction slash as flat as practicable on the completed slope. Place slash from the toe of the embankment to a point at least 600 mm below subgrade elevation. Priority for use of available slash is for: (1) through fills; (2) insides of curves; and (3) ditch relief outlets.

(13) Debris Mat. Use tree limbs, tops, cull logs, split stumps, wood chunks, and other debris to form a mat upon which construction equipment is operated. Place stumps upside down and blend stumps into the mat.

Measurement

201.06 Method. Use the method of measurement that is DESIGNATED IN THE SCHEDULE OF ITEMS.

Linear measurements are to be horizontal along the road centerline.

Area quantities are the number of hectares within the clearing limits.

Individual removal of trees is the number of trees of the various size designations removed. Measure tree diameters at a height of 300 mm above ground. Do not count trees less than 150 mm in diameter. Size designations are shown in table 201-1.

Table 201-1.—Size designations for trees removed.

	Size of Least Diameter at Height of 300 mm	
Pay Item Designation	Greater Than	Less Than
Small	150 mm	600 mm
Medium	600 mm	1 m
Large	1 m	–

Payment

201.07 Basis. The accepted quantities will be paid for at the contract unit price for each PAY ITEM DESIGNATED IN THE SCHEDULE OF ITEMS.

Payment will be made under:

Pay Item — Pay Unit

201 (01) Clearing and grubbing, slash treatment methods for tops and limbs ______ ______ ______ ______, logs ______ ______ ______, and stumps ______ ______ ______, utilization of timber __________ .... Hectare

201 (02) Clearing and grubbing, slash treatment methods for tops and limbs ______ ______ ______ ______, logs ______ ______ ______, and stumps ______ ______ ______, utilization of timber ______ .... Kilometer

201 (03) Clearing and grubbing, slash treatment methods for tops and limbs ______ ______ ______ ______, logs ______ ______ ______, and stumps ______ ______ ______, utilization of timber ______ .... Lump Sum

201 (04) Individual removal of trees, small; slash treatment methods for tops and limbs ______ ______ ______ ______ and logs ______ ______ ______ ______, utilization of timber ______ .... Each

201 (05) Individual removal of trees, medium; slash treatment methods for tops and limbs _____ _______ _____ ______ and logs _____ _____ ______ _____, utilization of timber ______ .... Each

201 (06) Individual removal of trees, large; slash treatment methods for tops and limbs _____ _____ ______ ______ and logs _____ ______ ______ _____, utilization of timber ______ .... Each

201 (07) Individual removal of trees, miscellaneous; slash treatment methods for tops and limbs _____ _____ _____ _____ and logs _____ _____ _____ _____, utilization of timber _____ .... Each

201 (08) Individual removal of trees; slash treatment methods for tops and limbs ____ ____ ____ ___ and logs ____ ____ ____ _____, utilization of timber _____ .... Kilometer

201 (09) Individual removal of stumps, slash treatment methods _____ .... Each

Section 202—Removal of Structures & Obstructions

Description

202.01 Work. Salvage, remove, and/or dispose of buildings, fences, structures, pavements, culverts, utilities, curbs, sidewalks, and other obstructions as SHOWN ON THE DRAWINGS. Salvage designated materials and backfill the resulting trenches, holes, pits, or as SHOWN ON THE DRAWINGS.

Construction

202.02 Salvaging Material. Use reasonable care to salvage all material designated to be salvaged. Salvage in readily transportable sections or pieces. Replace or repair all members, pins, nuts, plates, and related hardware damaged, lost, or destroyed during the salvage operations. Wire all loose parts to adjacent members or pack them in sturdy boxes with the contents clearly marked.

Carefully remove culvert, taking precautions to avoid damage. Store culverts to be relaid, when necessary, to prevent loss or damage before relaying. Replace without additional compensation all sections lost from storage or damaged by use of improper methods.

Matchmark members of salvaged structures. Furnish the CO with one set of drawings identifying the members and their respective matchmarks.

Stockpile salvaged material in a designated area.

202.03 Removing Material. Saw cut sidewalks, curbs, pavements, and structures when partial removal is required.

Raze and remove all buildings, foundations, pavements, sidewalks, curbs, fences, structures, and other obstructions that interfere with the work and are not designated to remain.

Existing culverts may be left in an embankment, provided that no portion of the culvert is within 600 mm of the subgrade, the embankment slope, or a new culvert or structure. Crush culvert ends.

Remove structures and obstructions in the roadbed to 300 mm below subgrade elevation. Remove structures and obstructions outside the roadbed to 300 mm below finished ground or to the natural stream bottom.

Prior to removal, place rock and soil material located on the bridge deck, or structure so that it does not enter a stream.

Remove the substructures of existing structures down to the natural stream bottom, and remove the parts outside of a stream down to at least 300 mm below natural ground surface or finished groundline, whichever is lower. Remove portions of existing structures that lie wholly or in part within the limits for a new structure to accommodate construction of the proposed structure.

Except in excavation areas, fill cavities left by structure removal with material to the level of the finished ground, and compact. Place and compact the type of backfill material that is SHOWN ON THE DRAWINGS, designated in the SPECIAL PROJECT SPECIFICATIONS, or approved by the CO.

202.04 Disposing of Material or Structures Not Designated for Salvage. Dispose of material and structures as SHOWN ON THE DRAWINGS or designated in the SPECIAL PROJECT SPECIFICATIONS, using one or more of the following methods:

(a) Removal From Project. Make necessary arrangements with property owners, and haul debris to suitable disposal locations as approved by the CO. Furnish a signed copy of the disposal agreement. Hazardous materials must be properly disposed of.

(b) Burning. Burn debris using high-intensity burning processes that produce few emissions. Examples include incinerating, high stacking, or pit and ditch burning. Provide a watchperson during burning operations.

When burning is complete, extinguish the fire so no smoldering debris remains. Dispose of unburned material in accordance with Subsection 202.04(c).

(c) Burying. Bury debris in trenches or pits in approved areas within the right-of-way. Do not bury debris inside the roadway prism limits, beneath drainage ditches, or in any riparian areas.

Place debris with earth material in alternating layers consisting of 1 m of debris covered by 600 mm of earth. Distribute stumps, logs, and other large pieces to form a compact mass and minimize air voids. Fill all voids. Cover the top layer of buried debris with at least 300 mm of compacted earth. Grade and shape the area.

Measurement

202.05 Method. Use the method of measurement that is DESIGNATED IN THE SCHEDULE OF ITEMS.

Payment

202.06 Basis. The accepted quantities will be paid for at the contract unit price for each PAY ITEM DESIGNATED IN THE SCHEDULE OF ITEMS.

Payment will be made under:

Pay Item		Pay Unit
202 (01)	Removal of structures and obstructions	Lump Sum
202 (02)	Removal of _______	Each
202 (03)	Removal of _______	Meter
202 (04)	Removal of _______	Square Meter
202 (05)	Removal of _______	Lump Sum

Section 203—Excavation, Embankment, & Haul

Description

203.01 Work. Excavate material and construct embankments. Furnish, haul, stockpile, place, dispose of, slope, shape, compact, and/or finish earthen and rocky material.

203.02 Excavation. Excavation consists of the excavation and placement of all excavated material that is not included under other PAY ITEMS listed in the SCHEDULE OF ITEMS.

203.03 Borrow Excavation. Excavate and utilize material from sources SHOWN ON THE DRAWINGS or described in the SPECIAL PROJECT SPECIFICATIONS. Additional sources of borrow excavation will be subject to advance approval by the CO. Develop sources in accordance with Section 611.

Construction

203.04 Clearing & Grubbing. Clear and grub in accordance with Section 201 before work under Section 203 begins. Road pioneering, slash disposal, and grubbing of stumps may proceed concurrently with excavation when approved by the CO. Conduct excavation and placement operations so material to be treated under Section 201 will not be incorporated in the roadway unless specifically included in the slash treatment method.

203.05 Pioneering. During pioneering operations, prevent undercutting of the final excavation slope. Avoid any restriction of drainages while pioneering the road. Keep all materials within the roadway limits unless otherwise SHOWN ON THE DRAWINGS.

203.06 Utilization of Excavated Materials. Use all suitable excavated material in the construction of embankments, subgrades, shoulders, slopes, bedding, and backfill for structures and other purposes, as SHOWN ON THE DRAWINGS.

(a) Excess Excavation. Place excess excavation as SHOWN ON THE DRAWINGS.

(b) Rock for Slope Protection. Conserve and use suitable excavated rock for protecting embankments.

(c) Conserving Material. Excavated material suitable for cushion, road finishing, or other purposes may be conserved and utilized instead of materials from designated sources. Field drain and dry excessively wet material that is otherwise suitable for embankment before placement.

(d) Excavation of Unsuitable Material & Backfill. Place unsuitable excavated material as SHOWN ON THE DRAWINGS. Backfill excavated areas with suitable material when necessary to complete the work. Do not place frozen material in embankments.

Break up rocks that are too large to be incorporated into the embankment or move them to locations approved by the CO. Broken pieces of rock may be placed on the face of the embankment and embedded so they will not roll or obstruct the use and maintenance of the roadbed. Immediately remove any excavated material that inadvertently reaches a stream course.

(e) Conservation of Topsoil. When SHOWN ON THE DRAWINGS, remove, transport, and deposit suitable topsoil in the designated stockpile areas.

(f) Abandoned Structures & Obstructions. Treat abandoned structures and obstructions in accordance with Section 202.

203.07 Drainage Excavation. Construct side ditches, minor channel changes, inlet and outlet ditches, furrow ditches, ditches along the road but beyond roadway limits, and other minor earth drainage structures as SHOWN ON THE DRAWINGS. Utilize excavated material in accordance with Subsection 203.06.

203.08 Sloping, Shaping, & Finishing. Complete slopes and ditches before placing aggregate courses. Slope, shape, and finish as follows:

(a) Sloping. Leave all earth slopes with uniform roughened surfaces, except as described in Subsection 203.08(b), with no noticeable break as viewed from the road. Except in solid rock, round the tops and bottoms of all slopes, including the slopes of drainage ditches, where SHOWN ON THE DRAWINGS. Round the material overlaying solid rock to the extent practical.

If a slide or slipout occurs on a cut or embankment slope, remove or replace the material, and repair or restore all damage to the work. Bench or key slope to stabilize the slide. Reshape the cut or embankment slope to an acceptable condition.

(b) Stepped Slopes. Where SHOWN ON THE DRAWINGS, construct steps on slopes of 1.3:1 to 1:2. Construct the steps about 500 mm high. Blend the steps into natural ground at the end of the cut. If the slope contains nonrippable rock outcrops, blend steps into the rock. Remove loose material found in transitional area. Except for removing large rocks that may fall, scaling stepped slopes is not required.

(c) ***Shaping.*** Shape the subgrade to a smooth surface and to the cross section required. Shape slopes to gradually transition into slope adjustments without noticeable breaks. At the ends of cuts and at intersections of cuts and embankments, adjust slopes in the horizontal and vertical planes to blend into each other or into the natural ground. For roads receiving base or surface course, rocks may remain in place if they do not protrude above the subgrade more than one-third of the depth of the base or surface course, or 75 mm, whichever is less.

(d) ***Finishing.*** Finish the road surface to be reasonably smooth, uniform, and shaped to conform to the typical sections as SHOWN ON THE DRAWINGS. Remove unsuitable material from the roadbed and replace it with suitable material. Finish roadbeds to the tolerance class shown in table 203-1 or as SHOWN ON THE DRAWINGS.

Ensure that the subgrade for both surfaced and unsurfaced roads is visibly moist during shaping and dressing. Bring low sections, holes, cracks, or depressions to grade with suitable material. Compact the subgrade as required by the designated embankment placing method.

Finish the roadbed for unsurfaced roads using one of the following methods, as SHOWN ON THE DRAWINGS:

(1) Method A. Ensure that the top 100 mm below the finished roadbed contains rocks no larger than 100 mm. Remove oversize material, reduce to acceptable size, or cover by importing suitable material approved by the CO.

(2) Method B. Roll the roadbed to break down rocky material. Roll a minimum of five full-width passes, or until visual displacement ceases, with a vibratory grid roller or equivalent weighing a minimum of 9 t.

(3) Method C. Tractor finish work by spreading the excavation for roads SHOWN ON THE DRAWINGS as Construction Tolerance Class K, L, or M, as shown in table 203-1. Eliminate rock berms that may form during embankment construction with a tractor finish.

203.09 Snow Removal. Remove snow and ice in advance of the work and deposit beyond the roadway limits in a manner that will not waste material. Snow and ice will not be incorporated into the embankment or be placed to cause damage.

203.10 Finishing Slopes. Ensure that finished slopes conform reasonably to the lines STAKED ON THE GROUND or SHOWN ON THE DRAWINGS. Finish slopes in a roughened condition to facilitate the establishment of vegetative growth. The finish associated with template and stringline or hand-raking methods will not be required. Remove rock, debris, and other loose material that are more than 150 mm in diameter, unless otherwise SHOWN ON THE DRAWINGS.

Table 203-1—Construction tolerences.

	Tolerance Class[a]												
	A	B	C	D	E	F	G	H	I	J	K	L	M
Roadbed width (mm)	150	150	300	300	300	300	300	450	300	600	600	600	600
Subgrade elevation (mm)	±30	±60	±60	±150	±150	±300	±300	±450	±600	±900	±600	±900	–[c]
Centerline alignment (mm)	60	60	150	150	300	300	300	450	600	900	900	1500	–[c]
Slopes, excavation, and embankment (% slope)[b]	±3	±5	±5	±5	±5	±5	±10	±10	±10	±10	±20	±20	±20

a. Maximum allowable deviation from construction stakes and drawings.

b. Maximum allowable deviation from staked slope measured from slope stakes or hinge points.

c. Unless otherwise SHOWN ON THE DRAWINGS, the centerline alignment and subgrade elevation, as built, have no horizontal curves with a radius of less than 26 m, and no vertical curves with a curve length of less than 25 m when the algebraic difference in the grade change is less than 10 percent, or a curve length of less than 30 m when the algebraic difference of the grade change is greater than or equal to 10 percent. The centerline grade is not to exceed 20 percent in 30 m of length.

In areas that require blasting, use blasting techniques in accordance with Section 220, as SHOWN ON THE DRAWINGS. Presplitting is not required, unless controlled backslope blasting is SHOWN ON THE DRAWINGS.

Perform test blasting in accordance with Subsection 220.06, unless directed otherwise by the CO.

203.11 Landscape & Stream Protection. Confine excavation, blasted material, and embankment material within the roadway limits, unless otherwise approved by the CO, to avoid overbuilding and to protect the landscape and streams. Retrieve and incorporate into designated areas all material deposited outside of the clearing limits.

203.12 Subgrade Treatments. Subgrade treatment consists of soil modification by admixing aggregates or placing geotextiles, fiber mat, wood corduroy, rock blanket, or other similar materials over areas of unsuitable embankment foundation materials that are SHOWN ON THE DRAWINGS. The construction and material requirements for the type of subgrade treatment will be specified in the SPECIAL PROJECT SPECIFICATIONS or SHOWN ON THE DRAWINGS.

203.13 Earth Berms. Construct permanent earth berms along the shoulder of traveled ways at locations SHOWN ON THE DRAWINGS. Use well-graded material that contains no rocks having a dimension greater than one-fourth the height of the berm in the construction. Acceptable material for the berm may be windrowed as the roadbed is constructed. When local material is not acceptable, import material from approved sources. Frozen material, roots, sod, or other deleterious material is unacceptable for berm construction. Do not waste materials over the embankment slope.

Accomplish compaction by operating the spreading equipment over the full section of the berm.

203.14 Water. Develop, haul, and apply water in accordance with Section 207.

203.15 Compaction Equipment. Use equipment capable of obtaining compaction requirements. The compacting units may be of any type, provided that they are capable of compacting each lift of material as specified, and that they meet the minimum requirements specified below. Heavier compacting units may be required to achieve the specified density of the embankment. Minimum requirements for rollers are as follows:

(a) Sheepsfoot, tamping, or grid rollers shall be capable of exerting a force of 4.5 kg/mm of width of roller drum.

(b) Steel-wheel rollers, other than the vibratory type, shall be capable of exerting a force of not less than 4.5 kg/mm of width of the compression roll or rolls.

(c) Vibratory steel-wheel rollers shall have a minimum weight of 5 t. The compactor shall be equipped with amplitude and frequency controls, and specifically designed to compact the material on which it is used.

(d) Pneumatic-tire rollers shall have smooth tread tires of equal size that will provide a uniform compacting pressure for the full width of the roller and capable of exerting a ground pressure of at least 550 kPa.

203.16 Embankment Placement. Place embankment in accordance with the following requirements:

(a) All Methods. Construct the lower part of the embankment in a single layer to the minimum depth necessary to support construction equipment when an embankment is to be placed across swampy ground and removal of unsuitable material or subgrade treatment is not required.

(b) Specific Methods. Place all embankments using one or more of the following methods, as SHOWN ON THE DRAWINGS and listed in the SCHEDULE OF ITEMS:

(1) Method 1—Side Casting & End Dumping. Embankment may be placed by side casting and end dumping. Build solid embankments by working smaller rocks and fines in with the larger rocks and fines to fill the voids.

(2) Method 2—Layer Placement. Roughen or step surfaces steeper than a ratio of 1 vertical to 3 horizontal (1:3) upon which embankment is to be placed, when SHOWN ON THE DRAWINGS, in order to provide permanent bonding of new and old materials.

Layer place embankment, except over rock surfaces. Over rock surfaces, material may be placed by end dumping to the minimum depth needed for operation of spreading equipment. Level and smooth each embankment layer before placement of subsequent layers. Operate hauling and spreading equipment uniformly over the full width of each layer.

Place suitable material in layers no more than 300 mm thick, except when the material contains rock more than 225 mm in diameter, in which case layers may be of sufficient thickness to accommodate the material involved. Ensure that no layer exceeds 600 mm before compaction.

Placing individual rocks or boulders greater than 600 mm in diameter will be permitted, provided that the embankment will accommodate them and that they are at least 150 mm below the subgrade. Carefully distribute rocks and fill the voids with finer material to form a dense and compacted mass.

Where material containing large amounts of rock is used to construct embankments, make layers of sufficient thickness to accommodate the material involved. Construct a solid embankment with adequate compaction by working smaller rock and fines in with the larger rocks to fill the voids, and by operating hauling and spreading equipment uniformly over the full width of each layer as the embankment is constructed.

Ensure that material is at a moisture content suitable to obtain a mass that will not visibly deflect under the load of the hauling and spreading equipment. Handle excessively wet material in accordance with Subsection 203.06(c).

(3) Method 3—Layer Placement (Roller Compaction). Place embankments as specified in method 2. Place in horizontal layers not exceeding 300 mm prior to compaction, except when the material contains rock more than 225 mm in diameter, in which case layers may be of sufficient thickness to accommodate the material involved. Obtain compaction using equipment listed in Subsection 203.15. Operate compaction equipment over the full width of each layer until visible deformation of the layer ceases or, in the case of the sheepsfoot roller, the roller "walks out" of the layer. Make at least three complete passes.

(4) Method 4—Controlled Compaction. Place embankments as specified in method 2; but place earth embankments in horizontal layers not exceeding 300 mm (loose measure), and compact them. Ensure that the moisture content of material is suitable for attaining the required compaction. Compact the embankments and the top 300 mm of excavation sections to at least 95 percent of the maximum density, as determined by AASHTO T 99, method C or D.

Determine the density of the embankment material during the progress of the work, in accordance with AASHTO T 191, T 205, or T 238; and T 217, T 239, or T 255. Correct for coarse particles in accordance with AASHTO T 224.

Density requirements will not apply to portions of rock embankments that cannot be tested in accordance with approved methods. When this condition exists, accomplish compaction by working smaller rocks and fines in with the larger rocks to fill the voids and by operating equipment over the embankment materials.

(5) Method 5—Special Project Controlled Compaction. Place and compact embankments to at least 90 percent of the maximum density determined by AASHTO T 180, method C or D, but obtain compaction of not less than 95 percent of AASHTO T 180, method C or D, for a minimum depth of 300 mm below subgrade for the width of the roadbed in both excavation and embankment sections.

Determine density during the work in accordance with AASHTO T 191, T 205, or T 238; T 217, T 239, or T 255; and T 224.

203.17 Construction Tolerances. Construct to the tolerance class as SHOWN ON THE DRAWINGS and in accordance with table 203-1. Construct roadway ditches to flow in the direction SHOWN ON THE DRAWINGS.

Ensure that deviations are uniform in the direction of change for a distance of 60 m or more along the project centerline.

203.18 Haul. Haul is incidental to excavation and borrow excavation, unless listed as a separate PAY ITEM in the SCHEDULE OF ITEMS.

Measurement

203.19 Method. Use the method of measurement that is DESIGNATED IN THE SCHEDULE OF ITEMS.

Quantities of excavation will include:

(a) Roadway excavation.

(b) Rock and unsuitable material below the required grade, and unsuitable material beneath embankment areas.

(c) Furrow ditches outside the roadway, except when furrow ditches are included in the SCHEDULE OF ITEMS.

(d) Topsoil and other material removed and stockpiled as directed.

(e) Borrow material used in the work, except when borrow is included in the SCHEDULE OF ITEMS.

(f) The volume of conserved materials taken from stockpiles and used in the work, except topsoil included under other PAY ITEMS.

(g) Slide material not attributable to negligence of the Contractor.

Quantities of excavation will not include:

(a) Material used for other than approved purposes.

(b) Unauthorized excavation or borrow.

(c) Quantity of material excavated from slope rounding or slope tapering.

(d) Overbreakage from the backslope in rock excavation requiring blasting.

(e) Material scarified in place to receive the first layer of embankment.

(f) Benching or stepping existing ground for embankment foundation.

(g) Stepping or scaling cut slopes.

(h) Oversize material removed when finishing unsurfaced roads.

When designed quantities are DESIGNATED IN THE SCHEDULE OF ITEMS as the method of measurement, estimate the quantities from design data based on undisturbed ground surface elevations.

When staked quantities are DESIGNATED IN THE SCHEDULE OF ITEMS as the method of measurement, determine excavation quantities by the average end area method using slope stake information taken prior to construction.

When actual quantities are DESIGNATED IN THE SCHEDULE OF ITEMS as the method of measurement, take preliminary cross sections, or comparable measurements, of the undisturbed ground surface; and measure final quantities in accordance with the following:

(a) When excavation is designated as a PAY ITEM in the SCHEDULE OF ITEMS, take final cross sections, or comparable measurements, of the completed and accepted work.

(b) When embankment is designated as a PAY ITEM in the SCHEDULE OF ITEMS, determine measurement in the final position.

(c) When borrow excavation is designated as a PAY ITEM in the SCHEDULE OF ITEMS, determine measurement in the original position.

Payment

203.20 Basis. The accepted quantities will be paid for at the contract unit price for each PAY ITEM DESIGNATED IN THE SCHEDULE OF ITEMS.

Payment will be made under:

Pay Item	Pay Unit
203 (01) Excavation, placement method 1	Cubic Meter
203 (02) Excavation, placement method 2	Cubic Meter
203 (03) Excavation, placement method 3	Cubic Meter

203 (04) Excavation, placement method 4 .... Cubic Meter

203 (05) Excavation, placement method 5 .... Cubic Meter

203 (06) Excavation, placement method ________ .... Kilometer

203 (07) Excavation, placement method ________ .... Lump Sum

203 (08) Borrow excavation, placement method _________ ... Cubic Meter

203 (09) Borrow excavation, placement method _________ ....Ton

203 (10) Unsuitable excavation .... Cubic Meter

203 (11) Embankment, placement method __________ .... Cubic Meter

203 (12) Embankment, placement method __________ .... Kilometer

203 (13) Subgrade treatment, type ______________............. Square Meter

203 (14) Rounding cut slopes .... Meter

203 (15) Drainage excavation, type _____________ .... Cubic Meter

203 (16) Drainage excavation, type _____________ .... Meter

203 (17) Drainage excavation, type _____________ .... Each

203 (18) Furrow ditches .... Meter

203 (19) Topsoil (stockpiled) .... Cubic Meter

203 (20) Earth berms .... Meter

203 (21) Haul .... Cubic Meter Kilometer

Section 204—Soil Erosion & Water Pollution Control

Description

204.01 Work. Furnish, construct, and maintain permanent and temporary erosion and sediment control measures.

Materials

204.02 Requirements. Ensure that materials meet the requirements specified in the following subsections:

Agricultural Limestone	713.02
Bales	713.13
Erosion Control Mats, Roving, & Geocell	713.07
Fertilizer	713.03
Geotextiles	714.01
Mulch	713.05
Sandbags	713.14
Seed	713.04
Stabilizing Emulsion Tackifiers	713.12

Ensure that all other materials are as SHOWN ON THE DRAWINGS or specified in the SPECIAL PROJECT SPECIFICATIONS.

Construction

204.03 Performance. Prior to the start of construction, submit a written plan that provides permanent and temporary erosion control measures to minimize erosion and sedimentation during and after construction. Do not begin work until the necessary controls for that particular phase of work have been implemented. Do not modify the type, size, or location of any control. An alternate erosion control plan with all necessary permits may be submitted 30 days before intended use.

Incorporate all permanent erosion control features into the project at the earliest practicable time, as outlined in the approved plan, to minimize the need for temporary erosion control.

Before grubbing and grading, construct all erosion controls around the perimeter of the project, including filter barriers, diversion, and settling structures. When required by the SPECIAL PROJECT SPECIFICATIONS, schedule clearing and grubbing so that grading operations and permanent erosion control measures can follow without interference.

Install any temporary erosion or pollution control measures that are required due to negligence or carelessness, without compensation.

204.04 Construction. Construct erosion control and sediment control measures as follows:

(a) Construct temporary erosion controls in incremental stages as construction proceeds.

(b) Construct temporary slope drains, diversion channels, and earth berms to protect disturbed areas and slopes.

(c) Apply permanent turf establishment to the finished slopes and ditches within 30 days, or as required in the SPECIAL PROJECT SPECIFICATION or SHOWN ON THE DRAWINGS.

(d) Apply temporary turf establishment on disturbed areas that will remain exposed for more than 30 days.

(e) Construct outlet protection as soon as culverts or other structures are complete.

(f) Construct permanent erosion controls, including waterway linings and slope treatments, as soon as practical or upon completion of the roadbed.

(g) Construct and maintain erosion controls on and around soil stockpiles to prevent soil loss.

(h) Following each day's grading operations, shape earthwork to minimize and control erosion from storm runoff.

204.05 Filter Barriers. Construct silt fence, straw bales, and brush barriers for filtering sediment from runoff and reducing the velocity of sheet flow.

204.06 Sediment Retention Structures. Construct sediment retention structures of the following types:

(a) Temporary Sediment Traps. Construct temporary sediment traps to detain runoff from disturbed areas and settle out sediment. Provide outlet protection.

(b) Sediment Basins. Construct sediment basins to store runoff and settle out sediment for large drainage areas. Construct sediment basins according to Section 203. Construct riser pipes according to Section 603A or 603B. Provide outlet protection.

204.07 Outlet Protection. Construct riprap aprons or basins to reduce water velocity and prevent scour at the outlet of permanent and temporary erosion control measures. Construct riprap according to Section 251.

204.08 Water Crossings. Construct temporary culvert pipe at temporary crossing where construction vehicles cross a live waterway.

204.09 Diversions. Construct temporary channels, temporary culverts, earth berms, or sandbags to divert water around disturbed areas and slopes. Use temporary channels, temporary culverts, pumps, sandbags, or other methods to divert the flow of live streams for permanent culvert installations and other work. Stabilize channels and provide outlet protection.

204.10 Waterway & Slope Protection & Stabilization. Use plastic lining, riprap, check dams, erosion control blankets and mats, and temporary slope drains as follows:

(a) Plastic Lining. Use plastic lining to protect underlying soil from erosion. Place the plastic lining loosely on a smooth soil surface free of projections or depressions that may cause the liner to puncture or tear. Lap transverse joints a minimum of 600 mm in the direction of flow. Do not use longitudinal joints. Anchor the lining in place using riprap.

(b) Riprap. Construct riprap for channel lining according to Section 251.

(c) Check Dams. Construct riprap, sandbags, or earth berms for temporary dams to reduce the velocity in ditches and swales.

(d) Temporary Slope Drains. Use drainpipe, riprap, or plastic lined waterway for temporary slope drains to channel runoff down slopes. Channel water into the slope drain with an earth berm constructed at the top of a cut or fill. Anchor slope drains to the slope. Provide outlet protection.

204.11 Temporary Turf Establishment. Apply seed, fertilizer, and mulch for soil erosion protection at the rates SHOWN ON THE DRAWINGS or in the SPECIAL PROJECT SPECIFICATIONS.

204.12 Inspection & Reporting. Inspect all erosion control facilities at least every 7 days, within 24 hours after more than 20 mm of rain in 24 hours, and as required in the contract permits.

Furnish inspection reports that include the following:

(a) Summary of the inspection.

(b) Names of personnel making the inspection.

(c) Date and time of inspection.

(d) Observations made.

(e) Corrective action necessary.

204.13 Maintenance & Cleanup. Maintain temporary erosion control measures in working condition until the project is complete or the measures are no longer needed. Clean erosion control measures when half full of sediment. Use the sediment in the work, if acceptable, or place it in accordance with Subsection 203.06.

Replace erosion control measures that cannot be maintained and those that are damaged by construction operations.

Remove and dispose of temporary erosion control measures when the turf is satisfactorily established, and when drainage and channels are lined and stabilized. Remove and dispose of erosion control measures according to Subsection 202.04(a).

Restore the ground to its natural or intended condition and provide permanent erosion control measures.

Measurement

204.14 Method. Use the method of measurement that is DESIGNATED IN THE SCHEDULE OF ITEMS.

Payment

204.15 Basis. The accepted quantities will be paid for at the contract unit price for each PAY ITEM DESIGNATED IN THE SCHEDULE OF ITEMS.

Payment will be made under:

Pay Item	Pay Unit
204 (01) Temporary seeding and fertilizing	Hectare
204 (02) Mulching	Ton
204 (03) Asphaltic material	Liter
204 (04) Temporary netting	Square Meter

204 (05) Straw/hay bales .... Each

204 (06) Gravel blanket .... Cubic Meter

204 (07) Silt fence .... Meter

204 (08) Brush barrier .... Meter

204 (09) Sediment basin .... Each

204 (10) Berm .... Meter

204 (11) Dike .... Meter

204 (12) Dam .... Each

204 (13) Temporary water bars .... Each

204 (14) ______ for soil erosion and pollution control .... Each

204 (15) ______ for soil erosion and pollution control .... Meter

204 (16) ______ for soil erosion and pollution control .... Square Meter

204 (17) ______ for soil erosion and pollution control .... Hectare

204 (18) ______ for soil erosion and pollution control .... Cubic Meter

204 (19) Soil erosion and pollution control .... Lump Sum

Section 206—Structural Excavation for Major Structures

Description

206.01 Work. Excavate, backfill, and dispose of material for the construction of structures. Preserve channels; shore and brace; construct cofferdams; seal foundations; dewater; excavate; prepare foundations; backfill; and subsequently remove safety features and cofferdams.

Materials

206.02 Requirements. Ensure that material conforms to specifications in the following sections and subsections:

Foundation Fill	704.01
Structural Backfill	704.04
Structural Concrete	552

Construction

206.03 Preparation for Structural Excavation. Clear the area of vegetation and obstructions according to Sections 201 and 202.

When structural excavation is to be measured and paid for by the cubic meter, notify the CO sufficiently before beginning any clearing, grubbing, or excavation so that cross-sectional measurements of the undisturbed ground may be taken. Do not disturb the natural ground adjacent to the structure until authorized by the CO.

206.04 General. Consider the elevations of the bottoms of footings or foundations when SHOWN ON THE DRAWINGS to be approximate elevations. The CO may order, in writing, changes in the elevations of footings and foundations when necessary to secure a satisfactory foundation.

Excavate trenches or foundation pits to a width and length that allows room for work. Provide a firm foundation of uniform density throughout its length and width. Do not place footings until the depth of excavation and the foundation material have been approved in writing.

Where necessary to blast rock, blast according to Section 220.

Follow Occupational Safety and Health Administration (OSHA) safety regulations (29 CFR, part 1926, subpart P, Excavation), or OSHA-approved State Plan requirements for sloping the sides of excavations and for using shoring, bracing, and other safety features. When sides of excavations are sloped for safety considerations, provide one copy of the design that demonstrates conformity with OSHA regulations. Submit working drawings and construction details when required by the SPECIAL PROJECT SPECIFICATIONS where support systems, shield systems, or other protective systems are used. Ensure that drawings demonstrate conformity with regulations.

Remove safety features when no longer necessary. Remove shoring and bracing to at least 300 mm below the surface of the finished ground.

Saw cut existing pavements or concrete structures that are adjacent to the area to be excavated and are designated to remain.

Conserve suitable material for structural backfill from excavated material. Do not deposit excavated material in or near a waterway. Do not stockpile excavated material closer than 1 m from the edge of the excavation.

Place unsuitable or excess material according to Subsection 203.06. If approved, suitable material may be used in embankment construction.

Remove all water as necessary to perform work.

206.05 Channel Preservation. Perform work in or next to a running waterway as follows:

(a) Excavate inside cofferdams, sheeting, or other approved separations such as dikes or sandbags.

(b) Do not disturb the natural bed of the waterway adjacent to the work.

(c) Backfill the excavation with structural backfill to original groundline.

(d) Do not pump water from foundation excavations directly into live streams. Pump water into settling areas as SHOWN ON THE DRAWINGS or as approved.

206.06 Cofferdams. Use cofferdams when excavating under water or when the excavation is affected by groundwater.

Submit three working copies of drawings and calculations 21 days prior to installation, showing proposed methods and construction details of cofferdams. Place seal and signature of a licensed professional engineer on the drawings and calculations.

Shore and construct cofferdams according to OSHA standards. Ensure that cofferdams:

(a) Extend below the bottom of the footing.

(b) Are braced to withstand expected pressures and loads without buckling, and are secured in place to prevent tipping or movement.

(c) Are as watertight as practicable.

(d) Provide sufficient clearance for the placement of forms and the inspection of their exteriors.

(e) Provide for dewatering.

(f) Protect fresh concrete against damage from sudden rises in water elevation.

(g) Prevent damage to the foundation by erosion.

When no longer required, remove all cofferdam material down to the natural bed of the waterway. Remove cofferdam material outside the waterway to a minimum of 300 mm below the surface of the finished ground.

Do not disturb, damage, or mar finished structure. Remove all timber or bracing in the cofferdam that extends into substructure masonry.

206.07 Foundation Seal. Construct a foundation seal of seal-concrete where a foundation area cannot be pumped reasonably free of water, and/or where the substructure concrete cannot be placed in accordance with Section 552.

While placing a foundation seal, maintain the water level inside the cofferdam at the same level as the water outside the cofferdam. Where a foundation seal is placed in tidal water or in a stream subject to sudden water level increases, vent or port the cofferdam at low water level.

Do not dewater a concrete-sealed cofferdam until the concrete strength is sufficient to withstand the hydrostatic pressure.

206.08 Dewatering. While placing concrete, locate and operate the pumps outside the foundation form. If pumping is permitted from the interior of any foundation enclosure, pump in a manner to avoid removal or disturbance of concrete material.

206.09 Foundation Preparation. Prepare footing foundations as follows:

(a) Footings Placed on Bedrock. Cut the bottom of the excavation to the specified elevations. Clean the foundation surface of loose or disintegrated material. Clean and grout all open seams and crevices that will remain beneath the footing.

(b) Footings Placed on an Excavated Surface Other Than Bedrock. Do not disturb the bottom of the foundation excavation. Remove material to foundation grade and compact the foundation immediately before concrete is placed. Treat material below the foundation grade that is disturbed as unstable material (see Subsection 206.09(d)).

(c) Footings Keyed Into Undisturbed Material. Excavate the foundation to the neat lines of the footing and compact the foundation. Where material does not stand vertically, fill all space between the neat lines of the footing and the remaining undisturbed material with concrete. If the top of the excavation is below the top of the footing, fill only to the top of the excavation; otherwise, fill to the top of the footing. Concrete placed against steel sheet piles in cofferdams is considered to be against undisturbed material.

(d) Unstable Material Below Footing Elevation. Excavate unstable material below foundation grade to the depth and lateral extent as approved, and replace it with foundation fill. Place foundation fill material in horizontal layers that, when compacted, do not exceed 150 mm in depth. Compact each layer according to Subsection 206.11.

(e) Foundations Using Piles. Excavate to the foundation elevation and drive the piles. Remove all loose and displaced material and reshape the bottom of the excavation to the foundation elevation. Smooth and compact the bed to receive the footing.

206.10 Backfill. Backfill structural excavation with structural backfill material.

Place structural backfill in horizontal layers that, when compacted, do not exceed 150 mm in depth. Compact each layer according to Subsection 206.11.

Do not place backfill or embankment behind the walls of concrete culverts or abutments of rigid frame structures until the top slab has been placed and cured. For all structures held at the top by the superstructure and behind the sidewalls of concrete culverts, bring backfill and embankment up evenly behind opposite abutments or sidewalls.

Do not place rock that is greater than 150 mm in its largest dimension within any backfill or embankment that is within 1 m of any structure.

Extend each layer to the limits of the excavation or to natural ground.

Do not place backfill against concrete that is less than 7 days old, or until 90 percent of the design strength is achieved.

206.11 Embankment. Construct all embankments, and backfill in horizontal layers adjacent to structures. Compact backfill in accordance with Subsection 203.16(b), method 4, except that mechanical tampers may be used for the required compaction. Use special care to prevent wedging action against the structure. Bench all slopes that bound or are within the areas to be backfilled to prevent wedging action. Extend compacted material horizontally for a distance at least equal to the height of the substructure or wall that is to be backfilled against, except where undisturbed material remains within the area.

Measurement

206.12 Method. Use the method of measurement that is DESIGNATED IN THE SCHEDULE OF ITEMS.

Measure structural excavation by the cubic meter that is in place in its original position. Do not include the following volumes in structural excavation:

(a) Material excavated outside vertical planes located 450 mm outside and parallel to the neat lines of footings or foundations. Use these vertical planes to determine pay quantities, regardless of the amount of material excavated inside or outside these planes.

(b) Any material included within the staked limits of the roadway excavation, such as contiguous channel changes and ditches, for which payment is otherwise provided in the contract.

(c) Water or other liquid material.

(d) Material excavated before the survey of elevations and measurements of the original ground.

(e) Material rehandled, except when the contract specifically requires excavation after embankment placement.

(f) Material excavated for footings or foundations at a depth more than 1.5 m below the lowest elevation for such footings or foundations, as shown on the plans.

Measure foundation fill, when DESIGNATED IN THE SCHEDULE OF ITEMS, by the cubic meter in place.

Measure structural backfill, and structural backfill for walls, by the cubic meter in place. Limit the volume of structural backfill measured to that placed inside vertical planes located 450 mm outside and parallel to the neat lines of footings or foundations. Use these vertical planes to determine pay quantities, regardless of the amount of backfill material placed outside these planes.

Measure work for shoring and bracing and for cofferdams on a lump-sum basis for all work needed to complete excavation to a depth of 1.5 m below the lowest elevation, as SHOWN ON THE DRAWINGS, for each foundation structure.

Payment

206.13 Basis. The accepted quantities will be paid for at the contract unit price for each PAY ITEM DESIGNATED IN THE SCHEDULE OF ITEMS.

Excavation for footings or foundations, shoring and bracing, and cofferdams at depths more than 1.5 m below the lowest elevation for such footing or foundation as SHOWN ON THE DRAWINGS will be paid for by design change.

Payment will be made under:

Pay Item		Pay Unit
206 (01)	Structural excavation	Cubic Meter
206 (02)	Foundation fill	Cubic Meter
206 (03)	Structural backfill	Cubic Meter
206 (04)	Structural backfill for walls	Cubic Meter
206 (05)	Shoring and bracing	Lump Sum
206 (06)	Cofferdams	Lump Sum
206 (07)	Structural excavation	Lump Sum

Section 206A—Structural Excavation for Minor Structures

Description

206A.01 Work. Excavate, backfill, and dispose of material for the construction of culverts and minor structures. Preserve channels; shore and brace; seal foundations; dewater; excavate; prepare foundations; bed; and backfill.

Materials

206A.02 Requirements. Ensure that material conforms to specifications in the following sections and subsections:

Backfill Material	704.03
Bedding	704.02
Foundation Fill	704.01
Minor Concrete Structures	602
Structural Concrete	552
Unclassified Borrow	704.06

Construction

206A.03 Preparation for Structural Excavation. Clear the area of vegetation and obstructions according to Sections 201 and 202.

206A.04 General. Excavate trenches or foundation pits according to Subsection 206.04 and the following:

(a) Minor Structures. Clean all loose material from all rock or other foundation material and cut to a firm surface that is level, stepped, or serrated. Remove all loose and disintegrated rock and thin strata. When the footing is to rest on material other than rock, complete the excavation just before the footing is to be placed.

(b) Culverts. Construct the width of trenches in natural ground to permit satisfactory joining of the culvert sections and thorough tamping of the bedding material under and around the culvert. Excavate trenches to a minimum width equal to the culvert diameter plus 600 mm.

Construct trenches for culverts being placed in embankments to a width of one diameter, plus one diameter on each side.

Excavate unsuitable foundation material below the invert of the culvert to an approximate depth of 600 mm and a width of at least the culvert diameter plus

1.25 m. Remove rock, hardpan, or other unyielding material below the foundation grade for a depth of at least 300 mm and a width of at least 600 mm greater than the outside width of the culvert.

Excavate to foundation grade without unduly disturbing the trench or foundation surface. Foundation grade is the elevation at the bottom of any bedding for the installation of the structure.

206A.05 Channel Preservation. Preserve channels according to Subsection 206.05, but excavate inside separations such as dikes or sandbags.

206A.06 Foundation Seal. Where necessary, construct foundation seal according to Subsection 206.07.

206A.07 Dewatering. Where necessary, dewater according to Subsection 206.08.

206A.08 Foundation Preparation. Excavate any unsuitable material present at foundation grade, and replace it with foundation fill. Place and compact the foundation fill material according to Subsection 206.09(d).

Where footing must be keyed into undisturbed material, prepare foundation and construct footing according to Subsection 206.09(c). Notify the CO when each excavation is completed, and receive written approval of the excavation and the foundation material before placing footings.

206A.09 Utilization of Excavated Materials. Utilize all suitable excavated material as backfill or embankment. Do not place excavated material in live streams.

Dispose of all surplus material as SHOWN ON THE DRAWINGS. Do not deposit excavated material in a manner that will endanger the partly finished structure.

206A.10 Backfill & Embankments for Minor Structures. Backfill excavated areas around minor structures to the level of the original ground surface. Backfill with selected material placed in horizontal layers not over 150 mm (loose measure) in depth. Use compactible material free of frozen lumps, chunks of highly plastic clay, or other objectionable material. Do not use rocks larger than 75 mm in diameter within 300 mm of the structure. Compact each layer in accordance with Subsection 203.16(b), method 4.

206A.11 Bedding, Backfill, & Embankment for Pipe Culverts. Install bedding, backfill, and embankment for pipe culverts in accordance with Sections 603, 603A, and 603B, unless otherwise SHOWN ON THE DRAWINGS or described in the SPECIAL PROJECT SPECIFICATIONS.

Measurement

206A.12 Method. Use the method of measurement that is DESIGNATED IN THE SCHEDULE OF ITEMS.

Measure bedding material by the cubic meter in place. Measure foundation fill under Section 206. Measure concrete under Section 552.

Payment

206A.13 Basis. The accepted quantities will be paid for at the contract unit price for each PAY ITEM DESIGNATED IN THE SCHEDULE OF ITEMS.

Payment will be made under:

Pay Item		Pay Unit
206A (01)	Minor structure excavation	Cubic Meter
206A (02)	Pipe culvert excavation	Cubic Meter
206A (03)	Bedding material	Cubic Meter
206A (04)	Foundation fill	Cubic Meter
206A (05)	Structural concrete class for ____________ *Description*	Cubic Meter
206A (06)	Minor structure excavation and backfill	Lump Sum
206A (07)	Pipe culvert excavation and backfill	Lump Sum

Section 207—Developing Water Supply & Watering

Description

207.01 Work. Furnish, haul, and apply water.

Materials

207.02 Requirements. In the planting or care of vegetation, use water that is free of substances injurious to plant life.

Ensure that water meets requirements of Subsection 725.01.

Water sources are SHOWN ON THE DRAWINGS. If other sources of water are used, obtain the right to use the water, and pay any royalty costs.

Construction

207.03 Development of Supply & Access. Develop water supplies and access as SHOWN ON THE DRAWINGS.

207.04 Equipment. Use watertight tanks of known capacity with mobile watering equipment. Provide uniform and controlled application of water without ponding or washing. Maintain positive control of water from the driver's position at all times.

Measurement

207.05 Method. Use the method of measurement that is DESIGNATED IN THE SCHEDULE OF ITEMS.

Furnish calibrated tanks, distributors, or accurate water meters for measurement when directed by the CO.

When the SCHEDULE OF ITEMS calls for developing water supply and water, the cost of developing the supply is included in the unit price for the quantity of water delivered.

Measure hauling of water along the shortest feasible route to the nearest water supply. Do not include the cost of developing the supply.

Payment

207.06 Basis. The accepted quantities will be paid for at the contract unit price for each PAY ITEM DESIGNATED IN THE SCHEDULE OF ITEMS.

Payment will be made under:

Pay Item		Pay Unit
207 (01)	Developing water supply	Lump Sum
207 (02)	Water	k Liters
207 (03)	Water	Lump Sum
207 (04)	Developing water supply and water	k Liters
207 (05)	Developing water supply and water	Lump Sum
207 (06)	Hauling water	k Liters-km

Section 210—Closing, Obliteration, or Treatment of Existing Roads

Description

210.01 Work. Treat roadways by removing rigid material, including culverts and bridges; constructing waterbars, overflow ditches, and earthen barriers; sloping and scarifying the roadbed; removing selected fill; hauling materials to designated disposal areas; and revegetating.

210.02 Performance. Break rigid material such as pavements, curbs, gutters, sidewalks, and other nonasphalt material into pieces with a maximum dimension of 300 mm.

Dispose of material by one of the methods listed in Subsection 202.04 or as SHOWN ON THE DRAWINGS.

Restore and maintain the natural drainage patterns.

210.03 Treatment of Roadway. Treat the roadway using one of the following methods, as specified in the SCHEDULE OF ITEMS:

*(a) **Method A.*** Fill ditches and restore the roadway to approximate original ground contour or shape to blend with the terrain. Loosen the roadbed by ripping, plowing, or scarifying to promote the establishment of vegetation. Scarify a representative area to determine the number of passes necessary to decompact the road surface. Apply this number to the entire project. Pull all major embankments, and use material to contour or fill ditches; or haul it to designated areas, as SHOWN ON THE DRAWINGS. Keep excavated material within the original roadway limits, unless otherwise directed by the CO or SHOWN ON THE DRAWINGS.

*(b) **Method B.*** Shape the roadway as SHOWN ON THE DRAWINGS to drain water. Fill ditches and outslope the roadbed when SHOWN ON THE DRAWINGS. Loosen the roadbed by ripping or scarifying to provide a seedbed, and promote establishment of vegetation.

*(c) **Method C.*** Treat the roadway as SHOWN ON THE DRAWINGS.

210.04 Waterbars & Barriers. Construct barriers to prevent vehicle access and waterbars as SHOWN ON THE DRAWINGS.

210.05 Establishing Vegetative Cover. Seed, fertilize, and mulch all disturbed areas as specified in Section 625 or SHOWN ON THE DRAWINGS.

Measurement

210.06 Method. Use the method of measurement that is DESIGNATED IN THE SCHEDULE OF ITEMS.

Payment

210.07 Basis. The quantities will be paid for at the contract unit price for each PAY ITEM DESIGNATED IN THE SCHEDULE OF ITEMS.

Seed, fertilizer, and mulch will be a separate PAY ITEM under Section 625.

Payment will be made under:

Pay Item		Pay Unit
210 (01)	Treatment of existing roadway, method _____	Square Meter
210 (02)	Treatment of existing roadway, method _____	Kilometer
210 (03)	Treatment of existing roadway, method _____	Lump Sum
210 (04)	Treatment of ______________________________	Each

Section 220—Rock Blasting

Description

220.01 Work. Fracture rock and construct stable final rock cut faces using controlled backslope blasting and production-blasting techniques, as SHOWN ON THE DRAWINGS. For controlled backslope blasting, use explosives to form a shear plane in the rock along a specified backslope. Controlled backslope blasting includes presplitting and cushion blasting. For production blasting, use explosives to fracture rock to produce slopes in reasonable conformity to the drawings and to minimize rock throw.

Materials

220.02 Requirements. Ensure that material conforms to the specifications in the following subsection:

Explosives & Blasting Accessories 725.24

Construction

220.03 Regulations. Furnish copies or other proof of all applicable permits and licenses. There are Federal, State, and local regulations on the purchase, transportation, storage, and use of explosive material. Federal regulations include the following:

(a) Safety & Health. OSHA, 29 CFR, part 1926, subpart U.

(b) Storage, Security, & Accountability. Bureau of Alcohol, Tobacco, and Firearms, 27 CFR, part 181.

(c) Shipment. U. S. Department of Transportation (DOT), 49 CFR, parts 171–179, 390–397.

220.04 Blasting Plan. Submit a blasting plan at least 14 days before drilling operations begin or whenever a change in drilling and blasting procedures is proposed. Include full details of drilling and blasting patterns and the techniques proposed for controlled and production blasting, including provisions for loading wet holes.

Ensure that the blasting plan contains at least the following:

(a) Maximum dimensions for width, length, and depth of shot.

(b) Typical plan and section view of the drill pattern for controlled backslope blast holes and production blast holes. Show the free face, burden, hole diameters, depths, spacings, inclinations, and depth of subdrilling, if any.

(c) Loading pattern diagram showing:

(1) Location of each hole.

(2) Location and amount of each type of explosive in each hole, including primer and initiators.

(3) Location, type, and depth of stemming.

(d) Initiation and delay methods, delay times, and overall powder factor.

(e) Manufacturer's data sheets for all explosives, primers, initiators, and other blasting devices.

(f) Working procedures and safety precautions for storing, transporting, and handling explosives.

(g) Working procedures and safety precautions for blasting.

The blasting plan is for quality control and recordkeeping purposes. The review of the blasting plan does not relieve the Contractor of the responsibility for using existing drilling and blasting technology, and for obtaining the required results.

220.05 Blaster-In-Charge. At least 10 days before the delivery or use of explosive material, designate in writing a blaster-in-charge.

220.06 Test Blasting. Drill, blast, and excavate one or more short test sections, as proposed in the blasting plan, before full-scale drilling and blasting. Test blasts may be made away from or at the final slope line.

Space blast holes according to the blasting plan for the initial test blast. Adjust the spacing as needed. A blast is unacceptable when it results in fragmentation beyond the final rock face; excessive fly rock; vibration; air blast; overbreak; damage to the final rock face; or a violation of other requirements. When a blast is determined to be unacceptable, revise the blasting plan and make an additional test blast.

220.07 Controlled Backslope Blasting. Perform controlled backslope blasting according to the following specifications:

(a) General. Drill and blast according to the blasting plan. Use controlled backslope blasting methods to form the final rock cut faces when the rock height is more than 3 m above ditch grade and slopes are staked 2:1 or steeper.

Use downhole angle or fan drill blast holes for pioneering the tops of rock cuts or preparing a working platform for controlled blasting. Use the blast hole diameter established for controlled backslope blasting and a hole spacing not exceeding 750 mm.

(b) Drilling. Remove overburden soil and loose rock along the top of the excavation for at least 10 m beyond the production hole drilling limits, or to the end of the cut.

Drill 75 ± 25-mm-diameter controlled backslope blast holes along the final rock face line. Drill controlled blast holes at least 10 m beyond the production holes to be detonated, or to the end of the cut.

Use drilling equipment with mechanical or electrical-mechanical devices that accurately control the angle at which the drill enters the rock. Select a lift height and conduct drilling operations so that the blast hole spacing and down-hole alignment do not vary more than 200 mm from the proposed spacing and alignment. When more than 5 percent of the holes exceed the variance, reduce the lift height and modify the drilling operations until the blast holes are within the allowable variance. Maximum lift height is limited to 20 m.

A 300-mm offset is allowed for a working bench at the bottom of each lift for drilling the next lower controlled blasting hole pattern.

Adjust the drill inclination angle or the initial drill collar location so that the required ditch cross section is obtained when the bench is used.

Drilling 500 mm below the ditch bottom is allowed for removing the toe.

(c) Blasting. Free blast holes of obstructions for their entire depth. Place charges with reasonable care to not cave in the blast hole walls.

Use the types of explosives and blasting accessories necessary to obtain the required results. A bottom charge may be larger than the line charges if no overbreak results. Do not use bulk ammonium nitrate and fuel oil for controlled blasting.

Stem the upper portion of all blast holes, preferably with dry sand or other granular material that passes a 9.5-mm sieve.

Where presplitting, delay the nearest production blast row at least 25 milliseconds after blasting the presplit line. Presplit a minimum of 10 m ahead of production blasting zone.

Where cushion (trim) blasting, delay the cushion blast row from 25 to 75 milliseconds after blasting the nearest production row.

220.08 Production Blasting. Perform production blasting according to the following specifications:

(a) General. Drill production holes and blast according to the blasting plan. Take all necessary precautions to minimize blast damage to the final rock face.

Following a blast, stop work in the entire blast area and check for misfires before allowing workers to return to excavate the rock.

Remove or stabilize all cut face rock that is loose, hanging, or potentially dangerous. Scale by hand using a standard steel mine scaling rod. Machine scale using hydraulic splitters or light blasting when necessary. Leave minor irregularities or surface variations in place if they do not create a hazard. Drill the next lift only after cleanup and stabilization work are completed.

If blasting operations cause fracturing of the final rock face, repair or stabilize it in an approved manner. Repair or stabilization may include removal, rock bolting, rock dowels, or other stabilization techniques.

Halt blasting operations if any of the following occur:

(1) Slopes are unstable.

(2) Slopes exceed tolerances.

(3) Backslope damage occurs.

(4) Public safety is jeopardized.

(5) Property or natural features are endangered.

(6) Fly rock is generated.

(b) Drilling. Drill the row of production blast holes closest to the controlled blast line parallel to the controlled blast line and no closer than 2 m to it. Do not drill production blast holes lower than the bottom of the controlled blast holes.

(c) Blasting. Use the types of explosives and blasting accessories that will obtain the desired fragmentation. Clean the blast holes, place the charges, and stem the holes according to Subsection 220.07(c). Detonate production holes on a delay sequence toward a free face.

220.09 Blasting Log. Submit a blasting log for each blast, including the following in the log:

(a) All actual dimensions of the shot, including blast hole depths, burden, spacing, subdrilling, stemming, powder loads, and timing.

(b) A drawing or sketch showing the direction of the face or faces, and the physical shot layout.

Measurement

220.10 Method. Use the method of measurement that is DESIGNATED IN THE SCHEDULE OF ITEMS.

Measure controlled backslope blast holes by the meter based on the actual length of drilling as recorded in the blasting log when it is included in the SCHEDULE OF ITEMS. All other rock blasting is subsidiary to other PAY ITEMS.

Payment

220.11 Basis. The accepted quantities will be paid for at the contract unit price for each PAY ITEM DESIGNATED IN THE SCHEDULE OF ITEMS.

Payment will be made under:

Pay Item	Pay Unit
220 (01) Controlled backslope blast hole	Meter

Section 221—Earthwork Geotextiles

Description

221.01 Work. Furnish and place a geotextile as a permeable separator or permanent erosion control measure.

Geotextile types are designated as follows:

Type II (A, B, or C) .... Separation
Type III (A or B) .... Stabilization
Type IV (A, B, C, D, E, or F) .... Permanent erosion control

Materials

221.02 Requirement. Ensure that material conforms to specifications in the following subsection:

Geotextiles 714.01

Construction

221.03 General. Where placing geotextiles on native ground, cut the trees and shrubs flush with the ground surface. Do not remove the topsoil and vegetation mat. Remove all sharp objects and large rocks. Fill depressions or holes with suitable material to provide a firm foundation.

Replace or repair all geotextile that is torn or punctured. When repairing, place a patch of the same type of geotextile overlapping 1 m beyond the damaged area.

221.04 Separation & Stabilization Applications. Where placing geotextiles on a subgrade, prepare the subgrade according to Subsections 203.08(c) and 203.08(d).

Place the geotextile smooth and free of tension, stress, or wrinkles. Fold or cut the geotextile to conform to curves. Overlap in the direction of construction. Overlap the geotextile a minimum of 500 mm at the ends and sides of adjoining sheets, or sew the geotextile joints according to the manufacturer's recommendations. Do not place longitudinal overlaps below anticipated wheel loads. Hold the geotextile in place with pins, staples, or piles of cover material.

End dump the cover material onto the geotextile from the edge of the geotextile or from previously placed cover material. Do not operate equipment directly on the geotextile. Spread the end-dumped pile of cover material, maintaining a minimum lift thickness as SHOWN ON THE DRAWINGS. Compact the cover material with rubber-tired or nonvibratory smooth drum rollers. Avoid sudden stops, starts, or turns of the construction equipment. Fill all ruts from construction equipment with additional cover material. Do not regrade ruts with placement equipment.

Place subsequent lifts of cover material in the same manner. Vibratory compactors may be used for compacting subsequent lifts. If foundation failures occur, repair the damaged areas and revert to the use of nonvibratory compaction equipment.

221.05 Erosion Control Applications. Place and anchor the geotextile on an approved smooth-graded surface. For slope or wave protection, place the long dimension of the geotextile down the slope. For stream bank protection, place the long dimension of the geotextile parallel to the centerline of the channel.

Overlap the geotextile a minimum of 300 mm at the ends and sides of adjoining sheets, or sew the geotextile joints according to the manufacturer's recommendations. Overlap the uphill or upstream sheet over the downhill or downstream sheet. Offset end joints of adjacent sheets a minimum of 1.5 m. Pins may be used to hold the geotextile sheets in place. Space pins along the overlaps at approximately 1 m between pins.

Place aggregate, slope protection, or riprap on the geotextile starting at the toe of the slope and proceed upward. Place riprap onto the geotextile from a height of less than 300 mm. Place slope protection rock or aggregate backfill onto the geotextile from a height less than 1 m. In underwater applications, place the geotextile and cover material on the same day.

Measurement

221.06 Method. Use the method of measurement that is DESIGNATED IN THE SCHEDULE OF ITEMS.

Measure earthwork geotextile by the square meter on a plane parallel to the ground surface, excluding overlaps.

Payment

221.07 Basis. The accepted quantities will be paid for at the contract unit price for each PAY ITEM DESIGNATED IN THE SCHEDULE OF ITEMS.

Payment will be made under:

Pay Item	Pay Unit
221 (01) Earthwork geotextile, type ____	Square Meter

Section 249—Composite Road Construction

Description

249.01 Work. Perform clearing and grubbing, excavation and embankment, and erosion control. During clearing and grubbing, treat merchantable timber and construction slash, including all trees designated for removal. During excavation and embankment, excavate and use borrow material; excavate drainage; shape the roadway, including approaches, turnarounds, ditches, and drainage dips; and place all excavated material, regardless of nature. Perform erosion control by furnishing and placing seed, fertilizer, mulch, and tackifier as SHOWN ON THE DRAWINGS. Construct the roadway in conformance with the dimensions SHOWN ON THE DRAWINGS or designated on the ground.

Materials

249.02 Requirement. Ensure that materials are as SHOWN ON THE DRAWINGS, and that they meet requirements specified in the following section and subsection:

Seeding & Mulching 625
Stabilizing Emulsion Tackifiers 713.12

Construction

249.03 Clearing & Disposal. Protect construction stakes and construction control markers. Remove or treat all trees, snags, downed timber, brush, and stumps within the clearing limits according to the following specifications:

(a) Merchantable Timber. Deck or remove timber meeting Utilization Standards as SHOWN ON THE DRAWINGS.

(b) Unmerchantable Timber. Treat unmerchantable timber as SHOWN ON THE DRAWINGS.

(c) Large Construction Slash. Treat construction slash larger than 75 mm in diameter and longer than 1 m by one or more of the following methods, as SHOWN ON THE DRAWINGS:

(1) Method A. Incorporate construction slash in the embankment.

(2) Method B. Windrow construction slash inside the clearing limits. When slash is windrowed, place it approximately parallel to the roadway outside the toe of the fill slope.

(3) Method C. Scatter construction slash outside the roadway without damaging trees. Limb all logs. Place logs and stumps away from trees, positioned so they will remain in place and are not on top of one another.

(4) Method D. Construct piles that are free of soil, with smaller slash well mixed with larger slash. Buck unmerchantable logs into lengths less than 6 m prior to placement in piles.

(5) Method E. Sidecast construction slash into the area below the roadway. Slash may be sidecast beyond the lower clearing limit for a distance not to exceed 3 m.

(6) Method F. Bury construction slash within the roadway limits. Construct mats in layers and cover the mats with at least 500 mm of rock and soil.

(7) Method G. Construct piles of construction slash in the areas SHOWN ON THE DRAWINGS or designated on the ground. Construct the piles so that burning does not damage standing trees. Burn the piles until all the material remaining in the pile is charred or ash.

(8) Method H. Bury the construction slash outside the roadway at the locations SHOWN ON THE DRAWINGS or designated on the ground. Construct mats in layers, and cover the mats with at least 500 mm of rock and soil. Slope the final surface to drain.

(9) Method J. Construct a debris mat of construction slash under the road subgrade. Use tree limbs, tops, cull logs, split stumps, wood chunks, and other debris to form a mat. Place stumps upside down and blended into the mat as SHOWN ON THE DRAWINGS.

(d) Small Construction Slash. Construction slash less than 75 mm in diameter and less than 1 m may be incorporated into embankments so long as the material is distributed so that it does not result in concentrations or matting.

Immediately remove slash deposited in stream courses.

Fell all dead trees outside the clearing limits that lean toward the road and are sufficiently tall to reach the roadbed. Fell hazard or unstable live trees designated on the ground outside the clearing limits before felling timber in the immediate clearing vicinity.

Leave stump heights less than 300 mm or one-third of the stump diameter, whichever is greater, measured on the side adjacent to the highest ground. Leave felled trees outside the clearing limits in place, and treat them no further unless otherwise SHOWN ON THE DRAWINGS.

249.04 Pioneering. Do not undercut the final back slope during pioneering operations. Deposit material inside the roadway limits. Do not restrict drainage.

249.05 Grubbing. Grub within the limits as SHOWN ON THE DRAWINGS. Stumps outside the grubbing limits may remain if cut no higher than 300 mm or one-third of the stump diameter, whichever is greater, above the original ground, measured on the uphill side, unless otherwise SHOWN ON THE DRAWINGS. Grub stumps that will protrude through the subgrade or have less than 150 mm of cover.

249.06 Excavation & Embankment. Construct the roadway to conform to the typical sections SHOWN ON THE DRAWINGS. Protect backslopes from being undercut. Embankment may be placed by side casting and end dumping.

Locate and use borrow material, and remove and treat unsuitable or excess material, as SHOWN ON THE DRAWINGS.

Place rocks that are too large to be incorporated in the embankment outside the traveled way on the downhill side such that they will not roll, obstruct drainage, or hinder roadbed use and maintenance.

Leave slopes that are to be seeded in a roughened condition.

Shape and finish the roadbed to the condition ordinarily accomplished by a crawler tractor with dozer blade to provide drainage of surface water, unless otherwise SHOWN ON THE DRAWINGS. Do not permit individual rocks to protrude more than 100 mm above the subgrade of the roadbed. A motor grader finish is not required.

Unless otherwise SHOWN ON THE DRAWINGS, observe a width tolerance for the traveled way of (+) 750 mm.

249.07 Erosion Control. Perform erosion control measures, including seeding, as SHOWN ON THE DRAWINGS. Use methods and rates of application, and types of seed, fertilizer, mulch, and tackifier, as specified in Section 625 and SHOWN ON THE DRAWINGS. Apply materials uniformly to the areas to be treated.

Measurement

249.08 Method. Use the method of measurement that is DESIGNATED IN THE SCHEDULE OF ITEMS.

Payment

249.09 Basis. The accepted quantities will be paid for at the contract unit price for each PAY ITEM DESIGNATED IN THE SCHEDULE OF ITEMS.

Payment will be made under:

Pay Item		Pay Unit
249 (01)	Composite road construction	Kilometer
249 (02)	Composite road construction	Lump Sum

DIVISION 250
Structural Embankments

Section 251—Riprap

Description

251.01 Work. Furnish and place riprap for bank protection, slope protection, drainage structures, and erosion control.

Riprap classes are designated as shown in table 705-1.

Materials

251.02 Requirements. Provide materials that conform to requirements in the following subsections:

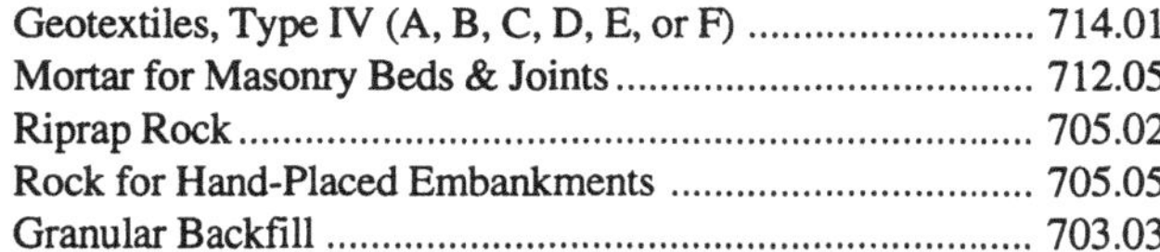

Geotextiles, Type IV (A, B, C, D, E, or F) 714.01
Mortar for Masonry Beds & Joints 712.05
Riprap Rock 705.02
Rock for Hand-Placed Embankments 705.05
Granular Backfill 703.03

Provide gravel cushion that meets the gradation requirements SHOWN ON THE DRAWINGS and the quality requirements specified in Subsection 703.06.

Construction

251.03 General. Minimize ground disturbance where practicable in preparing for placement of riprap. Prepare surfaces by removing logs, cutting brush and stumps flush with the ground, or as SHOWN ON THE DRAWINGS. Remove all soft or spongy material to the depths SHOWN ON THE DRAWINGS and replace it with approved material. Perform structural excavation and backfill as specified in Section 206A. Place geotextile as SHOWN ON THE DRAWINGS.

Control gradation by visual inspection. When SHOWN ON THE DRAWINGS, provide two samples of the specified class of rock. Each sample shall be at least 4.5 t or 10 percent of the total riprap weight, whichever is less. Provide one sample at the construction site, which may be a part of the finished riprap covering. Provide the other sample at the quarry. Use these samples as a frequent reference for judging the gradation of the riprap supplied. When specified in the SPECIAL PROJECT SPECIFICATIONS, provide mechanical equipment at the sorting site and the labor needed to assist in checking gradation.

251.04 Placed Riprap. Placed riprap is rock placed on a prepared surface to form a well-graded mass.

(a) Method A, Machine Placed. Place riprap to its full thickness in one operation to avoid displacing the underlying material. Do not place riprap material by methods that cause segregation or damage to the prepared surface. Place or rearrange individual rocks by mechanical or manual methods to obtain a compact uniform blanket with a reasonably smooth surface.

(b) Method B, End Dumped. Dump riprap to its full thickness in one operation. Avoid displacing the underlying material. Distribute larger rocks throughout the mass of stone. Obtain a uniformly thick blanket with a reasonably smooth surface.

251.05 Keyed Riprap. Keyed riprap is rock placed on a prepared surface and keyed into place by striking with a flat-faced weight.

Place rock for keyed riprap according to Subsection 251.04. Key the riprap into place by striking the surface with a 1.2 × 1.5-m flat-faced weight that weighs approximately 2,000 kg. Do not strike riprap below the water surface.

251.06 Mortared Riprap. Mortared riprap is rock placed on a prepared surface with the voids filled with Portland cement mortar.

Place rock for mortared riprap according to Subsection 251.04. Thoroughly moisten the rocks and wash any excess fines to the underside of the riprap. Place mortar only when the temperature is above 2 °C and rising. Place the mortar in a manner to prevent segregation. Fill all voids without unseating the rocks. Provide weep holes through the riprap as SHOWN ON THE DRAWINGS. Protect the mortared riprap from freezing and keep it moist for 3 days after the work is completed.

Where the depth SHOWN ON THE DRAWINGS for grouting is in excess of 300 mm, place the riprap in lifts of 300 mm or less. Grout each lift prior to placing the next lift. Construct and grout the succeeding lifts before the grout in the previous lift has hardened.

251.07 Sacked Concrete Riprap. Ensure that type A and type B sacked concrete riprap is prepared as described below.

(a) Type A. Prepare concrete containing at least 195 kg of cement per cubic meter; aggregate with a maximum size of 60 mm; and water limited to that necessary to ensure good workability without loss of cement by seepage through the sacks. Use reasonably clean and strong aggregate of appropriate size gradation. Use sacks that are at least 310 g/m^2 burlap, with a 1,016-mm width, or equivalent. Ensure that minimum weight of the filled sack is 25 kg. Place sacks while contents are moist. Premixed concrete that meets the requirements specified in this section is acceptable.

Loosely place the sacks, filled with concrete, to leave room for folding at the top. Make the fold just enough to retain the concrete at time of placing. Immediately after filling the sacks with concrete, place and lightly trample them to cause them to conform with the earth face and with adjacent sacks.

Remove all dirt and debris from the top of the sacks before the next course is laid thereon. Place stretchers so the folded ends will not be adjacent. Place headers with the folds toward the earth face. Do not place more than four vertical courses of sacks in any tier until initial set has taken place in the first course of any such tier.

(b) Type B. Provide type B (premixed) sacked concrete riprap containing commercially packaged dry combined materials for concrete. Ensure that each sack weighs at least 30 kg; is about 300 × 450 × 150 mm in size; and is strong enough for the mass of concrete it contains and free of tears and imperfections. Ensure that the concrete has not taken an initial set prior to placing.

Place and lightly compress sacks to cause them to conform with the earth surface and with adjacent sacks. When more than one layer of sacks is required, stagger joints one-half sack width. Do not place more than four vertical courses (one tier) of sacks until initial set has taken place in the first course of any such tier.

After placement, penetrate each sack at least six times from the top through the entire sack thickness, leaving at least a 13- to 25-mm-diameter void in the concrete mixture. Do not damage the sack through these penetrations to the extent that the concrete mixture is spilled or wasted.

When there will not be proper bearing or bond for the concrete because of delays in placing succeeding layers of sacks or because the work is hampered by storms, mud, or other causes, excavate a small trench behind the row of sacks already in place, and fill the trench with fresh concrete before laying the next layer of sacks.

Keep sacked concrete riprap moist and protected from freezing for a period of 4 days after placement.

251.08 Sacked Soil Cement Riprap. Sacked soil cement riprap may be composed of any combination of gravel, sand, silt, and clay with the following limitations: do not use topsoil; ensure that at least 55 percent of the mixed soil passes the 4.75-mm (no. 4) sieve, and that not more than 15 percent passes the 75-µm (no. 200) sieve; and ensure that the maximum size gravel passes the 37.5-mm sieve. Pulverize the soil so that no lumps exceed 13 mm in diameter. Thoroughly and uniformly mix the cement, soil, and water before placing in sacks. Limit moisture content to that necessary for good mixing without seepage. Provide sacks that are at least 310 g/m^2 burlap, with a 1,016-mm width, or equivalent. Ensure that the minimum weight of the filled sack is 25 kg. Place sacks while contents are moist.

The cement requirements in percent by volume for each soil group are shown below:

AASHTO Classification (M 145) Soil Group	Percent Cement by Volume
A-1-a	7
A-1-b	9
A-2	10
A-3	12
A-4	12
A-5	13
A-6	14
A-7	15

251.09 Hand-Placed Riprap. Securely bed the rock. Use spalls and small rocks to fill voids. Fill any spaces in back of the hand-placed riprap with compacted material.

251.10 Granular Filter Blanket. Place a sheathing material as specified in Subsection 703.04 where SHOWN ON THE DRAWINGS to the full specified thickness of each layer in one operation, using methods that will not cause segregation of particle sizes within the layer. Ensure that the surface of the finished layer is reasonably even and free of mounds or windrows. Place additional layers of filter material in a manner that will not cause mixture of the material in the different layers.

251.11 Geotextile. Place the geotextile as SHOWN ON THE DRAWINGS. Provide surfaces upon which the geotextile is to be placed with a uniform slope, and make them reasonably smooth and free of obstructions, depressions, and debris that could damage the geotextile. Have the surfaces approved before placing geotextile.

Loosely lay the geotextile without wrinkles or creases. Sew or overlap adjacent strips at joints. Insert securing pins through both strips of overlapped geotextile at maximum intervals of 900 mm, but not closer than 50 mm to each edge. Prevent the geotextile from being displaced.

Have the installed geotextile approved before covering with granular backfill or other materials. Carefully place the granular backfill on the geotextile to the depth SHOWN ON THE DRAWINGS by methods that will not damage the geotextile. Do not drop riprap placed on the granular backfill a distance greater than 900 mm.

Measurement

251.12 Method. Use the method of measurement that is DESIGNATED IN THE SCHEDULE OF ITEMS.

Payment

251.13 Basis. The accepted quantities will be paid for at the contract unit price for each PAY ITEM IN THE SCHEDULE OF ITEMS.

Payment will be made under:

Pay Item		Pay Unit
251 (01)	Placed riprap, class ______, method ______	Cubic Meter
251 (02)	Placed riprap, class ______, method ______	Ton
251 (03)	Placed riprap, class ______, method ______	Square Meter
251 (04)	Keyed riprap, class ______	Cubic Meter
251 (05)	Keyed riprap, class ______	Ton
251 (06)	Mortared riprap, class ______	Cubic Meter
251 (07)	Sacked concrete riprap, type ______	Square Meter
251 (08)	Sacked concrete riprap, type ______	Cubic Meter
251 (09)	Sacked soil cement riprap	Square Meter
251 (10)	Sacked soil cement riprap	Cubic Meter
251 (11)	Granular filter blanket	Cubic Meter
251 (12)	Hand-placed riprap	Square Meter
251 (13)	Geotextile, type IV, ______	Square Meter

Section 252—Special Rock Embankment & Rock Buttress

Description

252.01 Work. For special rock embankment, furnish rock and place it mechanically or by hand in fill sections. For rock buttress, furnish rock and place it mechanically or by hand in cut sections.

Special rock embankments and rock buttresses are designated as hand placed or mechanically placed.

Materials

252.02 Requirements. Furnish material that conforms to specifications in the following subsections:

Geotextiles, Type II (A, B, or C) 714.01
Rock for Hand-Placed Embankments 705.05
Rock for Mechanically Placed Embankments 705.04

Control gradation by visual inspection. When SHOWN ON THE DRAWINGS, provide two samples of the specified class of rock. Each sample shall be at least 4.5 t or 10 percent of the total rock weight, whichever is less. Provide one sample at the construction site, which may be a part of the finished rock covering. Provide the other sample at the quarry. Use these samples as a frequent reference for judging the gradation of the rock supplied. When specified in the SPECIAL PROJECT SPECIFICATIONS, provide mechanical equipment at the sorting site and the labor needed to assist in checking gradation.

Construction

252.03 Geotextile Installation. Place the geotextile as SHOWN ON THE DRAWINGS. Ensure that the surfaces upon which geotextile is to be placed have a uniform slope and are reasonably smooth and free of obstructions, depressions, and debris that could damage the geotextile. Have the surface approved by the CO prior to placing geotextile.

Lay the geotextile without tension, stress, wrinkles, or creases. Sew or overlap adjacent strips a minimum of 300 mm at joints. Insert securing pins through both strips of overlapped geotextile at maximum intervals of 900 mm, but no closer than 50 mm to each edge, to prevent the geotextile from being displaced.

Have the installed geotextile approved by the CO prior to covering.

252.04 Placing Rock. Perform the work specified in Section 203 or 206A, as required.

Place the rock in a stable orientation with minimal voids. Offset the rock to produce a random pattern. Use spalls smaller than the minimum rock size to chock the larger rock solidly in position and to fill voids between the large rocks.

Construct the exposed face of the rock mass to be reasonably uniform, with no projections beyond the neat line of the slope that are more than 500 mm for mechanically placed rock and 300 mm for hand-placed rock, or as SHOWN ON THE DRAWINGS.

Measurement

252.05 Method. Use the method of measurement that is DESIGNATED IN THE SCHEDULE OF ITEMS.

Payment

252.06 Basis. The accepted quantities will be paid for at the contract unit price for each PAY ITEM DESIGNATED IN THE SCHEDULE OF ITEMS.

Payment will be made under:

Pay Item	Pay Unit
252 (01) Special rock embankment, ______	Cubic Meter
252 (02) Special rock embankment, ______	Square Meter
252 (03) Special rock embankment, ______	Ton
252 (04) Rock buttress, ______	Cubic Meter
252 (05) Rock buttress, ______	Square Meter
252 (06) Rock buttress, ______	Ton
252 (07) Geotextile, type II, ______	Square Meter

Section 253—Gabions

Description

253.01 Work. Construct gabion structures and mattresses.

Materials

253.02 Requirements. Ensure that material conforms to specifications in the following subsections:

Gabion Material	720.02
Gabion Rock	705.01
Geotextiles, Type IV (A, B, C, D, E, or F)	714.01
Structural Backfill	704.04

Construction

253.03 General. Perform the work specified in Section 206A.

253.04 Basket Assembly. Do not damage wire coatings during basket assembly, structure erection, cell filling, or backfilling. Rotate the basket panels into position and join the vertical edges with fasteners according to Subsection 253.05. Temporary fasteners may be used for basket assembly if they are supplemented during structure erection with permanent fasteners according to Subsection 253.05.

Rotate the diaphragms into position and join the vertical edges according to Subsection 253.05.

253.05 Structure Erection. Place the empty gabion baskets on the foundation and interconnect the adjacent baskets along the top and vertical edges using fasteners.

Where lacing wire is used, wrap the wire with alternating single and double loops every other mesh opening and not more than 150 mm apart. Where spiral binders are used, crimp the ends to secure the binders in place. Where alternate fasteners are used, space the fasteners in every mesh opening and not more than 150 mm apart.

In the same manner, interconnect each vertical layer of baskets to the underlying layer of baskets along the front, back, and sides. Stagger the vertical joints between the baskets of adjacent rows and layers by at least one cell length.

253.06 Cell Filling. Remove all kinks and folds in the wire mesh, and properly align all the baskets. Place rock carefully in the basket cells to prevent the baskets from bulging and to minimize voids in the rock fill. Maintain the basket alignment.

Place internal connecting wires in each unrestrained exterior basket cell greater than 300 mm in height. This includes interior basket cells left temporarily unrestrained. Place internal connecting wires concurrently with rock placement.

Fill the cells in any row or layer so that no cell is filled more than 300 mm above an adjacent cell. Repeat this process until the basket is full and the lid bears on the final rock layer.

Secure the lid to the sides, ends, and diaphragms according to Subsection 253.05. Make all exposed basket surfaces smooth and neat, with no sharp rock edges projecting through the wire mesh.

253.07 Geotextile Installation. Place the geotextile according to Section 221 or as SHOWN ON THE DRAWINGS. Ensure that the surfaces upon which geotextile is to be placed have a uniform slope and are reasonably smooth and free of obstructions, depressions, and debris that could damage the geotextile. Have the surface approved by the CO prior to placing geotextile.

Lay the geotextile without tension, stress, wrinkles, or creases. Sew or overlap adjacent strips a minimum of 300 mm at joints. Insert securing pins through both strips of overlapped geotextile at maximum intervals of 900 mm, but no closer than 50 mm to each edge, to prevent the geotextile from being displaced.

Have the installed geotextile approved by the CO prior to backfilling with structural backfill or other approved material.

253.08 Backfilling. Backfill behind the gabion structure concurrently with the cell-filling operation. Backfill the area behind the gabion structure with structural backfill or acceptable roadway excavation, as approved by the CO. Compact each layer in accordance with Subsection 203.16(b), method 4.

253.09 Gabion Mattresses. Place a geotextile according to Section 221. Construct revet mattresses according to Subsections 253.04 through 253.06. Anchor the mattresses in place. Place geotextile against the vertical edges of the mattress, and backfill against the geotextile using structural backfill or other material, as approved by the CO. Overfill gabion mattresses by 30 to 50 mm.

Measurement

253.10 Method. Use the method of measurement that is DESIGNATED IN THE SCHEDULE OF ITEMS.

Payment

253.11 Basis. The accepted quantities will be paid for at the contract unit price for each PAY ITEM DESIGNATED IN THE SCHEDULE OF ITEMS.

Payment will be made under:

Pay Item		Pay Unit
253 (01)	Gabions, galvanized- or aluminized-coated	Square Meter
253 (02)	Gabions, epoxy- or polyvinylchloride (PVC)-coated	Square Meter
253 (03)	Gabions, galvanized- or aluminized-coated	Cubic Meter
253 (04)	Gabions, epoxy- or PVC-coated	Cubic Meter
253 (05)	Geotextile, type IV, ______	Square Meter

Section 254—Crib Walls

Description

254.01 Work. Construct concrete, metal, or timber crib retaining walls.

Materials

254.02 Requirements. Ensure that material conforms to specifications in the following subsections:

Bed Course	704.09
Crib Wall Backfill	704.12
Hardware & Structural Steel	716.02
Geotextiles, Type IV (A, B, C, D, E, or F)	714.01
Metal Bin Type Crib Walls	720.03
Precast Concrete Units	725.11
Structural Backfill	704.04
Treated Structural Timber & Lumber	716.03

Construction

254.03 General. Perform the work specified in Section 206. When the wall is set on a rocky foundation, place 200 mm of bed course under the wall base elements.

254.04 Erection. Furnish all necessary bolts, nuts, and hardware for complete assembly of the units into a continuous wall of connected units. Erect the crib wall according to the fabricator's or manufacturer's instructions. On curves, obtain the proper curvature for the face by using shorter stringers in the front or rear panels. Construct the wall to within 25 mm per 3 m from the lines and elevations, as SHOWN ON THE DRAWINGS.

(a) Concrete Crib Wall. Remove and replace all concrete members that are cracked or damaged.

(b) Metal Crib Wall. Torque bolts for metal crib walls according to manufacturer's recommendations.

(c) Timber Crib Wall. Construct timber cribs according to Section 557.

254.05 Geotextile Installation. Place the geotextile as SHOWN ON THE DRAWINGS. Ensure that the surfaces upon which geotextile is to be placed have a uniform slope and are reasonably smooth and free of obstructions, depressions, and debris that could damage the geotextile. Have the surface approved by the CO prior to placing geotextile.

Lay the geotextile without tension, stress, wrinkles, or creases. Sew or overlap adjacent strips a minimum of 300 mm at joints. Insert securing pins through both strips of overlapped geotextile at maximum intervals of 900 mm, but no closer than 50 mm to each edge, to prevent the geotextile from being displaced.

Have the installed geotextile approved by the CO prior to backfilling.

254.06 Backfilling. Backfill the inside of the cribs formed by the wall members with crib wall backfill in 150-mm layers. Place crib wall backfill, structural backfill, or acceptable roadway excavation behind cribs. Maintain an equal elevation of fill behind and inside cribs during backfilling operations. Compact each layer according to Subsection 203.16(b), method 4.

Measurement

254.07 Method. Use the method of measurement that is DESIGNATED IN THE SCHEDULE OF ITEMS.

Payment

254.08 Basis. The accepted quantities will be paid for at the contract unit price for each PAY ITEM DESIGNATED IN THE SCHEDULE OF ITEMS.

Payment will be made under:

Pay Item		Pay Unit
254 (01)	Concrete crib retaining wall	Square Meter
254 (02)	Metal crib retaining wall	Square Meter
254 (03)	Treated timber crib retaining wall	Square Meter
254 (04)	Crib wall backfill	Cubic Meter
254 (05)	Geotextile, type IV, ______	Square Meter

Section 255—Mechanically Stabilized Earth Walls

Description

255.01 Work. Construct mechanically stabilized earth walls.

Materials

255.02 Requirements. Ensure that material conforms to specifications in the following section and subsections:

Geotextiles, Type IV (A, B, C, D, E, or F)	714.01
Mechanically Stabilized Earth Wall Material	720.01
Minor Concrete Structures	602
Select Granular Backfill	704.10
Structural Backfill	704.04

Construction

255.03 General. Perform the work specified in Section 206. Grade the foundation for a width equal to the length of reinforcing mesh or strips plus 500 mm. Where the wall is set on a rocky foundation, place 150 mm of select granular backfill under the reinforcing mesh or strips.

For concrete-faced walls, provide a precast reinforced or a nonreinforced cast-in-place concrete leveling pad. Cure cast-in-place leveling pads a minimum of 12 hours before placing wall panels.

255.04 Wall Erection. Erect the wall as SHOWN ON THE DRAWINGS and according to the manufacturer's recommendations.

(a) Concrete-Faced Walls. Erect panels by means of lifting devices connected to the upper edge of the panel. Align precast facing panels within 19 mm vertically and horizontally when measured with a 3-m straightedge.

Make the joint openings 19 mm ± 6 mm wide. Install joint material according to the drawings. Cover all joints on the backside of the panels with a 300-mm-wide strip of geotextile. Overlap geotextile splices a minimum of 100 mm.

Hold the panels in position with temporary wedges or bracing during backfilling operations. Erect the wall so the overall vertical tolerance (top to bottom) does not exceed 13 mm per 3 m of wall height.

(b) Wire-Faced Walls. Place backing mats and 6-mm hardware cloth in successive horizontal lifts as backfill placement proceeds. Connect, tighten, and anchor soil reinforcement elements to the wall facing units before placing backfill. Do not exceed an individual lift vertical tolerance and an overall-wall (top-to-bottom) vertical tolerance of 25 mm per 3 m of wall height. Place reinforcement elements within 25 mm vertically of the corresponding connection elevation at the wall face. Do not deviate from the designed batter of the wall by more than 25 mm per 3 m of wall height. Do not deviate more than 50 mm at any point in the wall from a 3-m straightedge placed horizontally on the theoretical plane of the design face.

(c) Gabion-Faced Walls. Furnish and assemble gabion baskets according to Subsection 253.04. Lay reinforcement mesh horizontally on compacted fill. Place the soil reinforcement mesh normal to the face of the wall. Connect the gabion facing unit to reinforcement mesh with spiral binders or tie wire at 100 mm nominal spacing with alternating single and double locked loops. Join adjacent baskets along vertical edges according to Subsection 253.05. Fill gabion basket cells according to Subsection 253.06. Pull and anchor the reinforcement mesh taut before placing backfill. Place gabion baskets in successive horizontal lifts in the sequence shown on the drawings as backfill placement proceeds.

255.05 Backfilling. Backfill and compact the stabilized volume with select granular backfill in accordance with Subsection 203.16(b), method 4. Ensure that no voids exist beneath the reinforcing mesh or strips. Where spread footings used to support bridge or other structural loads are supported by the stabilized volume, compact the top 1.5 m below the footing elevation to at least 100 percent of the maximum density, as determined by AASHTO T 99, method C or D.

Do not damage or disturb the facing, reinforcing mesh, or strips. Compact within 1 m of the wall face with an approved lightweight mechanical tamper, roller, or vibratory system. Correct all damaged, misaligned, or distorted wall elements.

Backfill and compact behind the stabilized volume with structural backfill or suitable roadway excavation.

At the end of each day's operation, slope the last lift of backfill away from the wall face to rapidly direct runoff away from the wall. Do not allow surface runoff from adjacent areas to enter the wall construction site.

Measurement

255.06 Method. Use the method of measurement that is DESIGNATED IN THE SCHEDULE OF ITEMS.

Payment

255.07 Basis. The accepted quantities will be paid for at the contract unit price for each PAY ITEM DESIGNATED IN THE SCHEDULE OF ITEMS.

Payment will be made under:

Pay Item	Pay Unit
255 (01) Mechanically stabilized earth wall	Square Meter

Section 257—Alternate Retaining Walls

Description

257.01 Work. Ensure that various types of retaining walls are constructed at the Contractor's option. The alternate wall types are gabions, crib walls, mechanically stabilized earth walls, permanent ground anchor walls, reinforced concrete retaining walls, and reinforced soil embankments.

Materials

257.02 Requirements. Ensure that material conforms to specifications in the following sections:

Crib Walls	254
Driven Piles	551
Gabions	253
Mechanically Stabilized Earth Walls	255
Reinforced Soil Embankments	259
Reinforcing Steel	554
Structural Concrete	552

Construction

257.03 General. The designer/supplier furnishing the proposed wall is responsible for the stability of the wall. Do not qualify the responsibility for the design or restrict the use of the drawings or calculations for the proposed alternate. Indemnify the Government from all claims for infringement of proprietary rights by others without the consent of the patent holders or licensees.

257.04 Submittal. Submit a proposal using any of the wall types listed. Submit wall type proposals on a site-by-site basis. Different types may be used at individual sites on the project.

Provide drawings of the proposed wall within 120 days of the notice to proceed and at least 90 days before starting wall construction. Verify the limits of the wall before preparing drawings.

All drawings shall be signed by a licensed professional engineer.

Include all details, dimensions, quantities, ground profiles, and cross sections necessary to construct the wall. Submit design calculations on sheets about

200 × 300 mm in size with the project number, wall location, designation, date of preparation, initials of designer and checker, and page number at the top of the page.

Provide an index page with the design calculations. Ensure that the drawings include, but are not limited to, the following items:

(a) Plan and elevation drawings for each wall, containing the following:

(1) A plan view of the wall, identifying:

(a) The offset from the construction centerline to the face of the wall at its base at all changes in horizontal alignment.

(b) The limit of widest module, mesh, strip, or anchor.

(c) The centerline of any drainage structure or drainage pipe behind, passing through, or passing under the wall.

(2) An elevation view of the wall identifying:

(a) The elevation at the top of the wall, at all horizontal and vertical break points, and at least every 10 m along the wall.

(b) Elevations at the wall base, the top of leveling pads and footings, or the bottom of soldier piles.

(c) The wall batter.

(d) The distance along the face of the wall to all steps in the wall base, footings, leveling pads, or lagging.

(e) The type of panel or depth of module or lagging.

(f) The length and type of mesh, strips, or anchors.

(g) The distance along the face of the wall to where changes in length of the mesh, strips, or anchors occur.

(h) The original and final ground line.

(3) General notes for constructing the wall.

(4) Horizontal and vertical curve data affecting the wall. Match lines or other details to relate wall stationing to centerline stationing.

(5) A listing of the summary of quantities on the elevation drawing of each wall.

(b) Dimensions and schedules of all reinforcing steel, including reinforcing bar bending details, dowels, and/or studs for attaching the facing.

(c) Details and dimensions for foundations and leveling pads, including steps in the footings or leveling pads.

(d) Details and dimensions for:

(1) All panels, modules, soldier piles, and lagging necessary to construct the element.

(2) All reinforcing steel in the element

(3) The location of mesh, strip attachment, or anchor devices embedded in the panels.

(4) All anchors and soldier piling, including the spacing and size of piles and the spacing and angle of anchors.

(e) Details for constructing walls around drainage facilities.

(f) Details for terminating walls and adjacent slope construction.

(g) Architectural treatment details.

(h) Design notes including an explanation of any symbols and computer programs used in the design of the walls. Specify the factors of safety for sliding, pullout, and overturning. Specify the bearing pressure beneath the wall footing, stabilized earth mass, or soldier piles.

(i) Verification of the design criteria for the site-specific wall locations with test procedures, results, and interpretations. Include results from creep, durability, construction-induced damage, and junction strength tests.

(j) Other design calculations.

Process all submissions unless written permission is given for the wall designer/ supplier and the CO to communicate directly.

Submit three sets of the wall drawings with the initial submission. One set will be returned with any indicated corrections. If revisions are necessary, make the necessary corrections and resubmit three revised sets.

When the drawings are approved, furnish five sets and a mylar sepia set of the drawings.

257.05 Construction. Construct the wall according to the approved drawings and the following sections, as applicable:

(a) Gabions—Section 253.

(b) Crib walls—Section 254.

(c) Mechanically stabilized earth walls—Section 255.

(d) Permanent ground anchor walls—Sections 551 and 552.

(e) Reinforced soil embankments—Section 259.

Revise the drawings when plan dimensions are revised due to field conditions or for other reasons.

Measurement

257.06 Method. Use the method of measurement that is DESIGNATED IN THE SCHEDULE OF ITEMS.

Measure alternate retaining walls by the lump sum.

Payment

257.07 Basis. The accepted quantities will be paid for at the contract unit price for each PAY ITEM DESIGNATED IN THE SCHEDULE OF ITEMS.

When plan dimensions are changed by the CO during construction to account for field conditions, the lump sum price of the wall will be adjusted by applying a calculated per-square-meter cost adjustment factor to the added or decreased wall front face area resulting from the change. The adjustment factor will be determined by dividing the lump sum price bid for each wall by its estimated area shown in the PAY ITEM.

The alternate retaining wall lump sum will be paid based on the progress of the work under this section.

Payment will be made under:

Pay Item	Pay Unit
257 (01) Alternate retaining wall ____________ *Location* (__________ square meters) *Estimated*	Lump Sum

Section 259—Reinforced Soil Embankments

Description

259.01 Work. Construct reinforced soil embankments. Geogrid soil reinforcement material categories are designated as shown in table 714-7.

Materials

259.02 Requirements. Ensure that material conforms to specifications in the following subsections:

Geogrids, Categories 1, 2, 3, 4, 5, or 6 .... 714.03
Select Granular Backfill 704.10

Construction

259.03 General. Lay soil reinforcements at the proper elevation and alignment, as SHOWN ON THE DRAWINGS.

Orient soil reinforcements such that the maximum tensile strength available is in the direction of primary reinforcement, as SHOWN ON THE DRAWINGS.

Do not splice reinforcements in the primary direction. Geogrids may be overlapped three ribs (approximately 25 mm) in the direction transverse to the primary direction, and may be held together with hog rings or other approved devices.

Install soil reinforcements in accordance with manufacturer's recommendations for their intended purpose. Include a copy of those recommendations with project submittals.

Exercise care to prevent wrinkle development and/or slippage of reinforcement during fill placement and spreading. Prior to beginning work, submit a plan indicating how the stretching and staking will be accomplished.

Do not permit construction equipment to operate directly upon the reinforcement material. A minimum fill thickness of 150 mm is required prior to operation of tracked vehicles over the reinforced material. Keep turning of tracked vehicles to a minimum to prevent tracks from displacing the fill and damaging the material.

Ensure that the following tolerances apply to the elevations and dimension as SHOWN ON THE DRAWINGS: install the base of reinforcement material within ± 100 mm of that specified; and do not permit any layer to exceed 300 mm in thickness.

259.04 Field Adjustments. The final limits and configuration of the fills or reinforcement sections may vary, depending on the foundation materials encountered during excavation. Ensure that final foundation limits are approved by the CO in writing before placing any backfill material.

At least 48 hours before beginning work, notify the CO to inspect the foundation.

259.05 Excavation. Dispose of unsuitable material as SHOWN ON THE DRAWINGS.

259.06 Performance. Maintain all construction stakes to control the work.

259.07 Backfilling. Provide select granular backfill material that meets requirements specified in Subsection 704.10, unless otherwise SHOWN ON THE DRAWINGS. Place and compact backfill in accordance with Subsection 203.16(b), method 4. Do not use sheepsfoot rollers to obtain compaction. On geogrid fills, place backfill so as to keep tension in the geogrid.

259.08 Slope Face Treatment. Treat the face of the reinforced slope for erosion control in accordance with Section 204 and as SHOWN ON THE DRAWINGS.

Measurement

259.09 Method. Use the method of measurement that is DESIGNATED IN THE SCHEDULE OF ITEMS.

Payment will be based on the contract unit price of reinforcing material placed and accepted. This will be exclusive of overlap and wastage.

Payment

259.10 Basis. The accepted quantities will be paid for at the contract unit price for each PAY ITEM DESIGNATED IN THE SCHEDULE OF ITEMS.

Payment will be made under:

Pay Item	Pay Unit
259 (01) Geogrid category__________	Square Meter

DIVISION 300
Bases

Section 304—Aggregate Base or Surface Course

Description

304.01 Work. Furnish, haul, and place aggregate base or surface course on the subgrade or base or stockpile site approved by the CO. Work may include additive mineral filler, or binder as specified in the SPECIAL PROJECT SPECIFICATIONS. Produce aggregate by pit-run, grid-rolling, screening, or crushing methods, or procure Government-furnished aggregate, as DESIGNATED IN THE SCHEDULE OF ITEMS.

Materials

304.02 Source. Obtain materials from sources or stockpiles SHOWN ON THE DRAWINGS, or from other approved sources.

Develop and utilize Government-furnished sources in accordance with Section 611.

304.03 Gradation. Ensure that grading requirements for crushing or screening operations meet the requirements specified in Subsection 703.05.

After a representative quantity of aggregate has been produced, and before incorporating the aggregate into the work, set target values (TV's) for the required gradation within the gradation ranges shown in table 703-2 or table 703-3.

No gradation other than maximum size will be required for pit-run or grid-rolled material. For grid rolling, utilize all suitable material that can be reduced to maximum size as DESIGNATED IN THE SCHEDULE OF ITEMS. After processing on the road, remove all oversize material from the road and dispose of it as SHOWN ON THE DRAWINGS.

304.04 Quality. Ensure that all aggregate meets the quality requirements specified in Subsection 703.05, unless otherwise required in the SPECIAL PROJECT SPECIFICATIONS.

304.05 Additives. Provide additives, if required, that meet specifications in the following subsections:

Bentonite 725.20
Calcium Chloride Flake 723.02
Hydrated Lime 725.03
Lignin Sulfonate 723.03
Magnesium or Calcium Chloride Brine 723.01

304.06 Water. Develop, haul, and apply water in accordance with Section 207.

304.07 Mineral Filler. Add mineral filler to meet quality and/or gradation requirements as specified in the SPECIAL PROJECT SPECIFICATIONS.

Construction

304.08 Preparation of Roadbed. Complete the roadbed in accordance with Section 203, 306, or 309, and have it approved in writing by the CO before placing base or surface course.

304.09 Mixing & Placing. Ensure that aggregate and any required additives, water, mineral filler, and binder are mixed by method(s) as SHOWN ON THE DRAWINGS; except, if crushed aggregate products are being produced and mineral filler, binder, or additives are required, uniformly blend during crushing.

(a) Stationary Plant Method. Mix the aggregate with other required materials in an approved mixer. Add water during the mixing operation in the amount necessary to provide the moisture content for compacting to the specified density. After mixing, transport the aggregate to the jobsite while it contains the proper moisture content, and place it on the subgrade or base course using an aggregate spreader.

(b) Travel Plant Method. After the aggregate for each layer has been placed with an aggregate spreader or windrow-sizing device, uniformly mix it with other required materials using a traveling mixing plant. During mixing, add water to provide the necessary moisture content for compacting.

(c) Road Mix Method. After the aggregate for each layer has been placed, mix it with other required materials at the required moisture content until the mixture is uniform throughout. Mix aggregate and all other materials until a uniform distribution is obtained.

Spread the aggregate in a uniform layer, with no segregation of size, and to a loose depth that has the required thickness when compacted.

If the required compacted depth of any aggregate base or surface course exceeds 150 mm, place the aggregate base or surface course in two or more layers of approximately equal thickness. If the nominal maximum particle size exceeds 75 mm, place the aggregate in layers that do not exceed twice the maximum size of the specified aggregate size.

During placement of aggregate over geotextile, place aggregate in a single lift to the full depth specified, unless otherwise SHOWN ON THE DRAWINGS or in the SPECIAL PROJECT SPECIFICATIONS.

Operate hauling equipment over the surface of the previously constructed layer in a dispersed manner to minimize rutting or uneven compaction.

304.10 Compaction. Compact the aggregate using one of the following methods, as specified in the SCHEDULE OF ITEMS:

(a) Compaction A. Compact the aggregate by operating spreading and hauling equipment over the full width of each layer of the aggregate.

(b) Compaction B. Moisten or dry the aggregate to a uniform moisture content suitable for compaction. Operate rollers that meet the requirements specified in Subsection 203.15(b), (c), or (d) over the full width of each layer until visual displacement ceases, making no fewer than three complete passes.

(c) Compaction C. Compact each layer of aggregate to a density of at least 95 percent of the maximum density, as determined by AASHTO T 99, method C or D.

(1) Compaction C-1. Compact each layer of aggregate to a density of at least 96 percent of the maximum density, as determined by the Modified Marshall Hammer Compaction Method (available upon request from USDA Forest Service, Regional Materials Engineering Center, P.O. Box 7669, Missoula, MT 59807).

(d) Compaction D. Compact each layer of aggregate to a density of at least 95 percent of the maximum density, as determined by AASHTO T 180, method C or D.

(1) Compaction D-1. Compact each layer of aggregate to a density of at least 100 percent of the maximum density as determined by the Modified Marshall Hammer Compaction Method (available upon request from USDA Forest Service, Regional Materials Engineering Center, P.O. Box 7669, Missoula, MT 59807).

(e) Compaction E. Ensure that materials produced by pit-run and grid-rolling are visually moist and compacted using operating compaction equipment defined in Subsection 203.15(b), (c), and (d) over the full width of each layer until visual displacement ceases.

For all compaction methods, blade the surface of each layer during the compaction operations to remove irregularities and produce a smooth, even surface. When a density requirement is specified, determine the density of each layer in accordance with AASHTO T 191, T 205, T 238, T 217, T 239, T 255, or T 224.

304.11 Stockpiling. If DESIGNATED IN THE SCHEDULE OF ITEMS or if the Contractor elects to produce and stockpile aggregates prior to placement, the aggregates shall be handled and stockpiled in accordance with Section 305. Establish stockpile sites at locations SHOWN ON THE DRAWINGS or approved by the CO.

Clear and grub stockpile sites, if required, in accordance with Section 201.

304.12 Thickness & Width Requirements. Ensure that the thickness and width of the compacted aggregate conform to the dimensions SHOWN ON THE DRAWINGS, and that measurements on the compacted aggregate meet the following criteria:

(a) The maximum variation from the specified thickness is 25 mm.

(b) The compacted thickness is not consistently above or below the specified thickness, and the average thickness of 4 or more measurements for any 1 km of road segment is within ± 5 mm of the specified thickness.

(c) The compacted width has a (+) 300 mm tolerance.

Measurement

304.13 Method. Use the method of measurement that is DESIGNATED IN THE SCHEDULE OF ITEMS.

Aggregate quantities include mineral filler, binder, and water.

Payment

304.14 Basis. The accepted quantities will be paid for at the contract unit price for each PAY ITEM DESIGNATED IN THE SCHEDULE OF ITEMS.

Payment will be made under:

Pay Item		Pay Unit
304 (01)	Pit run aggregate, maximum size _____, compaction_____	Cubic Meter
304 (02)	Pit run aggregate, maximum size _____, compaction _____	Ton
304 (03)	Pit run aggregate, maximum size _____, compaction______	Lump Sum
304 (04)	Grid-rolled aggregate, maximum size _____, compaction______	Cubic Meter

304 (05) Grid-rolled aggregate, maximum size _____,
compaction ______ .... Ton

304 (06) Grid-rolled aggregate, maximum size _____,
compaction ______ .... Lump Sum

304 (07) Screened aggregate, grading ________,
compaction ________ .... Cubic Meter

304 (08) Screened aggregate, grading ________,
compaction ________ .... Ton

304 (09) Screened aggregate, grading ________,
compaction ________ .... Lump Sum

304 (10) Crushed aggregate, type ____, grading ___,
compaction ________ .... Cubic Meter

304 (11) Crushed aggregate, type ____, grading ___,
compaction ________ .... Ton

304 (12) Crushed aggregate, type ____, grading ___,
compaction ________ .... Lump Sum

304 (13) Placing aggregate, compaction ___________ .... Cubic Meter

304 (14) Placing aggregate, compaction ___________ .... Ton

304 (15) Placing aggregate, compaction ___________ .... Lump Sum

304 (16) Stockpiled aggregate, type _____,
grading ___________ .... Cubic Meter

304 (17) Stockpiled aggregate, type _____,
grading ___________ .... Ton

304 (18) Furnishing and applying magnesium/calcium
chloride brine .... Ton

304 (19) Furnishing and applying calcium chloride flake .... Ton

304 (20) Furnishing and applying lignin sulfonate .... Ton

304 (21) Furnishing and applying hydrated lime .... Ton

304 (22) Furnishing and applying bentonite .... Ton

Section 305—Stockpiled Aggregates

Description

305.01 Work. Furnish and place aggregate in a stockpile location as SHOWN ON THE DRAWINGS.

Materials

305.02 Requirements. Furnish material that conforms to specifications in the following section:

Aggregate 703

Furnish stockpiled aggregates that conform to the gradation and quality requirements specified in the SCHEDULE OF ITEMS.

After a representative quantity of aggregate has been produced, and before incorporating the aggregate into the stockpile, set TV's for the required gradation within the gradation ranges.

Construction

305.03 Stockpile Site. Prepare existing sites as necessary to accommodate the quantity of aggregate to be stockpiled and as SHOWN ON THE DRAWINGS.

Prepare new sites as follows:

(a) Clear and grub in accordance with Section 201.

(b) Grade and shape the site to a uniform cross section that drains.

(c) Compact the floor of the site with at least three passes using compaction equipment conforming to Subsection 203.15.

(d) Place, compact, and maintain a minimum 150-mm layer of crushed aggregate over the stockpile site and access roads for stabilization and to prevent contamination of the stockpiles.

305.04 Stockpile. Obtain site approval before stockpiling aggregates. Make the stockpiles neat and regular in shape. Make the side slopes no flatter than 1:1.5.

Build the stockpiles in layers not to exceed 2 m in thickness. Complete each layer before depositing aggregates in the next layer.

Construct stockpile layers by spreading aggregates with trucks or other approved pneumatic-tired equipment. Do not push aggregates into piles.

Do not dump the aggregate such that any part of it runs down and over the lower layers in the stockpile. Do not drop aggregates from a bucket or spout in one location to form a cone-shaped pile.

When operating trucks on stockpiles, avoid tracking dirt or other foreign matter onto the stockpiled material.

Space stockpiles far enough apart to prevent different aggregate gradations from mixing, or use suitable walls or partitions to separate stockpiles.

Measurement

305.05 Method. Use the method of measurement that is DESIGNATED IN THE SCHEDULE OF ITEMS.

Payment

305.06 Basis. The accepted quantities will be paid for at the contract unit price for each PAY ITEM DESIGNATED IN THE SCHEDULE OF ITEMS.

Payment will be made under:

Pay Item		Pay Unit
305 (01)	Stockpiled aggregate, section ___, grading ___	Ton
305 (02)	Stockpiled aggregate, section ___, grading ___	Cubic Meter
305 (03)	Preparation of stockpile site	Hectare

Section 306—Reconditioning Existing Road

Description

306.01 Work. Recondition existing road; clean ditches, cattleguards, and culverts, including inlets and outlets; remove slide material; scarify where SHOWN ON THE DRAWINGS; and shape and compact the traveled way and shoulders, including parking areas, turnouts, and approach road intersections.

Construction

306.02 Blading & Shaping. Unless otherwise SHOWN ON THE DRAWINGS, blade and shape the existing traveled way and shoulders, including turnouts, to remove minor surface irregularities. Maintain the existing cross slope or crown, unless otherwise SHOWN ON THE DRAWINGS. Establish a blading pattern that will retain the surfacing on the roadbed and provide a thorough mixing of the materials within the completed surface width.

Scarify and shape the existing traveled way and shoulders at locations and to the depth and width SHOWN ON THE DRAWINGS. Remove any rock larger than 100 mm in its greatest dimension that is brought to the surface during scarification, except as provided below.

When a base or surface course is required, remove to at least 150 mm below existing surface any rocks that protrude above the existing surface more than one-third of the depth of the base or surface course, and any rocks with exposed surface area exceeding 0.2 m^2. Remove all unsuitable materials and place them in areas as SHOWN ON THE DRAWINGS.

Scarify and pulverize existing bituminous surfaces, as SHOWN ON THE DRAWINGS, until all lumps are reduced to the maximum size SHOWN ON THE DRAWINGS. Incorporate the bituminous pulverized aggregate into the traveled way and shoulders.

Similarly treat the traveled way and shoulders of intersecting roads to provide a smooth transition for the distance SHOWN ON THE DRAWINGS.

306.03 Rock Reduction. Where SHOWN ON THE DRAWINGS, roll the traveled way to break down rocky material. Continue rolling until visual breakdown or visual displacement of rocky material ceases, up to a maximum of five full-width passes. A pass is defined as one complete coverage of the surface.

Use rollers that are specifically designed to break down rocky material and that meet the requirements specified in Subsection 203.15(a) or (c), unless otherwise approved by the CO.

306.04 Drainage. Grade the ditches to the typical sections and at the locations SHOWN ON THE DRAWINGS. Clean culverts to drain.

Remove excess materials from the roadbed, culverts, and ditches, and place material as SHOWN ON THE DRAWINGS.

306.05 Cattleguards. Remove the cattleguard deck prior to cleaning, and reinstall it upon completion. Clean the area beneath the cattleguard of soil and other material to the bottom of the original foundation, or as SHOWN ON THE DRAWINGS, over the entire width of the installation.

Never leave the cattleguard opening unattended or open without an adequate barricade.

306.06 Compaction. Shape and compact the traveled way and shoulders using one of the methods described in Subsection 304.10, as DESIGNATED IN THE SCHEDULE OF ITEMS.

Measurement

306.07 Method. Use the method of measurement that is DESIGNATED IN THE SCHEDULE OF ITEMS.

Payment

306.08 Basis. The accepted quantities will be paid for at the contract unit price for each PAY ITEM DESIGNATED IN THE SCHEDULE OF ITEMS.

Payment will be made under:

Pay Item		Pay Unit
306 (01)	Reconditioning of roadbed, compaction _______	Kilometer
306 (02)	Reconditioning of roadbed, compaction _______	Lump Sum
306 (03)	Cleaning cattleguard	Each
306 (04)	Cleaning culvert	Each

Section 307—Portland Cement-Treated Base—Central Plant Mix

Description

307.01 Work. On a prepared roadbed, construct one or more courses of a mixture of aggregate, water, and cement, or a mixture of aggregate, water, fly ash, lime, and/or cement.

Treated aggregate courses are designated as cement or aggregate, water, fly ash, lime, and/or cement.

Materials

307.02 Requirements. Ensure that materials meet the requirements specified in the following subsections:

Blotter	703.12
Cement	701.01
Chemical Admixtures (Set-Retarding)	711.03
Emulsified Asphalt	702.03
Fly Ash	725.04
Fog Seal	410.10
Hydrated Lime	725.03
Subbase, Base, and Surface Course Aggregate	703.05
Water	725.01

Construction

307.03 General. Store all chemical additives in closed, weatherproof containers. Prepare the surface that the Portland cement-treated base is to be placed upon in accordance with Section 203 or 306, as applicable.

Thirty days before producing the treated aggregate mixture, submit the mix design for approval. Include the following, as applicable:

(1) Source of each component.

(2) Report showing the results of the following tests:

(a) Moisture density relationship at selected cement content or selected aggregate, water, fly ash, lime, and/or cement content.

(b) Compressive strength data for all specimens tested, at each cement content or each aggregate, water, fly ash, lime, and/or cement content.

(3) Target values for each aggregate sieve size specified.

(4) 90-kg sample of aggregate.

(5) 25-kg sample of fly ash.

(6) 10-kg sample of hydrated lime.

(7) 10-kg sample of Portland cement.

(8) 2-kg sample of the retarder or other additives.

Furnish a new mix design if there is a change in the source of material. Begin production only after the mix design is approved.

At least 10 days prior to construction, submit a quality control plan for review and approval. Include all tolerance items, field tests, and measurements shown in the specification. Ensure that the plan verifies compliance with these requirements for each 500-t or smaller sublot produced. Perform field testing and measurement at statistically random points within each sublot.

307.04 Mix Design Requirements. Provide test samples with a minimum average compressive strength of 3.4 MPa, with no single test lower than 2.8 MPa.

(a) Aggregate/Cement-Treated Course. Provide a mix design, based on dry weight, within the limits shown in table 307-1.

Table 307-1.—Range of aggregate/cement mix design parameters.

Material	Percent[a]
Aggregate	90–96
Portland cement	4–10

a. By weight of total dry mix.

(b) Aggregate/Water/Fly Ash/Lime/Cement-Treated Course. Provide an aggregate, water, fly ash, lime, and/or cement mix design, based on dry weight, within the limits shown in table 307-2.

Table 307-2.—Range of aggregate, water, fly ash, lime, and/or cement mix design parameters.

Material	Percent[a]
Aggregate	75–92
Fly ash	6–20
Hydrated lime and/or Portland cement	2–5

a. By weight of total dry mix.

Mold, cure, and test samples of the aggregate, water, fly ash, lime, and/or cement mixture in accordance with ASTM C 593, parts 10 and 11, but revise the curing period from 7 to 28 days at 38 ± 2 °C.

307.05 Mixing. Do not begin mixing operations when the atmospheric temperature is expected to fall below 4 °C within 48 hours. Do not place a treated aggregate course when the underlying surface is frozen or muddy, or when it is raining or snowing. Protect the finished course from freezing for at least 7 days after mixing.

Mix the components with a stationary pugmill-type central mixing plant until a uniform mixture is obtained. The CO may require additional mixing to insure uniformity. During mixing, add sufficient water to obtain the optimum moisture content for compaction, plus 2 percent.

Equip the mixer with batching or metering devices for proportioning the components either by weight or by volume. Maintain the accuracy of the amounts of aggregate, chemical additives, and water (based on total dry weight of mixture) within the following tolerances:

Aggregate .... ± 2.0% by weight
Fly ash .... ± 1.5% by weight
Hydrated lime or cement .... ± 0.5% by weight
Water .... ± 2.0% by weight

A retarder may be used to slow initial set for a maximum of 2 hours. Dissolve retarder in water and uniformly add the solution to the mixture.

Transport the mixture in vehicles that maintain moisture content, and that prevent segregation and loss of the fine material.

307.06 Placing, Compacting, & Finishing. Maintain the moisture content (± 2 percent of optimum) during placing, compacting, and finishing.

Spread the mixture on the prepared surface in a uniform layer. Do not place the mixture in a layer exceeding 150 mm in compacted thickness. When more than one layer is necessary, shape and compact each layer before the succeeding layer is placed. Route hauling equipment uniformly over the full width of the surface to minimize rutting or uneven compaction. Shape the final layer to line, grade, and cross section.

Compact each layer full width. Roll from the sides to the center, parallel to the centerline of the road. Compact each layer of aggregate to not less than 95 percent of maximum density.

Use AASHTO T 99 to determine the maximum density. Determine in-place density using any applicable AASHTO procedure.

Do not leave any treated aggregate that has not been compacted undisturbed for more than 30 minutes. Complete the compaction and finishing within 1 hour (up to 2 hours for mixtures with a retarder) from the time water is added to the mixture. Make the compacted surface smooth, dense, and free of compaction planes, ridges, or loose material.

The finished course thickness must be within a tolerance of 13 mm from that specified. Ensure that the finished thickness is not consistently above or below the specified thickness. Finish the surface to within ± 10 mm from the staked line and grade elevation. Correct all defective areas by loosening the material, adding or removing material, reshaping, and compacting.

If the time between placing adjacent partial widths exceeds 30 minutes, provide a construction joint.

307.07 Construction Joints. For lime and fly ash mixtures, tie each day's operation into the completed work of the previous day by remixing approximately 500 mm of the completed course before processing additional sections. Add 50 percent of the original amount of lime or fly ash to the remixed material.

For cement mixtures, or when a lime or fly ash mixture remains undisturbed for more than 24 hours, make a transverse construction joint by cutting back into the completed work to form an approximately vertical face.

307.08 Curing. Minimize hauling over the unprotected treated aggregate course. Limit traffic over the treated surface to prohibit any visible deflection, rutting, raveling, or wear of the surface. Keep the completed layer or course continuously moist until the subsequent layer or course is placed. Apply all curing water under pressure through a spray bar equipped with nozzles that produce a fine, uniform spray. Place and compact the subsequent layer or course within 2 days after the treated aggregate course has been compacted and finished. Placement of the subsequent layer or course may be deferred up to 21 days by placing a fog seal over the treated course. Do not place fog seals on intermediate layers of a course.

When a fog seal is used, keep the surface continuously moist for a minimum of 7 days after compacting and finishing. After the 7-day period, apply a diluted slow-setting emulsified asphalt at a rate of 0.45 to 1.15 L/m^2 to provide a continuous film over the course. Apply the fog seal in accordance with Subsection 410.10. When necessary for the maintenance of vehicular traffic, furnish and apply blotter material.

Reapply the fog seal where it has been determined that the "continuous film" has been damaged.

If the material loses the required stability, density, or finish before placement of the next course or acceptance of the work, reprocess, recompact, and add additives as necessary to restore the strength of the damaged material to that specified in the design.

Measurement

307.09 Method. Use the method of measurement that is DESIGNATED IN THE SCHEDULE OF ITEMS.

Measure fog seal and blotter under Section 410.

Payment

307.10 Basis. The accepted quantities will be paid for at the contract unit price for each PAY ITEM DESIGNATED IN THE SCHEDULE OF ITEMS.

Payment will be made under:

Pay Item		Pay Unit
307 (01)	______ Portland cement-treated base, grading ___	Ton
307 (02)	______ Portland cement-treated base, grading ___, ______ -mm depth	Square Meter
307 (03)	Portland cement	Ton
307 (04)	Fly ash	Ton
307 (05)	Hydrated lime	Ton

Section 308—Portland Cement-Treated Base—In-Place Stabilization

Description

308.01 Work. Process and incorporate chemical additives (fly ash, hydrated lime, or Portland cement) and water to a depth as SHOWN ON THE DRAWINGS.

Materials

308.02 Requirements. Ensure that materials meet the requirements specified in the following subsections:

Blotter	703.12
Cement	701.01
Chemical Admixtures (Set-Retarding)	711.03
Crushed Aggregate	703.06
Emulsified Asphalt	702.03
Fly Ash	725.04
Fog Seal	410.10
Hydrated Lime	725.03
Water	725.01

Construction

308.03 General. Scarify the road surface to the depth SHOWN ON THE DRAWINGS. Windrow the scarified materials or spread them to a uniform thickness for mixing. Remove any material that would be retained on a 50-mm sieve, along with other unsuitable materials. If additional aggregate is required, blend it with the existing material. If necessary, clean and trim all butt joints at existing pavement or structures prior to mixing.

Ensure that the subgrade supports all equipment required in the construction of the base. Prior to mixing, subexcavate soft or yielding areas and replace them with suitable materials, in accordance with Subsection 203.06(d).

At least 10 days prior to construction, submit a quality control plan for review and approval. Include all tolerance items, field tests, and measurements shown in this specification. Ensure that the plan verifies compliance with these requirements for each sublot produced that is 500 m^3 or less. Perform field testing and measurement at statistically random points within each sublot.

308.04 Application. Apply additives to a dry, unfrozen surface with a surface temperature of at least 4 °C. Do not apply when there is excessive loss of additive due to washing or blowing, or when the air temperature is expected to fall below 4 °C within 48 hours.

Apply additives at the required rates using one of the following methods:

(a) Dry Method. Uniformly apply the additives using an approved spreader. A motor grader is not an approved spreader. Apply water using approved methods to obtain the proper moisture content for mixing and compaction.

(b) Slurry Method. Mix additives with water and apply as a thin water suspension or slurry using either trucks with approved distributors or rotary mixers. Equip the distributor truck or rotary mixer tank with an agitator to keep the additives and water in suspension. Make successive passes over the aggregate course until the moisture and additive content for mixing and compaction are obtained.

Maintain the application of components within the following tolerances based on total dry weight of mixture:

Fly ash	± 2%
Hydrated lime or cement	± 1%
Water	± 2%

308.05 Mixing. Perform initial and final mixing as follows:

(a) Initial Mixing. Mix the material until a homogeneous friable mixture is obtained.

For lime or fly ash mixtures, add sufficient water and thoroughly mix to obtain the optimum moisture content for compaction, in accordance with AASHTO T 99, plus necessary hydration moisture. Hydration moisture is 1-1/2 percent for each percent of additive in the mixture. Complete the mixing within 6 hours of additive application.

For cement mixtures, add sufficient water and thoroughly mix to obtain the optimum moisture content for compaction, in accordance with AASHTO T 99, plus 2 percent. Complete the mixing within 2 hours of cement application.

Moist cure the mixture for 2 to 4 days.

(b) Final Mixing. For lime or fly ash mixtures, remix the material using an approved road mixer. Retarder may be added as desired. If the remixed material contains clods, rescarify in accordance with Subsection 306.02. Do not remix Portland cement mixtures.

308.06 Compaction & Finishing. Immediately after final mixing, shape and compact each layer full width. Roll from the sides to the center, parallel to the centerline of the road. Compact each layer of aggregate to not less than 95 percent of maximum density.

Use AASHTO T 99 to determine the maximum density. Use AASHTO T 224 to correct for coarse particles. Determine in-place density using any applicable AASHTO procedure.

Do not leave any treated aggregate that has not been compacted undisturbed for more than 30 minutes. Complete the compaction and finishing within 1 hour (up to 2 hours for mixtures with a retarder) from the time water is added to the mixture. Make the compacted surface smooth, dense, and free of compaction planes, ridges, or loose material.

The finished course thickness must be within a tolerance of 25 mm from that specified. Ensure that the finished thickness is not consistently above or below the specified thickness. Finish the surface to within ± 20 mm from the staked line and grade elevation. Correct all defective areas by loosening the material, adding or removing material, reshaping, and compacting.

If the time between placing adjacent partial widths exceeds 30 minutes, provide a construction joint.

308.07 Construction Joints. For lime and fly ash mixtures, tie each day's operation into the completed work of the previous day by remixing approximately 0.5 m of the completed course before processing additional sections. Add 50 percent of the original amount of lime or fly ash to the remixed material.

For cement mixtures or when a lime or fly ash mixture remains undisturbed for more than 24 hours, make a transverse construction joint by cutting back into the completed work to form an approximately vertical face. Use crushed aggregate meeting the requirements of Subsection 703.06 to repair transverse construction joints.

308.08 Curing. Minimize hauling over the unprotected treated aggregate course. Limit traffic over the treated surface to prohibit any visible deflection, rutting, raveling, or wear of the surface. Keep the completed layer or course continuously moist until the subsequent layer or course is placed. Apply all curing water under pressure through a spray bar equipped with nozzles that produce a fine, uniform spray. Place and compact the subsequent layer or course within 2 days after the treated aggregate course has been compacted and finished. Placement of the subsequent layer or course may be deferred up to 21 days by placing a fog seal over the treated course. Do not place fog seals on intermediate layers of a course.

When a fog seal is used, keep the surface continuously moist for a minimum of 7 days after compacting and finishing. After the 7-day period, apply a diluted slow-setting emulsified asphalt at a rate of 0.45 to 1.15 L/m^2 to provide a continuous film over the course. Apply the fog seal in accordance with Subsection 410.10. When necessary for the maintenance of vehicular traffic, furnish and apply blotter material.

Reapply the fog seal where it has been determined that the "continuous film" has been damaged.

If the material loses the required stability, density, or finish before placement of the next course or acceptance of the work, reprocess, recompact, and add additives as necessary to restore the strength of the damaged material to that specified in the design.

Measurement

308.09 Method. Use the method of measurement that is DESIGNATED IN THE SCHEDULE OF ITEMS.

Measure fog seal and blotter under Section 410.

Payment

308.10 Basis. The accepted quantities will be paid for at the contract unit price for each PAY ITEM DESIGNATED IN THE SCHEDULE OF ITEMS.

Payment will be made under:

Pay Item		Pay Unit
308 (01)	Portland cement-treated base	Kilometer
308 (02)	Portland cement-treated base	Square Meter
308 (03)	Hydrated lime	Ton
308 (04)	Portland cement	Ton
308 (05)	Fly ash	Ton

Section 309—Subgrade Stabilization

Description

309.01 Work. Process and incorporate additives into the upper layer of a subgrade.

Additives are designated as fly ash, hydrated lime, or Portland cement.

On existing subgrade, remove and dispose of slide material, vegetation, or other debris from roadbed shoulders, ditches, and culvert inlets and outlets, and reshape the roadbed shoulders, ditches, and culvert inlets and outlets.

Materials

309.02 Requirements. Ensure that materials meet the requirements specified in the following subsections:

Blotter 703.12
Cement 701.01
Chemical Admixtures (Retarder) 711.03
Emulsified Asphalt 702.03
Fly Ash 725.04
Fog Seal 410.10
Hydrated Lime 725.03
Water 725.01

Construction

309.03 General. Apply stabilization additives at the rates SHOWN ON THE DRAWINGS.

Store chemical additives and admixtures in closed, weatherproof containers. Prepare the subgrade in accordance with applicable requirements specified in Section 306. Scarify and pulverize the roadbed to a depth of 150 mm. Size and shape the material for the addition of additives. Limit the quantity of material being processed to a volume that is suitable for the mixing machine.

309.04 Mixing. Do not apply additives when the surface is wet, the surface temperature is below 5 °C, or the air temperature is expected to fall below 5 °C within 48 hours after mixing. Do not apply when additive is lost due to washing or blowing.

Mix the additives with the prepared material in either a dry or a slurry state, in accordance with Subsection 308.04. Continue mixing or remixing until a homogeneous mixture is obtained. Adjust the moisture content of the mixture in accordance with Subsection 308.05.

Except for the mixing equipment, keep all traffic off the spread material until mixing is completed.

For lime and hydrated lime/fly ash mixtures, complete the first mixing within 6 hours of application, and cure the mixture for 2 to 4 days by keeping the mixture moist. Perform final mixing by uniformly remixing the material. Mixing is complete when 95 percent of all material except rock passes a 45-mm sieve, and at least 50 percent of the material passes a 4.75-mm sieve when tested in accordance with AASHTO T 27, in a nondried condition. Retarders may be added.

For Portland cement mixtures, complete the mixing within 2 hours of application.

309.05 Compacting & Finishing. Immediately after final mixing, shape and finish the mixture in accordance with Subsection 203.08(c) and (d). Within 2 days after final mixing, finish the surface to make it smooth and suitable for placing a base or surface course.

Compact treated material in accordance with Subsection 203.16(b), method 4. During compaction and until final finishing, aerate or water as necessary to maintain the moisture content within ± 2 percent of the optimum moisture content necessary for compaction. If the time between placing adjacent partial widths exceeds 30 minutes, or when tying into the previous work, provide a construction joint in accordance with Subsection 307.07.

309.06 Curing. Cure in accordance with Subsection 307.08.

If the material loses the required stability, density, or finish before placement of the next course or acceptance of the work, reprocess or recompact as necessary to restore the strength of the damaged material to that specified in the mix design.

Measurement

309.07 Method. Use the method of measurement that is DESIGNATED IN THE SCHEDULE OF ITEMS.

Measure fog seal and blotter material under Section 410.

Payment

309.08 Basis. The accepted quantities will be paid for at the contract unit price for each PAY ITEM DESIGNATED IN THE SCHEDULE OF ITEMS.

Payment will be made under:

Pay Item		Pay Unit
309 (01)	Subgrade stabilization with ____________, _____ mm depth	Square Meter
309 (02)	Hydrated lime	Ton
309 (03)	Portland cement	Ton
309 (04)	Fly ash	Ton

Section 310—Dust Abatement

Description

310.01 Work. Furnish and apply dust palliative and blotter material, if necessary, to a road surface.

Materials

310.02 Requirements. Furnish material of the type and grade DESIGNATED IN THE SCHEDULE OF ITEMS that meet the requirements specified in the following subsections:

Calcium Chloride Flake	723.02
Lignin Sulfonate	723.03
Magnesium or Calcium Chloride Brine	723.01

Apply product at temperatures as stated in Section 723. Use the preparation method DESIGNATED IN THE SCHEDULE OF ITEMS.

310.03 Blotter Material. Furnish blotter material that meets the requirements specified in Subsection 703.12.

Construction

310.04 Weather Limitations. Application during a light rain is acceptable, provided the dust palliative penetrates the road surface and does not flow to low areas or off of the road surface. Apply chloride brines and lignin sulfonate only when the ambient temperature is 5 °C or higher and the ground is not frozen.

310.05 Equipment. Ensure that distribution equipment applies dust palliative uniformly on variable widths of road surface. The maximum allowable variation from the specified application rate is ± 10 percent of the specified rate for individual distributor loads, and ± 2 percent of the specified rate over the total project.

For liquid products, meet the following requirements:

(a) Application at controlled rates from 0.5 to 2.6 L/m^2 with uniform pressure and application.

(b) Spray pattern from each nozzle on the spray bar is uniform across the spray bar.

(c) Provide distribution equipment that includes accurate volume-measuring devices or a calibrated tank; a thermometer for measuring temperatures of tank contents; and a hose-and-nozzle attachment for applying material to areas inaccessible to the spray bar.

When compaction is included in the SCHEDULE OF ITEMS, provide rollers that meet the requirements specified in Subsection 203.15(b), (c), or (d).

310.06 Preparation of Road Surface. Use one or more of the following preparation and application methods, as SHOWN ON THE DRAWINGS or in the SPECIAL PROJECT SPECIFICATIONS. For all methods, apply water to the road surface or to the windrow at the times and in the quantities directed by the CO to ensure a near-optimum moisture content that will allow adequate penetration or mixing of the dust palliatives and surface materials. To accelerate penetration and absorption of calcium chloride flake materials, the road surface may be dampened prior to or after flake application.

(a) Method 1. Apply the dust palliative directly to the prepared and compacted surface.

(b) Method 2. Develop a layer of loose cushion material approximately 25 mm in depth for the full width of traveled way, and keep it in a loose condition prior to applying dust palliative.

(c) Method 3. Blade approximately 25 mm of the surface material into a berm on the shoulder. Then make the initial application on the exposed surface. As soon as practicable, but no more than 4 hours after application, blade the material in the berm(s) to a uniform depth across the previously treated surface, and water it if necessary to meet required water content. Then apply the second application.

For methods 2 and 3, process the road surface by blading below the elevation of raveling, washboarding, and potholes. After processing, shape the surface by blading to the required cross section as SHOWN ON THE DRAWINGS. Have the prepared surface approved in writing by the CO prior to treatment.

310.07 Application of Dust Palliative Treatment. Ensure that the dust palliative application rate and width of road surface to be covered are as SHOWN ON THE DRAWINGS or in the SPECIAL PROJECT SPECIFICATIONS. Adjust the application rate and apply the product in as many stages as necessary to assure thorough penetration and to prevent product runoff. If more than one application is required, do not apply the second application until the previous application has sufficiently penetrated the road surface.

Obtain uniform distribution at all points. Correct any overlapping or skipping between spread sections. Use blotter material to cover areas with excess dust palliative. Protect the surface of adjacent structures and trees from spattering or marring. Discharge dust palliative material only in approved areas, and do not allow it to flow into ditches or stream courses.

310.08 Compaction. Compaction is required if included in the SCHEDULE OF ITEMS. Begin compaction as soon as the dust palliative has penetrated enough to prevent pickup of material. Operate rollers over the full width of each layer until visual displacement ceases, but make no fewer than three complete passes.

310.09 Maintenance & Opening to Traffic. Do not permit traffic on dust palliative treatment until the treatment has penetrated and cured enough to prevent excessive pickup under traffic. If it becomes necessary to permit traffic prior to that time, apply blotter material where necessary.

Measurement

310.10 Method. Use the method of measurement that is DESIGNATED IN THE SCHEDULE OF ITEMS.

Payment

310.11 Basis. The accepted quantities will be paid for at the contract unit price for each PAY ITEM DESIGNATED IN THE SCHEDULE OF ITEMS, with the following exceptions:

(a) If laboratory quality assurance tests indicate that the minimum magnesium or calcium chloride concentrations applied to the road surface were not as specified in Subsection 723.01, the CO may reduce payment by multiplying the pay factor as calculated below, times the contract unit price for the appropriate PAY ITEMS, times the accepted quantity. No payment will be made for brine concentrations below 20 percent.

Magnesium Chloride Brine Pay Factor =
1.0 – ((28% – Concentration Applied)/8%)

Calcium Chloride Brine Pay Factor =
1.0 – ((36% – Concentration Applied)/16%)

(b) If laboratory quality assurance tests indicate that the minimum lignin sulfonate concentrations were not as specified in Subsection 723.03, the CO may reduce payment by multiplying the pay factor as calculated below times the contract unit price for the appropriate PAY ITEMS, times the accepted quantity. No payment will be made for concentrations below 24 percent.

Lignin Pay Factor =
1.0 – ((48% – Concentration Applied)/24%)

Payment will be made under:

Pay Item		Pay Unit
310 (01)	Magnesium chloride brine at 28 percent minimum concentration, preparation method ____________	Ton
310 (02)	Lignin sulfonate solution at 48 percent minimum concentration, preparation method____________	Ton
310 (03)	Calcium chloride brine at 36 percent minimum concentration, preparation method____________	Ton
310 (04)	Magnesium chloride brine at 28 percent minimum concentration, or calcium chloride brine at 36 percent minimum concentration, or calcium chloride flake at 77 percent minimum concentration, preparation method____________	Lump Sum
310 (05)	Calcium chloride flake at 77 percent minimum concentration, preparation method____________	Ton
310 (06)	Compaction	Kilometer

DIVISION 400
Bituminous Pavements

Section 401—Major Hot Asphalt Concrete Pavement With Pay Factor

Description

401.01 Work. Construct one or more courses of hot asphalt concrete pavement as SHOWN ON THE DRAWINGS. Have the surface approved in writing by the CO before placing the hot asphalt concrete pavement.

Hot asphalt concrete pavement classes are designated as shown in table 401-1. Hot asphalt concrete pavement aggregate grading is designated as shown in table 703-4. State asphalt concrete pavement classes are designated by local State department of transportation designations. Superpave asphalt concrete pavement nominal size and grading are designated as shown in table 703-10, 703-11, or 703-12.

Table 401-1.—Asphalt concrete mixture requirements.

Design Parameters	Class of Mixture A	Class of Mixture B	Class of Mixture C
(a) Hveem (AASHTO T 246 and T 247):			
(1) Stabilometer, min.	37	35	30
(2) Air voids, %[a]	3–5	3–5	3–5
(3) Voids in Mineral Aggregate (VMA), min. %	See table 401-2		
(b) Marshall (AASHTO T 245):[b]			
(1) Stability, kN min.	8.0	5.3	4.4
(2) Flow, 0.25 mm	8–14	8–16	8–20
(3) Air voids, %[a]	3–5	3–5	3–5
(4) VMA, min. %	See table 401-2		
(5) Compaction, number of blows each end of test specimen	75	50	50
(c) Immersion-Compression (AASHTO T 165 and T 167):			
(1) Compressive strength, MPa min.	2.1	1.7	1.4
(2) Retained strength, min. %	70	70	70
(d) Root–Tunnicliff (ASTM D 4867):			
(1) Tensile strength ratio, min. %	70	70	70
(e) Dust/asphalt ratio[c]	0.6–1.3	0.6–1.3	0.6–1.3

a. The percent of air voids is based on AASHTO T 166, AASHTO T 209, and AASHTO T 269. Maximum specific gravity will be based on AASHTO T 209.

b. Following mixing, asphalt cement mixtures will be cured in an oven maintained at 12 °C to 18 °C above the compaction temperature for 90 ± 10 minutes.

c. Dust/asphalt ratio is defined as the percent of material, including nonliquid antistrip and mineral filler, that passes the 75-µm sieve, divided by the percent of asphalt (calculated by weight of mix).

Table 401-2.—VMA for Marshall, Hveem, and Superpave mix design.

Sieve Size[a]	Minimum Voids[b,c] (%)		
	Marshall	Hveem	Superpave
2.36 mm	21	19	–
4.75 mm	18	16	–
9.5 mm	16	14	15
12.5 mm	15	13	14
19 mm	14	12	13
25 mm	13	11	12
37.5 mm	12	10	11
50 mm	11.5	9.5	10.5

a. The largest sieve size listed in the applicable specification upon which any material is permitted to be retained.

b. VMA to be determined in accordance with Asphalt Institute (AI) Manual Series number 2 (MS–2).

c. When a mineral filler or nonliquid antistrip is used, include the percentage specified in the calculation for compliance with the VMA.

Asphalt cement grade is designated as shown in AASHTO M 20, M 226, or MP 1, or in applicable State department of transportation specifications for asphalt materials for the grade specified.

A prepaving conference will be held at least 10 working days prior to the beginning of paving operations. At that time, the Contractor and the CO will discuss methods of accomplishing all phases of the paving work, including laydown operations, work schedules, work force, quality control systems, spill prevention and contingency plans, and asphalt concrete mix delivery.

Materials

401.02 Requirements. Ensure that material conforms to specifications in the following subsections:

Antistrip Additive 702.07
Asphalt Cement 702.01
Hot Asphalt Concrete Pavement Aggregate 703.07
Mineral Filler 725.05
Recycling Agent 702.05
Superpave Asphalt Concrete Pavement Aggregate 703.14

Ensure that reclaimed asphalt pavement material conforms to the following:

(a) 100 percent passes the 50-mm screen.

(b) The material consists of asphalt cement and asphalt cement-coated aggregate.

Construction

401.03 Composition of Mixture (Job-Mix Formula). Up to 20 percent reclaimed asphalt pavement material may be used, subject to approval of a Contractor quality control plan and submission of test data demonstrating that the mixture will meet the requirements specified in this section.

Furnish the appropriate mixture as follows:

(a) Hot Asphalt Concrete Pavement Mixture. Furnish aggregate, asphalt, additives, and, when applicable, reclaimed asphalt pavement material that meet the applicable aggregate gradation requirement shown in table 703-4, and design parameters (a) or (b); (c) or (d); and (e) shown in table 401-1.

(b) Superpave Asphalt Concrete Pavement Mixture. Furnish aggregate, asphalt, and additives that meet applicable gradation and material requirements specified in Subsection 703.14, and the appropriate design parameters shown in tables 401-2 and 401-2A. Compact specimens with the gyratory compactive effort specified in table 401-2B for the specified air temperature as SHOWN ON THE DRAWINGS.

Table 401-2A.—Superpave asphalt concrete mixture requirements.

Design Parameters	Requirement
Percent air voids at design gyrations, N_{des}	4.0
Percent maximum density at initial gyrations, N_{init}	89 max.
Percent maximum density at maximum gyrations, N_{max}	98 max.
Tensile strength ratio (AASHTO T 283)	80 min.
Voids Filled With Asphalt	70–80%
Dust/asphalt ratio[a]	0.6–1.2

a. Dust/asphalt ratio is defined as the percent of material passing the 75-µm sieve, divided by the effective asphalt content as calculated by weight of mix.

Table 401-2B.—Gyratory compactive effort.

Average Design High Air Temperature	N_{init}	N_{des}	N_{max}
< 39 °C	7	68	104
39–40 °C	7	74	114
41–42 °C	7	78	121
43–44 °C	7	82	127

(c) State Asphalt Concrete Pavement Mixture. Furnish aggregate, asphalt, and additives that meet the applicable aggregate gradation and aggregate quality specified by the local State department of transportation, and design parameters (a) or (b); (c) or (d); and (e) shown in table 401-1. Local State department of transportation design parameters in lieu of those shown in table 401-1 may be used if approved by the CO.

Submit written job-mix formulas for approval at least 21 days before production. For each job-mix formula, submit the following:

(1) Aggregate and mineral filler, including:

(a) TV for percent passing each sieve size for the aggregate blend. Ensure that the gradation of the blended aggregate and reclaimed asphalt pavement material falls within the gradation band for each sieve size designated in the specified grading.

(b) Source and percentage of each aggregate stockpile to be used.

(c) Average gradation of each aggregate stockpile.

(d) Results of aggregate quality tests.

(e) Samples, when SHOWN ON THE DRAWINGS.

(2) Asphalt cement, including:

(a) TV for percent of asphalt cement based on total weight of mix.

(b) Recent quality test results from the manufacturer for the asphalt cement, including a temperature/viscosity curve.

(c) Material safety data sheets.

(d) Samples, when SHOWN ON THE DRAWINGS.

(3) Antistrip additives. When applicable, furnish:

(a) Type and TV for percent of antistrip additive.

(b) Material safety data sheet.

(c) Samples, when SHOWN ON THE DRAWINGS.

(4) Mix temperatures, including:

(a) Temperature leaving the mixer.

(b) Temperature immediately preceding initial compaction.

(5) Maximum specific gravity, determined according to AASHTO T 209 at the asphalt cement TV.

(6) Reclaimed asphalt pavement material; when applicable, the percent reclaimed asphalt pavement material and the type and percent of recycling agent.

(7) Asphalt mixtures; when applicable, the location of all commercial mixing plants to be used. A job-mix formula is needed for each plant.

The CO will evaluate the suitability of the material and the proposed job-mix formula. After reviewing the proposed job-mix formula, the CO will develop a TV for the asphalt cement content and determine the need for antistrip additive, the specific gravity in accordance with AASHTO T 209, and the discharge temperature range.

If a job-mix formula is rejected, submit a new job-mix formula as described above.

Changes to an approved job-mix formula require approval before production. Allow up to 14 days to evaluate a change. Approved changes in TV's will not be applied retroactively for payment.

401.04 Mixing Plant. Use mixing plants that conform to ASTM D 995, unless producing approved material for a local State department of transportation. Supplement mixing plant as follows:

(a) All Plants. For all plants, use:

(1) Automated Controls. Automatically control the proportioning, mixing, and discharging of the mixture.

(2) Emission Controls. If a wet scrubber is used, circulate the collected material though sludge pits or settling tanks. Remove the resultant sediment from the project or bury according to Subsection 202.04.

(b) Drum Dryer-Mixer Plants. For drum dryer-mixer plants, use:

(1) Bins. Provide a separate bin in the cold aggregate feeder for each individual aggregate stockpile in the mixture. Use bins of sufficient size to keep the plant in continuous operation, and of proper design to prevent overflow of material from one bin to another.

(2) Stockpiling Procedures. Separate aggregate into at least two stockpiles with different gradations. At a minimum, designate one stockpile to contain mostly coarse material, and one stockpile to contain mostly fine material. Stockpile material according to Subsections 305.03 and 305.04.

(3) Reclaimed Asphalt Pavement Material. Modify drum dryer-mixer plants to prevent direct contact of the reclaimed asphalt pavement material with the burner flame and to prevent overheating of the reclaimed asphalt pavement material. Stockpile the material according to Subsection 305.03 and 305.04.

(c) ***Batch & Continuous Mix Plants.*** For batch and continuous mix plants, use:

(1) A Hot Aggregate Bin. Provide a bin with three or more separate compartments for storage of the screened aggregate fractions to be combined for the mixture. Make the partitions between the compartments tight and of sufficient height to prevent spillage of aggregate from one compartment into another.

(2) Load Cells. Calibrated load cells instead of scales may be used in batch plants.

(3) Reclaimed Asphalt Pavement Material. Modify batch plants to allow the introduction of reclaimed asphalt pavement material into the mixture using methods that bypass the dryer. Design the cold feed bin, conveyor system, and special bin adjacent to the weigh hopper, if used, to avoid segregation and sticking of the reclaimed asphalt pavement material. Heat the new aggregate and/or reclaimed aggregate material to a temperature that will transfer sufficient heat to the reclaimed asphalt pavement material to produce a mix of uniform temperature within the range specified in the approved job-mix formula.

401.05 Pavers. Use pavers that are:

(a) Self-contained, power-propelled units with adjustable vibratory screeds with full-width screw augers.

(b) Heated for the full width of the screed.

(c) Capable of spreading and finishing courses of asphalt mixture in widths at least 300 mm more than the width of one lane.

(d) Equipped with a receiving hopper with sufficient capacity to ensure a uniform spreading operation.

(e) Equipped with automatic feed controls that are properly adjusted to maintain a uniform depth of material ahead of the screed.

(f) Capable of being operated at forward speeds consistent with satisfactory laying of the mixture.

(g) Capable of producing a finished surface with the required smoothness and texture without segregating, tearing, shoving, or gouging the mixture.

(h) Equipped with automatic screed controls with sensors capable of sensing grade from an outside reference line, sensing the transverse slope of the screed, and providing the automatic signals that operate the screed to maintain grade and transverse slope.

401.06 Surface Preparation. Prepare the surface in accordance with Section 304, 306, 307, or 308, as applicable. Apply an asphalt tack coat to contact surfaces of curbing, gutters, manholes, and other structures, in accordance with Section 407.

401.07 Weather Limitations. Place hot asphalt concrete pavement on a dry, unfrozen surface when the air temperature in the shade is above 2 °C and rising, and when the temperature of the road surface in the shade, the lift thickness, and the minimum laydown temperature are as shown in table 401-3.

Table 401-3.—Minimum laydown temperature[a] for hot asphalt concrete mixture placement (°C).

	Lift Thickness		
Road Surface Temperature (°C)	≤ 50 mm	50–75 mm	≥ 75 mm
< 2	–[b]	–[b]	–[b]
2–3.9	–[b]	–[b]	138
4–9.9	–[b]	141	135
10–14.9	146	138	132
15–19.9	141	135	129
20–24.9	138	132	129
25–29.9	132	129	127
≥ 30	129	127	124

a. Never heat the asphalt concrete mixture above the temperature specified in the approved mix design.
b. Paving not allowed.

401.08 Asphalt Preparation. Uniformly heat the asphalt cement to provide a continuous supply of the heated asphalt cement from storage to the mixer. Do not heat asphalt cement above 175 °C.

If the job-mix formula requires a liquid heat-stable antistrip additive, meter it into the asphalt cement transfer lines at a bulk terminal or mixing plant. Inject the additive for at least 80 percent of the transfer or mixing time to obtain uniformity.

401.09 Aggregate Preparation. If nonliquid antistrip is used, adjust the aggregate moisture to at least 4 percent by weight of aggregate. Mix the antistrip uniformly with the aggregate before introducing the aggregate into the dryer or dryer drum. Use calibrated weighing or metering devices to measure the amount of antistrip and moisture added to the aggregate.

For batch plants, heat, dry, and deliver aggregate for pugmill mixing at a temperature sufficient to produce a mixture temperature within the approved range. Adjust flames used for drying and heating to prevent damage to and contamination of the aggregate.

Control plant operations so the moisture content of the mixture behind the paver is 0.5 percent or less, in accordance with AASHTO T 110.

401.10 Mixing. Measure the aggregate and asphalt into the mixer in accordance with the approved job-mix formula. Mix until all the particles are completely and uniformly coated with asphalt, in accordance with ASTM D 995. Maintain the discharge temperature within the approved range.

401.11 Hauling. Use vehicles with tight, clean, and smooth metal beds for hauling asphalt concrete mixtures.

Thinly coat the beds with an approved material to prevent the mixture from adhering to the beds. Do not use petroleum derivatives or other coating material that contaminates or alters the characteristics of the mixture. Drain the bed before loading.

Equip each truck with a canvas cover or other suitable material of sufficient size to protect the mixture from the weather. When necessary to maintain temperature, use insulated truck beds and securely fastened covers. Provide access ports or holes for checking the temperature of the asphalt mixture in the truck.

401.12 Placing & Finishing. Do not use mixtures produced from different plants unless the mixtures are produced in accordance with the same job-mix formula, contain material from the same sources, and are approved.

Place asphalt concrete mixture at a temperature conforming to table 401-3. Measure temperature of the mixture in the hauling vehicle just before dumping into spreader, or measure it in the windrow immediately before pickup.

Place the mixture with a paver that conforms to specifications in Subsection 401.05. Control horizontal alignment using a reference line. Automatically control the grade and slope from reference lines, a ski and slope control device, dual skis. Use skis with a minimum length of 6 m.

Limit the compacted thickness to 75 mm, unless otherwise SHOWN ON THE DRAWINGS.

On areas where mechanical spreading and finishing is impractical, place and finish the mixture with other equipment to produce a uniform surface closely matching the surface obtained when using a mechanical paver.

Offset the longitudinal joint of one layer at least 150 mm from the joint in the layer immediately below. Make the longitudinal joint in the top layer along the centerline of two-lane roadways, or at the lane lines of roadways with more than two lanes. Offset transverse joints in adjacent lanes and in multiple lifts by at least 3 m.

The CO will designate the job-mix formula to be used for wedge and leveling courses at each location unless DESIGNATED IN THE SCHEDULE OF ITEMS. Place wedge and leveling courses in maximum 75-mm lifts and compact with a pneumatic-tire roller meeting the requirements of Subsection 203.15(d). Complete the wedge and leveling before starting normal paving operations.

401.13 Compacting. Furnish at least three rollers, one each for breakdown, intermediate, and finish rolling. Furnish at least one roller with pneumatic tires. Size the rollers to achieve the required results. Operate rollers in accordance with manufacturer's recommendations.

Thoroughly and uniformly compact the asphalt surface by rolling. Do not cause undue displacement, cracking, or shoving. Continue rolling until all roller ridges, ruts, and humps are eliminated and the required compaction is obtained. Do not vibratory roll the mixture after its surface cools below 80 °C.

Along forms, curbs, headers, walls, and other places not accessible to the rollers, compact the mixture with other equipment to obtain the minimum compaction.

401.14 Joints, Trimming Edges, & Cleanup. At connections to existing pavements and previously placed lifts, make the transverse joints vertical to the depth of the new pavement. Form transverse joints by cutting back on the previous run to expose the full depth course. Dispose of trimmed asphalt material in accordance with Subsection 202.04 (a).

Apply an asphalt tack coat to the edge of the joint for both transverse and longitudinal joints, and where SHOWN ON THE DRAWINGS, in accordance with Section 407.

Place the asphalt concrete mixture as continuously as possible. Do not pass rollers over the unprotected end of a freshly laid mixture.

401.15 Acceptance. Provide a quality control plan and then sample, test, and maintain records according to Section 160. See table 401-4 for minimum sampling and testing requirements. Sample to ensure that:

Table 401-4.—Sampling and testing.

Type of Acceptance	Material or Product	Property or Characteristic	Test Method or Specification	Frequency	Sampling Point
Production certification (Subsection105.04)	Asphalt cement	Contract requirements	AASHTO M 20, M 226, or MP 1, as applicable	Daily	–
Tested conformance	Material source	Los Angeles abrasion	AASHTO T 96	Three times for each undeveloped source,[b] or once for all other sources	Material source
		Sodium sulfate soundness loss	AASHTO T 104		
		Durability index (coarse and fine)	AASHTO T 210		
	Aggregate	Fractured faces (coarse)[a]	FLH T 507	Three times for each undeveloped source,[b] or once for all other sources	Cold feed prior to entering dryer
		Sand equivalent value (fine)	AASHTO T 176, alternate method number 2 (referee method)		
	Asphalt cement	Sample	Subsection 105.04(b)	Once for each 500 t of mix, and not more than three times per day	At point of shipment delivery
	Job-mix formula	Contract requirements	Subsection 401.03	Once for each product or material change	–
Mix evaluation	Hot asphalt concrete pavement	Asphalt content	AASHTO T 164, method B or E	Once for each 500 t, and not more than three times per day	At plant, in hauling units, or behind laydown machine before rolling
		Gradation	AASHTO T 30		
		Compaction	ASTM D 2950, "Procedure"	Five times for each 500 t, and not less than five times per day	In-place after compaction
		Maximum specific gravity	AASHTO T 209	Once for each 1,000 t	At plant, in hauling units, or behind laydown machine before rolling

a. Use only for gravel sources.

b. An undeveloped source is a source that has not supplied aggregate for asphalt concrete within 365 days of the start of producing asphalt concrete for this particular project.

- The sample size is adequate to provide a duplicate to the CO and to meet potential need for retesting as specified in Subsection 401.18.
- Samples are prepared according to AASHTO T 248 or other procedures applicable to the item being sampled.
- The sample is adequately identified and placed in CO-approved containers provided by the Contractor.

The CO may perform quality assurance testing, and these tests will be made available to the Contractor upon request.

A lot is defined as the number of tons of material or work produced and/or placed under one set of TV's. The lot will be represented by randomly selected samples tested for acceptance. Plant and equipment operators will not be advised ahead of time when samples are taken.

Acceptance will be evaluated as follows:

(a) Asphalt Cement. Asphalt cement will be evaluated for acceptance under Subsection 106.05.

(b) Pavement Smoothness. Use a 3-m metal straightedge to measure at right angles and parallel to the centerline at designated sites. Defective areas are surface deviations in excess of 10 mm between any two contacts of the straightedge with the surface. Correct defective areas using approved methods.

(c) Thickness & Width. Ensure that the thickness and width of the compacted mixture conform to the dimensions SHOWN ON THE DRAWINGS and meet the following requirements:

(1) The maximum variation from the specified thickness is less than 6 mm for the wearing course or 12 mm for the base course.

(2) The compacted width has a +150-mm tolerance.

(3) The compacted thickness and width are not consistently above or below the specified dimensions.

(d) Asphalt Concrete Mixture Gradation and Asphalt Content. Gradation and asphalt content will be evaluated for acceptance under Subsection 401.16.

(e) Asphalt Concrete Pavement Compaction. Compaction will be evaluated for acceptance under Subsection 401.17.

401.16 Acceptance Sampling & Testing of Asphalt Concrete Mixture Gradation & Asphalt Content. Take statistically random samples in accordance with the tests specified in table 401-4. Take a minimum of three tests per lot. Acceptance or rejection of completed work will be on a lot basis. If the Contractor's quality control tests required in table 401-4 are validated by the CO in accordance with Subsection 401.18 (Test Result Validation Procedure), then the Contractor's tests will be used for acceptance tests.

Obtain samples of the mixture at the plant in approved State department of transportation sampling devices, or after the mixture has been discharged into hauling units or placed on the road in accordance with AASHTO T 168. Test samples for asphalt content by means of AASHTO T 164, method B (Reflux Method) or method E (Vacuum Extraction). Other methods, including nuclear, require approval in writing by the CO and may require an increased sampling and testing frequency. Report the asphalt content to the nearest 0.01 percent. Determine gradation of the entire quantity of extracted material in accordance with AASHTO T 30, except that results shall be reported to the nearest 0.1 percent for all sieves except the 75-µm sieve. Report this sieve to the nearest 0.01 percent. Determine the percent moisture in the asphalt mixture in accordance with AASHTO T 110.

If samples are tested for asphalt content by means of AASHTO T 164, determine an Extraction Retention Factor based on the average difference between at least three samples of known asphalt content, and corresponding asphalt content by the same procedure that will be used for acceptance. Prepare the samples in accordance with table 401-1, (b) Marshall, unless otherwise approved by the CO.

If areas of isolated defect are identified by the CO, treat these areas as a separate lot.

The mix tolerance, also referred to as the upper and lower specification limits (USL's and LSL's), is as shown in table 401-5. For Superpave asphalt concrete pavement, the sieve tolerances will follow the allowable deviations in Subsection 703.14 for the designated nominal size.

Table 401-5.—Mix tolerances.

Mixture Characteristic	Tolerances
Bitumen content	TV ± 0.5
Sieve size:	
4.75 mm and larger	TV ± 6.0
600 µm to 4.74 mm	TV ± 4.0
300 to 599 µm	TV ± 3.0
75 to 299 µm	TV ± 2.0

The Contractor may request a change in TV's subject to the provisions in Subsection 401.03. If the TV's are changed, evaluate all of the material produced up to the time of the change as a lot, and begin a new lot.

The lot will be accepted with respect to gradation and asphalt content using statistical evaluation procedures in accordance with Subsection 401.19.

401.17 Acceptance Sampling & Testing of Asphalt Concrete Pavement Compaction. Take statistically random samples in accordance with the tests specified in table 401-4. Take a minimum of five tests per lot. Acceptance or rejection of completed work will be on a lot basis. If the Contractor's quality control tests required in table 401-4 are validated by the CO in accordance with Subsection 401.18 (Test Result Validation Procedure), then the Contractor's tests will be used for acceptance tests.

Use the nuclear gauge for acceptance. Calibrate the nuclear gauge in accordance with ASTM D 2950, Calibration section, within 6 months prior to use on this project, and check the standard and reference on each day of use in accordance with ASTM D 2950, Standardization and Reference Check sections. Do not take acceptance samples within 0.3 m from the edges of the panel. Determine the LSL for compaction using either the control strip method or the maximum density method as follows:

(a) Control Strip Method. Use a control strip to establish the LSL. To determine the LSL, construct a control strip at the beginning of work on each type of material to be compacted. Leave each control strip, constructed to acceptable density and surface tolerances, in place to become a section of the completed roadway. Correct or remove unacceptable control strips and replace them at the Contractor's expense. Construct a control strip at least 100 m long and one lane wide, and of the designated lift thickness SHOWN ON THE DRAWINGS.

Ensure that the materials used in the construction of the control strip meet the specification requirements. Furnish them from the same source and of the same type and asphalt content used in the remainder of the course represented by the control strip.

Use equipment in the construction of the control strip that meets the requirements of specified in Subsections 401.05 and 401.13, and that is of the same type and weight as that used on the remainder of the course represented by the control strip.

Begin compaction of the control strips immediately after the course has been placed to the specified thickness. Ensure that compaction is continuous and uniform over the entire surface. Continue compaction of the control strip until no discernible increase in density can be obtained by additional compactive effort.

Upon completion of the compaction, determine the mean density of the control strip by averaging the results of 10 consecutive nuclear density tests taken at randomly selected sites within the control strip. The mean density of the control strip must equal or exceed the density shown in table 401-6.

Table 401-6.—Compaction requirements.

Road Grade (%)	% of AASHTO T 209 converted to density
≤ 8	90
8–12	89
> 12	88

Cease paving if three consecutive control strips fail to achieve the specified minimum density. Take all necessary actions to resolve compaction problems. Do not resume paving without approval of the CO. The LSL shall then be 98 percent of the mean density of the control strip. Construct a new control strip if any of the following occur:

(1) A change in the properties of the material.

(2) A change in the rollers.

(3) A new layer.

(4) Changes in grade as indicated in table 401-6.

(b) Maximum Density Method. The LSL shall be as shown in table 401-6.

After the bituminous mixture has been placed and compacted, the lot will be accepted with respect to compaction using statistical evaluation procedures in accordance with Subsection 401.19. The maximum pay factor for compaction will be 1.00 for the control strip method and 1.05 for the maximum density method. If areas of isolated defect are identified by the CO, these areas will be treated as a separate lot.

401.18 Test Result Validation Procedure. Provide the CO with a duplicate of all required samples, specified in table 401-4. If the Government decides to run assurance tests on the duplicate samples, the CO will determine the number to be run. Normally, the first three samples submitted will be tested, and 10 percent thereafter.

As testing is completed, the CO will evaluate all the Contractor testing. If Contractor testing is verified by Government testing, the Contractor's test results may be used by the Government to evaluate work for acceptance. If Contractor testing is not verified by Government testing, the Contractor has the option of either retesting or having the Government test the duplicate sample. The Contractor or the CO may witness the testing of the remaining sample portions. If the Contractor retests the sample, the test results will again be evaluated based on Government verification testing. If the test results are not valid, the Government test results will then be used for acceptance.

If it becomes necessary for the Government to test all of the samples for a work item due to the Contractor's tests being declared invalid, a payment deduction equal to the total cost of performing all of the testing for the applicable item will be made.

If the Contractor's test results are shown to be valid, but significant differences or shifts make the test results questionable, the CO will review the Contractor's equipment and test procedures.

If any deficiencies are identified that would account for the significant differences or shifts, the CO will suspend acceptance of all material until the deficiencies have been corrected. If no deficiencies that would account for the significant differences or shifts in test results can be identified, continue testing. In order to identify the deficiencies, the CO will increase testing frequency of sample portions.

401.19 Statistical Evaluation of Materials for Acceptance. Analyze all test results for a lot collectively and statistically by the Quality Level Analysis—Standard Deviation Method, using the procedures listed to determine the total estimated percent of the lot that is within specification limits. Quality Level Analysis is a statistical procedure for estimating the percent compliance to a specification. This procedure is affected by shifts in the arithmetic mean (X) and by the sample standard deviation(s). The analysis of each test parameter is based on an Acceptable Quality Level (AQL) of 95.0 and a producer's risk of 5 percent. The AQL may be viewed as the lowest percent of material inside the specification limit that is acceptable at the contract price. The producer's risk is the probability that when the Contractor is producing material at exactly the AQL, the material will receive less than a 1.00 pay factor. As an incentive to produce uniform quality material, payment of up to 5 percent more than the contract unit price may be obtained.

Quality Level Analysis—Standard Deviation Method Procedures are as follows:

(a) Determine the arithmetic mean (X) of each component tested:

$$X = \sum x / n$$

where

$\sum x$ = summation of individual test values
n = total number test values

(b) Compute the sample standard deviation (s):

$$s = \sqrt{\left(\frac{\sum x^2 - nX^2}{n-1}\right)}$$

where

$\sum x^2$ = summation of the squares of individual test values
X^2 = arithmetic mean squared
n = total number test values

(c) Compute the upper quality index (Q_U):

$$Q_U = \frac{USL - X}{s}$$

where

USL = TV plus allowable deviation
s = sample standard deviation

(d) Compute the lower quality index (Q_L):

$$Q_L = \frac{X - LSL}{s}$$

where

LSL = TV minus allowable deviation
s = sample standard deviation

(e) Determine P_U (the percent within the USL that corresponds to a given Q_U) from table 401-7. Note that if a USL is not specified, P_U will be 100.

Table 401-7.—Quality Level Analysis by the Standard Deviation Method.

Estimated % Inside Specification Limits (P_U and/or P_L)	Upper Quality Index (Q_U) or Lower Quality Index (Q_L)														
	$n = 3$	$n = 4$	$n = 5$	$n = 6$	$n = 7$	$n = 8$	$n = 9$	$n = 10$ to $n = 11$	$n = 12$ to $n = 14$	$n = 15$ to $n = 18$	$n = 19$ to $n = 25$	$n = 26$ to $n = 37$	$n = 38$ to $n = 69$	$n = 70$ to $n = 200$	$n = 200$ to 999
100	1.16	1.05	1.79	2.03	2.23	2.39	2.53	2.65	2.83	3.03	3.20	3.38	3.54	3.70	3.83
99	–	1.47	1.67	1.80	1.89	1.95	2.00	2.04	2.09	2.14	2.18	2.22	2.26	2.29	2.31
98	1.15	1.44	1.60	1.70	1.76	1.81	1.84	1.86	1.91	1.93	1.96	1.99	2.01	2.03	2.05
97	–	1.41	1.54	1.62	1.67	1.70	1.72	1.74	1.77	1.79	1.81	1.83	1.85	1.86	1.87
96	1.14	1.38	1.49	1.55	1.59	1.61	1.63	1.65	1.67	1.68	1.70	1.71	1.73	1.74	1.75
95	–	1.35	1.44	1.49	1.52	1.54	1.55	1.56	1.58	1.59	1.61	1.62	1.63	1.63	1.64
94	1.13	1.32	1.39	1.43	1.46	1.47	1.48	1.49	1.50	1.51	1.52	1.53	1.54	1.55	1.55
93	–	1.29	1.35	1.38	1.40	1.41	1.42	1.43	1.44	1.44	1.45	1.46	1.46	1.47	1.47
92	1.12	1.26	1.31	1.33	1.35	1.36	1.36	1.37	1.37	1.38	1.39	1.39	1.40	1.40	1.40
91	1.11	1.23	1.27	1.29	1.30	1.30	1.31	1.31	1.32	1.32	1.33	1.33	1.33	1.34	1.34
90	1.10	1.20	1.23	1.24	1.25	1.25	1.26	1.26	1.26	1.27	1.27	1.27	1.28	1.28	1.28
89	1.09	1.17	1.19	1.20	1.20	1.21	1.21	1.21	1.21	1.22	1.22	1.22	1.22	1.22	1.23
88	1.07	1.14	1.15	1.16	1.16	1.16	1.16	1.17	1.17	1.17	1.17	1.17	1.17	1.17	1.17
87	1.06	1.11	1.12	1.12	1.12	1.12	1.12	1.12	1.12	1.12	1.12	1.12	1.12	1.13	1.13
86	1.04	1.08	1.08	1.08	1.08	1.08	1.08	1.08	1.08	1.08	1.08	1.08	1.08	1.08	1.08
85	1.03	1.05	1.05	1.04	1.04	1.04	1.04	1.04	1.04	1.04	1.04	1.04	1.04	1.04	1.04
84	1.01	1.02	1.01	1.01	1.00	1.00	1.00	1.00	1.00	1.00	1.00	1.00	0.99	0.99	0.99
83	1.00	0.99	0.98	0.97	0.97	0.96	0.96	0.96	0.96	0.96	0.96	0.96	0.95	0.95	0.95
82	0.97	0.96	0.95	0.94	0.93	0.93	0.93	0.92	0.92	0.92	0.92	0.92	0.92	0.92	0.92
81	0.96	0.93	0.91	0.90	0.90	0.89	0.89	0.89	0.89	0.88	0.88	0.88	0.88	0.88	0.88
80	0.93	0.90	0.88	0.87	0.86	0.86	0.86	0.85	0.85	0.85	0.85	0.84	0.84	0.84	0.84
79	0.91	0.87	0.85	0.84	0.83	0.82	0.82	0.82	0.82	0.81	0.81	0.81	0.81	0.81	0.81
78	0.89	0.84	0.82	0.80	0.80	0.79	0.79	0.79	0.78	0.78	0.78	0.78	0.78	0.78	0.78
77	0.87	0.81	0.78	0.77	0.76	0.76	0.76	0.75	0.75	0.75	0.75	0.75	0.75	0.75	0.75

Table 401-7.—Quality Level Analysis by the Standard Deviation Method (cont.).

76	0.84	0.78	0.75	0.74	0.73	0.73	0.72	0.72	0.72	0.71	0.71	0.71	0.71	0.71	0.71
75	0.82	0.75	0.72	0.71	0.70	0.70	0.69	0.69	0.69	0.68	0.68	0.68	0.68	0.68	0.68
74	0.79	0.76	0.69	0.68	0.67	0.66	0.66	0.66	0.66	0.65	0.65	0.65	0.65	0.64	0.64
73	0.76	0.69	0.66	0.65	0.64	0.63	0.63	0.63	0.62	0.62	0.62	0.62	0.62	0.61	0.61
72	0.74	0.66	0.63	0.62	0.61	0.60	0.60	0.60	0.59	0.59	0.59	0.59	0.59	0.58	0.58
71	0.71	0.63	0.60	0.59	0.58	0.57	0.57	0.57	0.57	0.56	0.56	0.56	0.56	0.55	0.55
70	0.68	0.60	0.57	0.56	0.55	0.55	0.54	0.54	0.54	0.53	0.53	0.53	0.53	0.53	0.52
69	0.65	0.57	0.54	0.53	0.52	0.52	0.51	0.51	0.51	0.50	0.50	0.50	0.50	0.50	0.50
68	0.62	0.54	0.51	0.50	0.49	0.49	0.48	0.48	0.48	0.48	0.47	0.47	0.47	0.47	0.47
67	0.59	0.51	0.47	0.47	0.46	0.46	0.46	0.45	0.45	0.45	0.45	0.44	0.44	0.44	0.44
66	0.56	0.48	0.45	0.44	0.44	0.43	0.43	0.43	0.42	0.42	0.42	0.42	0.41	0.41	0.41
65	0.52	0.45	0.43	0.41	0.41	0.40	0.40	0.40	0.40	0.39	0.39	0.39	0.39	0.39	0.39
64	0.49	0.42	0.40	0.39	0.38	0.38	0.37	0.37	0.37	0.37	0.36	0.36	0.36	0.36	0.36
63	0.46	0.39	0.37	0.36	0.35	0.35	0.35	0.34	0.34	0.34	0.34	0.34	0.33	0.33	0.33
62	0.43	0.36	0.34	0.33	0.35	0.32	0.32	0.32	0.31	0.31	0.31	0.31	0.31	0.31	0.31
61	0.39	0.33	0.31	0.30	0.32	0.29	0.29	0.29	0.29	0.29	0.28	0.28	0.28	0.28	0.28
60	0.36	0.30	0.28	0.27	0.30	0.27	0.26	0.26	0.26	0.26	0.26	0.26	0.26	0.25	0.25
59	0.32	0.27	0.25	0.25	0.24	0.24	0.24	0.24	0.23	0.23	0.23	0.23	0.23	0.23	0.23
58	0.29	0.24	0.23	0.22	0.21	0.21	0.21	0.21	0.21	0.21	0.20	0.20	0.20	0.20	0.20
57	0.25	0.21	0.20	0.19	0.19	0.19	0.19	0.19	0.18	0.18	0.18	0.18	0.18	0.18	0.18
56	0.22	0.18	0.17	0.16	0.16	0.16	0.16	0.16	0.16	0.15	0.15	0.15	0.15	0.15	0.15
55	0.18	0.15	0.14	0.14	0.13	0.13	0.13	0.13	0.13	0.13	0.13	0.13	0.13	0.13	0.13
54	0.14	0.12	0.11	0.11	0.11	0.11	0.10	0.10	0.10	0.10	0.10	0.10	0.10	0.10	0.10
53	0.11	0.09	0.08	0.08	0.08	0.08	0.08	0.08	0.08	0.08	0.08	0.08	0.08	0.08	0.08
52	0.07	0.06	0.06	0.05	0.05	0.05	0.05	0.05	0.05	0.05	0.05	0.05	0.05	0.05	0.05
51	0.04	0.03	0.03	0.03	0.03	0.03	0.03	0.03	0.03	0.03	0.03	0.03	0.03	0.03	0.03
50	0.00	0.00	0.00	0.00	0.00	0.00	0.00	0.00	0.00	0.00	0.00	0.00	0.00	0.00	0.00

Note: If the value of Q_U or Q_L does not correspond to a value in the table, use the next lower value. If Q_U or Q_L are negative values, P_U or P_L is equal to 100 minus the table value for P_U or P_L. If the value of Q_U or Q_L does not correspond exactly to a figure in the table, use the next higher figure.

(f) Determine P_L (the percent within the *LSL* that corresponds to a given Q_L) from table 401-7. Note that if an *LSL* is not specified, P_L will be 100.

(g) Determine the Quality Level (the total percent within specification limits) as follows:

$$\text{Quality Level} = (P_U + P_L) - 100$$

(h) Using the Quality Level from step (g), determine the pay factor (*PFi*) from table 401-8 for each constituent tested.

The contract unit price will be paid for any lot for which at least three samples have been obtained and all of the test results meet the appropriate criteria listed below:

- All test results are within the allowable deviations specified for the item, or
- All test results are greater than or equal to a minimum specification limit, or
- All test results are less than or equal to a maximum specification limit.

Compute the Quality Level and composite pay factor (CPF) in these instances to determine the amount of any bonus that might be warranted.

If less than three samples have been obtained at the time a lot is terminated, include the material in this shortened lot as part of an adjacent lot at the pay factor computed for the revised lot.

If the lot does not meet the criteria for payment at the contract unit price, the lot will be accepted if the CPF is greater than 0.75, provided there are no isolated defects identified by the CO. If a lot contains a CPF less than 0.75, the lot will be rejected. The CO may permit one or more of the following:

- Require complete removal and replacement with specification material at no cost to the Government.
- At the Contractor's written request, allow corrective work at no additional cost to the Government, and then apply an appropriate price reduction that may range from no reduction to no payment.
- At the Contractor's written request, allow material to remain in place with an appropriate price reduction that may range from a designated percentage reduction to no payment.

Table 401-8.—Required quality level for a given sample size (n) and a given pay factor (PFi).

PFi	$n = 3$	$n = 4$	$n = 5$	$n = 6$	$n = 7$	$n = 8$	$n = 9$	$n = 10$ to $n = 11$	$n = 12$ to $n = 14$	$n = 15$ to $n = 18$	$n = 19$ to $n = 25$	$n = 26$ to $n = 37$	$n = 38$ to $n = 69$	$n = 70$ to $n = 200$	$n = 201$ to 999
1.05	100	100	100	100	100	100	100	100	100	100	100	100	100	100	100
1.04	90	91	92	93	93	93	94	94	95	95	96	96	97	97	99
1.03	80	85	87	88	89	90	91	91	92	93	93	94	95	96	97
1.02	75	80	83	85	86	87	88	88	89	90	91	92	93	94	95
1.01	71	77	80	82	84	85	85	86	87	88	89	90	91	93	94
1.00	68	74	78	80	81	82	83	84	85	86	87	89	90	91	93
0.99	66	72	75	77	79	80	81	82	83	85	86	87	88	90	92
0.98	64	70	73	75	77	78	79	80	81	83	84	85	87	88	90
0.97	62	68	71	74	75	77	78	78	80	81	83	84	85	87	89
0.96	60	66	69	72	73	75	76	77	78	80	81	83	84	86	88
0.95	59	64	68	70	72	73	74	75	77	78	80	81	83	85	87
0.94	57	63	66	68	70	72	73	74	75	77	78	80	81	83	86
0.93	56	61	65	67	69	70	71	72	74	75	77	78	80	82	84
0.92	55	60	63	65	67	69	70	71	72	74	75	77	79	81	83
0.91	53	58	62	64	66	67	68	69	71	73	74	76	78	80	82
0.90	52	57	60	63	64	66	67	68	70	71	73	75	76	79	81
0.89	51	55	59	61	63	64	66	67	68	70	72	73	75	77	80
0.88	50	54	57	60	62	63	64	65	67	69	70	72	74	76	79
0.87	48	53	56	58	60	62	63	64	66	67	69	71	73	75	78
0.86	47	51	55	57	59	60	62	63	64	66	68	70	72	74	77
0.85	46	50	53	56	58	59	60	61	63	65	67	69	71	73	76
0.84	45	49	52	55	56	58	59	60	62	64	65	67	69	72	75
0.83	44	48	51	53	55	57	58	59	61	63	64	66	68	71	74
0.82	42	46	50	52	54	55	57	58	60	61	63	65	67	70	72
0.81	41	45	48	51	53	54	56	57	58	60	62	64	66	69	71
0.80	40	44	47	50	52	53	54	55	57	59	61	63	65	67	70
0.79	38	43	46	48	50	52	53	54	56	58	60	62	64	66	69
0.78	37	41	45	47	49	51	52	53	55	57	59	61	63	65	68
0.77	36	40	43	46	48	50	51	52	54	56	57	60	62	64	67
0.76	34	39	42	45	47	48	50	51	53	55	56	58	61	63	66
0.75[a]	33	38	41	44	46	47	49	50	51	53	55	57	59	62	65

Note: If the computed Quality Level does not correspond exactly to a figure in the table, use the next lower value.

a. Reject quality levels less than those specified for a 0.75 pay factor.

Determine the CPF for each lot as follows:

$$CPF = \frac{[f1(PF1) + f2(PF2) + ... + fi(PFi)]}{\Sigma fi}$$

where

PFi = pay factor for each constituent tested
fi = weighting factor listed below for the applicable material:

Item	Weighting factor (fi)
9.5-mm and larger material	1
4.75-mm to 9.49-mm material	3
2.36-mm to 4.74-mm material	5
74-µm to 2.35-mm material	3
75-µm material	12
Asphalt content	29
Compaction	45

Measurement

401.20 Method. Use the method of measurement that is DESIGNATED IN THE SCHEDULE OF ITEMS.

Calculate tonnage as the weight used in the accepted pavement. No deduction will be made for the weight of asphalt cement in the mixture.

Payment

401.21 Basis. The accepted quantities will be paid for at the contract unit price for each PAY ITEM listed below that is DESIGNATED IN THE SCHEDULE OF ITEMS, with the following exceptions:

(a) Payment for hot asphalt concrete pavement and asphalt cement will be made at a price determined by multiplying contract unit bid price by the CPF, as determined in Subsection 401.19.

(b) Payment for sampling and testing will be made as follows:

(1) Twenty-five percent of the lump sum, not to exceed 0.5 percent of the original contract amount, will be paid after all the testing facilities are in place, qualified sampling and testing personnel are identified, and the work being tested has started.

(2) Payment for the remaining portion of the lump sum will be prorated based on the total work completed.

(3) Payment for all or part of this PAY ITEM may be retained if the Government assurance tests invalidate the Contractor's testing.

Payment will be made under:

Pay Item		Pay Unit
401 (01)	Hot asphalt concrete pavement, class ___, grading ___	Ton
401 (02)	Superpave asphalt concrete pavement, nominal size ___	Ton
401 (03)	State asphalt concrete pavement, class ___	Ton
401 (04)	Hot asphalt concrete pavement, class ___, grading ___, wedge and leveling course	Ton
401 (05)	Asphalt cement, grade ___	Ton
401 (06)	Asphalt cement, State department of transportation grade ___	Ton
401 (07)	Sampling and testing	Lump Sum

Section 402—Major Hot Asphalt Concrete Pavement

Description

402.01 Work. Construct one or more courses of hot asphalt concrete pavement as SHOWN ON THE DRAWINGS. Have the surface approved by the CO in writing before placing the hot asphalt concrete pavement.

Hot asphalt concrete pavement classes are designated as shown in table 402-1. Hot asphalt concrete pavement aggregate grading is designated as shown in table 703-4. State asphalt concrete pavement classes are designated by local State department of transportation designations.

Asphalt cement grade is designated as shown in AASHTO M 20, M 226, or MP 1, or in applicable State department of transportation specifications for asphalt materials for the grade specified.

A prepaving conference will be held at least 10 working days prior to the beginning of paving operations. At that time, the Contractor and the CO will discuss methods of accomplishing all phases of the paving work, including laydown operations, work schedules, work force, quality control systems, spill prevention and contingency plans, and asphalt concrete mix delivery.

Materials

402.02 Requirements. Ensure that material conforms to the requirements specified in the following subsections:

Antistrip Additive 702.07
Asphalt Cement 702.01
Hot Asphalt Concrete Pavement Aggregate 703.07
Mineral Filler 725.05
Recycling Agent 702.05

Ensure that reclaimed asphalt pavement material conforms to the following:

(a) 100 percent passes the 50-mm screen.

(b) The material consists of asphalt cement and asphalt cement-coated aggregate.

Table 402-1—Asphalt concrete mixture requirements.

Design Parameters[a]	Class of Mixture A	B	C
(a) Hveem (AASHTO T 246 and T 247):			
(1) Stabilometer, min.	37	35	30
(2) Air voids, %[a]	3–5	3–5	3–5
(3) VMA, min. %	See table 402-2		
(b) Marshall (AASHTO T 245):[b]			
(1) Stability, kN min.	8.0	5.3	4.4
(2) Flow, 0.25 mm	8–14	8–16	8–20
(3) Air voids, %[a]	3–5	3–5	3–5
(4) VMA, min. %	See table 402-2		
(5) Compaction, number of blows each end of test specimen	75	50	50
(c) Immersion-Compression (AASHTO T 165 and T 167):			
(1) Compressive strength, MPa min.	2.1	1.7	1.4
(2) Retained strength, min. %	70	70	70
(d) Root–Tunnicliff (ASTM D 4867):			
(1) Tensile strength ratio, min. %	70	70	70
(e) Dust/asphalt ratio[c]	0.6–1.3	0.6–1.3	0.6–1.3

a. Percent of air voids is based on AASHTO T 166, AASHTO T 209, and AASHTO T 269. Maximum specific gravity will be based on AASHTO T 209.

b. Following mixing, asphalt cement mixtures will be cured in an oven maintained at 12 °C to 18 °C above the compaction temperature for 90 ± 10 minutes.

c. Dust/asphalt ratio is defined as the percent of material, including nonliquid antistrip and mineral filler, that passes the 75-µm sieve, divided by the percent of asphalt (calculated by weight of mix).

Construction

402.03 Composition of Mixture (Job-Mix Formula). Up to 20 percent reclaimed asphalt pavement material may be used, subject to approval of a Contractor quality control plan and submission of test data demonstrating that the mixture will meet the requirements specified in this section.

Furnish the appropriate mixture as follows:

(a) Hot Asphalt Concrete Pavement Mixture. Furnish aggregate, asphalt additives, and, when applicable, reclaimed asphalt pavement material that meet the applicable aggregate gradation requirement shown in table 703-4, and design parameters (a) or (b); (c) or (d); and (e) shown in table 402-1.

Table 402-2.—VMA for Marshall or Hveem mix design.

Sieve Size[a]	Minimum Voids[b,c] (%)	
	Marshall	Hveem
2.36 mm	21	19
4.75 mm	18	16
9.5 mm	16	14
12.5 mm	15	13
19 mm	14	12
25 mm	13	11
37.5 mm	12	10
50 mm	11.5	9.5

a. The largest sieve size listed in the applicable specification upon which any material is permitted to be retained.

b. VMA to be determined in accordance with AI Manual Series number 2 (MS–2).

c. When a mineral filler or nonliquid antistrip is used, include the percentage specified in the calculation for compliance with the VMA.

(b) State Asphalt Concrete Pavement Mixture. Furnish aggregate, asphalt, and additives that meet applicable aggregate gradation and aggregate quality requirements of the local State department of transportation, and design parameters (a) or (b); (c) or (d); and (e) shown in table 402-1. Local State department of transportation design parameters in lieu of those in table 402-1 may be used if approved by the CO.

Submit written job-mix formulas for approval at least 21 days before production. For each job-mix formula, submit the following:

(1) Aggregate and mineral filler, including:

(a) TV for percent passing each sieve size for the aggregate blend. Ensure that the gradation of the blended aggregate and reclaimed asphalt pavement falls within the gradation band for each sieve size designated in the specified grading.

(b) Source and percentage of each aggregate stockpile to be used.

(c) Average gradation of each aggregate stockpile.

(d) Results of aggregate quality tests.

(e) Samples, when SHOWN ON THE DRAWINGS.

(2) Asphalt cement, including:

(a) TV for percent of asphalt cement based on total weight of mix.

(b) Recent quality test results from the manufacturer for the asphalt cement, including a temperature-viscosity curve.

(c) Material safety data sheets.

(d) Samples, when SHOWN ON THE DRAWINGS.

(3) Antistrip additives. When applicable, furnish:

(a) Type and TV for percent of antistrip additive.

(b) Material safety data sheet.

(c) Samples, when SHOWN ON THE DRAWINGS.

(4) Mix temperatures, including:

(a) Temperatures leaving the mixer.

(b) Temperature immediately preceding initial compaction.

(5) Maximum specific gravity, determined according to AASHTO T 209 at the asphalt cement TV.

(6) Reclaimed asphalt pavement material; when applicable, the percent reclaimed asphalt pavement material and type and percent of recycling agent.

(7) Asphalt mixtures; when applicable, the location of all commercial mixing plants to be used. A job-mix formula is needed for each plant.

The CO will evaluate the suitability of the material and the proposed job-mix formula. After reviewing the proposed job-mix formula, the CO will develop a TV for the asphalt cement content and determine the need for antistrip additive, the specific gravity in accordance with AASHTO T 209, and the discharge temperature range.

If a job-mix formula is rejected, submit a new job-mix formula as described above.

Changes to an approved job-mix formula require approval before production. Allow up to 14 days to evaluate a change. Approved changes in TV's will not be applied retroactively for payment.

402.04 Mixing Plant. Use mixing plants that conform to ASTM D 995, unless producing approved materials for a local State department of transportation. Supplement mixing plant as follows:

(a) All Plants. For all plants, use:

(1) Automated Controls. Automatically control the proportioning, mixing, and discharging of the mixture.

(2) Emission Controls. If a wet scrubber is used, circulate the collected material through sludge pits or settling tanks. Remove the resultant sediment from the project or bury according to Subsection 202.04.

(b) Drum Dryer-Mixer Plants. For drum dryer-mixer plants, use:

(1) Bins. Provide a separate bin in the cold aggregate feeder for each individual aggregate stockpile in the mixture. Use bins of sufficient size to keep the plant in continuous operation, and of proper design to prevent overflow of material from one bin to another.

(2) Stockpiling Procedures. Separate aggregate into at least two stockpiles with different gradations. As a minimum, one stockpile shall contain mostly coarse material, and one stockpile shall contain mostly fine material.

(3) Reclaimed Asphalt Pavement Material. Modify drum dryer-mixer plants to prevent direct contact of the reclaimed asphalt pavement material with the burner flame and to prevent overheating of the reclaimed asphalt pavement material.

(c) Batch & Continuous Mix Plants. For batch and continuous mix plants, use:

(1) A Hot Aggregate Bin. Provide a bin with three or more separate compartments for storage of the screened aggregate fractions to be combined for the mixture. Make the partitions between the compartments tight and of sufficient height to prevent spillage of aggregate from one compartment into another.

(2) Load Cells. Calibrated load cells instead of scales may be used in batch plants.

(3) Reclaimed Asphalt Pavement Material. Modify batch plants to allow the introduction of reclaimed asphalt pavement material into the mixture using methods that bypass the dryer. Design the cold feed bin, conveyor system, and special bin adjacent to the weigh hopper, if used, to avoid segregation and sticking of the reclaimed asphalt pavement material. Heat the new aggregate and/or reclaimed

aggregate material to a temperature that will transfer sufficient heat to the reclaimed asphalt pavement material to produce a mix of uniform temperature within the range specified in the approved job-mix formula.

402.05 Pavers. Use pavers that are:

(a) Self-contained, power-propelled units with adjustable vibratory screeds with full-width screw augers.

(b) Heated for the full width of the screed.

(c) Capable of spreading and finishing courses of asphalt mixture in widths at least 300 mm more than the width of one lane.

(d) Equipped with a receiving hopper with sufficient capacity to ensure a uniform spreading operation.

(e) Equipped with automatic feed controls that are properly adjusted to maintain a uniform depth of material ahead of the screed.

(f) Capable of being operated at forward speeds consistent with satisfactory laying of the mixture.

(g) Capable of producing a finished surface of the required smoothness and texture without segregating, tearing, shoving, or gouging the mixture.

(h) Equipped with automatic screed controls with sensors capable of sensing grade from an outside reference line, sensing the transverse slope of the screed, and providing the automatic signals that operate the screed to maintain grade and transverse slope.

402.06 Surface Preparation. Prepare the surface in accordance with Section 304, 306, 307, or 308, as applicable. Apply an asphalt tack coat to contact surfaces of curbing, gutters, manholes, and other structures in accordance with Section 407.

402.07 Weather Limitations. Place hot asphalt concrete pavement on a dry, unfrozen surface when the air temperature in the shade is above 2 °C and rising, and the temperature of the road surface in the shade, the lift thickness, and the minimum laydown temperature are as shown in table 402-3.

402.08 Asphalt Preparation. Uniformly heat the asphalt cement to provide a continuous supply of the heated asphalt cement from storage to the mixer. Do not heat asphalt cement above 175 °C.

Table 402-3.—Minimum laydown temperature[a] for hot asphalt concrete mixture placement (°C).

Road Surface Temperature (°C)	Lift Thickness		
	≤ 50 mm	50–75 mm	≥ 75 mm
< 2	–[b]	–[b]	–[b]
2–3.9	–[b]	–[b]	138
4–9.9	–[b]	141	135
10–14.9	146	138	132
15–19.9	141	135	129
20–24.9	138	132	129
25–29.9	132	129	127
≥ 30	129	127	124

a. Never heat the asphalt concrete mixture above the temperature specified in the approved mix design.
b. Paving not allowed.

If the job-mix formula requires a liquid heat-stable antistrip additive, meter it into the asphalt cement transfer lines at a bulk terminal or mixing plant. Inject the additive for at least 80 percent of the transfer or mixing time to obtain uniformity.

402.09 Aggregate Preparation. If nonliquid antistrip is used, adjust the aggregate moisture to at least 4 percent by weight of aggregate. Mix the antistrip uniformly with the aggregate before introducing the aggregate into the dryer or dryer drum. Use calibrated weighing or metering devices to measure the amount of antistrip and moisture added to the aggregate.

For batch plants, heat, dry, and deliver aggregate for pugmill mixing at a temperature sufficient to produce a mixture temperature within the approved range. Adjust flames used for drying and heating to prevent damage to, and contamination of, the aggregate.

Control plant operations so the moisture content of the mixture behind the paver is 0.5 percent or less, in accordance with AASHTO T 110.

402.10 Mixing. Measure the aggregate and asphalt into the mixer in accordance with the approved job-mix formula. Mix until all the particles are completely and uniformly coated with asphalt in accordance with ASTM D 995. Maintain the discharge temperature within the approved range.

402.11 Hauling. Use vehicles with tight, clean, smooth metal beds for hauling asphalt concrete mixtures.

Thinly coat the beds with an approved material to prevent the mixture from adhering to the beds. Do not use petroleum derivatives or other coating material that contaminates or alters the characteristics of the mixture. Drain the bed before loading.

Equip each truck with a canvas cover or other suitable material of sufficient size to protect the mixture from the weather. When necessary to maintain temperature, use insulated truck beds and securely fastened covers. Provide access ports or holes for checking the temperature of the asphalt mixture in the truck.

402.12 Placing & Finishing. Do not use mixtures produced from different plants unless the mixtures are produced in accordance with the same job-mix formula, use material from the same sources, and are approved.

Place asphalt concrete mixture at a temperature that conforms to table 402-3. Measure temperature of the mixture in the hauling vehicle just before dumping into spreader, or measure it in the windrow immediately before pickup.

Place the mixture with a paver that conforms to specifications in Subsection 402.05. Control horizontal alignment using a reference line. Automatically control the grade and slope from reference lines, a ski and slope control device, dual skis. Use skis with a minimum length of 6 m.

Limit the compacted thickness to 75 mm, unless otherwise SHOWN ON THE DRAWINGS.

On areas where mechanical spreading and finishing is impractical, place and finish the mixture with alternate equipment to produce a uniform surface closely matching the surface obtained when using a mechanical paver.

Offset the longitudinal joint of one layer at least 150 mm from the joint in the layer immediately below. Make the longitudinal joint in the top layer along the centerline of two-lane roadways or at the lane lines of roadways with more than two lanes. Offset transverse joint in adjacent lanes and in multiple lifts at least 3 m.

The CO will designate the job-mix formula to be used for wedge and leveling courses at each location unless DESIGNATED IN THE SCHEDULE OF ITEMS. Place wedge and leveling courses in maximum 75-mm lifts and compact with a pneumatic-tire roller meeting the requirements of Subsection 203.15(d). Complete the wedge and leveling before starting normal paving operations.

402.13 Compacting. Furnish at least three rollers, one each for breakdown, intermediate, and finish rolling. Furnish at least one roller with pneumatic tires. Size the rollers to achieve the required results. Operate rollers in accordance with manufacturer's recommendations.

Thoroughly and uniformly compact the asphalt surface by rolling. Do not cause undue displacement, cracking, or shoving. Continue rolling until all roller marks are eliminated and the required compaction is obtained. Do not vibratory roll the mixture after its surface cools below 80 °C.

Along forms, curbs, headers, walls, and other places not accessible to the rollers, use other equipment to obtain the minimum compaction of the mixture.

402.14 Joints, Trimming Edges, & Cleanup. At connections to existing pavements and previously placed lifts, make the transverse joints vertical to the depth of the new pavement. Form transverse joints by cutting back on the previous run to expose the full depth course. Dispose of trimmed asphalt material in accordance with Subsection 210.02(a).

Apply an asphalt tack coat to the edge of the joint for both transverse and longitudinal joints, and where SHOWN ON THE DRAWINGS, in accordance with Section 407.

Place the asphalt concrete mixture as continuously as possible. Do not pass rollers over the unprotected end of a freshly laid mixture.

402.15 Acceptance. Provide a quality control plan, and then sample, test, and maintain records according to Section 160. See table 402-4 for minimum sampling and testing requirements. Sample to ensure that:

(a) The sample size is adequate to provide a duplicate to the CO and to meet potential need for retesting as specified in Subsection 402.18.

(b) Samples are prepared according to AASHTO T 248 or other procedures applicable to the item being sampled.

(c) The sample is adequately identified and placed in CO-approved containers provided by the Contractor.

The CO may perform quality assurances testing, and these tests will be made available to the Contractor upon request.

A lot is defined as the number of tons of material or work produced, and/or placed under one set of TV's. The lot will be represented by randomly selected samples tested for acceptance. Plant and equipment operators will not be advised ahead of time when samples are taken.

Acceptance will be evaluated as follows:

(a) Asphalt Cement. Asphalt cement will be evaluated for acceptance under Subsection 106.05.

Table 402-4.—Sampling and testing.

Type of Acceptance	Material or Product	Property or Characteristic	Test Method or Specification	Frequency	Sampling Point
Production certification (Subsection 105.04)	Asphalt cement	Contract requirements	AASHTO M 20, M 226, or MP 1, as applicable	Daily	–
Tested conformance	Material source	Los Angeles abrasion	AASHTO T 96	Three times for each undeveloped source, [b] or once for all other sources	Material source
		Sodium sulfate soundness loss	AASHTO T 104		
		Durability index (coarse and fine)	AASHTO T 210		
	Aggregate	Fractured faces (coarse)[a]	FLH T 507	Three times for each undeveloped source,[b] or once for all other sources	Cold feed prior to entering dryer
		Sand equivalent value (fine)	AASHTO T 176, alternate method number 2 (referee method)		
	Asphalt cement	Sample	Subsection 105.04(b)	Once for each 500 t of mix, and not more than three times per day	At point of shipment delivery
	Job-mix formula	Contract requirements	Subsection 402.03	Once for each product or material change	–
Mix evaluation	Hot asphalt concrete pavement	Asphalt content	AASHTO T 164, method B or E	Once for each 500 t, and not more than three times per day	At plant, in hauling units, or behind laydown machine before rolling
		Gradation	AASHTO T 30		
		Compaction	ASTM D 2950, "Procedure"	Five times for each 500 t, and not less than five times per day	In-place after compaction
		Maximum specific gravity	AASHTO T 209	Once for each 1,000 t	At plant, in hauling units, or behind laydown machine before rolling

a. Use only for gravel sources.

b. An undeveloped source is a source that has not supplied aggregate for asphalt concrete within 365 days of the start of producing asphalt concrete for this particular project.

(b) Pavement Smoothness. Use a 3-m metal straightedge to measure at right angles and parallel to the centerline at designated sites. Surface deviations in excess of 10 mm between any two contacts of the straightedge with the surface are defective areas. Correct these areas using approved methods.

(c) Thickness & Width. Ensure that the thickness and width of the compacted mixture conform to the dimensions SHOWN ON THE DRAWINGS and meet the following requirements:

(1) The maximum variation from the specified thickness is less than 6 mm for the wearing course or 12 mm for the base course.

(2) The compacted width has a +150-mm tolerance.

(3) The compacted thickness and width are not consistently above or below the specified dimension.

(d) Asphalt Concrete Mixture Gradation and Asphalt Content. Gradation and asphalt content will be evaluated for acceptance under Subsection 402.16.

(e) Asphalt Concrete Pavement Compaction. Compaction will be evaluated for acceptance under Subsection 402.17.

402.16 Acceptance Sampling & Testing of Asphalt Concrete Mixture Gradation & Asphalt Content. Take statistically random samples in accordance with the tests specified in table 402-4. Take a minimum of three tests per lot. Acceptance or rejection of completed work will be on a lot basis. If the Contractor quality control tests required in table 402-4 are validated by the CO in accordance with Subsection 402.18 (Test Result Validation Procedure), then the Contractor tests shall be used for acceptance tests. Take samples of the mixture at the plant in approved State department of transportation sampling devices, or after the mixture has been discharged into hauling units or placed on the road in accordance with AASHTO T 168. Test the samples for asphalt content by means of AASHTO T 164, method B (Reflux Method) or method E (Vacuum Extraction). Other methods, including nuclear, require approval in writing by the CO, and may require an increased sampling and testing frequency. Report the asphalt content to the nearest 0.01 percent. Determine gradation of the entire quantity of extracted material in accordance with AASHTO T 30, but report results to the nearest 0.1 percent for all sieves except the 75-µm sieve; report this sieve to the nearest 0.01 percent. Determine the percent moisture in the asphalt concrete mixture in accordance with AASHTO T 110.

If samples are tested for asphalt content by means of AASHTO T 164, determine an Extraction Retention Factor based on the average difference between at least three samples of known asphalt content and corresponding asphalt content by the same procedure that will be used for acceptance. Prepare the samples in accordance with table 402-1, (b) Marshall, unless otherwise approved by the CO.

If areas of isolated defect are identified by the CO, treat these areas as a separate lot.

The Contractor may request a change in TV's subject to the provisions in Subsection 402.03. If the TV's are changed, evaluate all of the material produced up to the time of the change as a lot, and begin a new lot.

The lot will be accepted with respect to gradation and asphalt if the average of all test results fall within the tolerances shown in table 402-5.

Table 402-5.—Mix tolerances.

Mixture Characteristic	Tolerances
Bitumen content	TV ± 0.5
Sieve size:	
4.75 mm and larger	TV ± 6.0
600 µm to 4.74 mm	TV ± 4.0
300 to 599 µm	TV ± 3.0
75 to 299 µm	TV ± 2.0
Temperature:	
Leaving the mixture	TV ± 6 °C
Placed on the road	TV ± 8 °C

402.17 Acceptance Sampling & Testing of Asphalt Concrete Pavement Compaction. Take statistically random samples in accordance with the tests specified in table 402-4. Take a minimum of five tests per lot. Acceptance or rejection of completed work will be on a lot basis. If the Contractor's quality control tests required in table 402-4 are validated by the CO in accordance with Subsection 402.18 (Test Result Validation Procedure), then the Contractor's tests will be used for acceptance tests.

Use the nuclear gauge for acceptance. Calibrate the nuclear gauge in accordance with ASTM D 2950, Calibration section, within 6 months prior to use on this project, and check the standard and reference on each day of use in accordance with ASTM D 2950, Standardization and Reference Check sections. Do not take acceptance samples within 0.3 m from the edges of the panel. Determine the TV for compaction using either the Control Strip Method or the Maximum Density Method as follows:

(a) Control Strip Method. Construct a control strip at the beginning of work on each type of material to be compacted. Construct each control strip to acceptable density and surface tolerances, and leave it in place to become a section of the completed roadway. Correct or remove unacceptable control strips, and replace them at the Contractor's expense. Construct a control strip at least 100 m long and one lane wide, and at the compacted lift thickness SHOWN ON THE DRAWINGS.

Ensure that the materials used in the construction of the control strip meet the specification requirements. Furnish them from the same source and of the same type and asphalt content used in the remainder of the course represented by the control strip.

Use equipment in the construction of the control strip that meets the requirements specified in Subsections 402.05 and 402.13, and is of the same type and weight as that to be used on the remainder of the course represented by the control strip.

Begin compacting the control strips immediately after the course has been placed to the specified thickness. Ensure that compaction is continuous and uniform over the entire surface. Continue compaction of the control strip until no discernible increase in density can be obtained by additional compactive effort.

Upon completion of the compaction, determine the mean density of the control strip by averaging the results of 10 consecutive nuclear density tests taken at randomly selected sites within the control strip. The mean density of the control strip must equal or exceed the density shown in table 402-6. The TV shall then be 98 percent of the mean density of the control strip.

Table 402-6.—Compaction requirements.

Road Grade (%)	% of AASHTO T 209 converted to density
$\leq$ 8	90
8–12	89
> 12	88

Cease paving if three consecutive control strips fail to achieve the specified minimum density. Take all necessary actions to resolve compaction problems. Do not resume paving without approval by the CO. Construct a new control strip in case of any of the following:

(1) Any change in the properties of the material.

(2) Any change in the rollers.

(3) A new layer.

(4) Or changes in grade as indicated in table 402-6.

(b) Maximum Density Method. The TV shall be as shown in table 402-6.

After the bituminous mixture has been placed and compacted, compaction for the lot will be accepted if both of the following apply:

(1) All individual test results equal or exceed 98 percent of the TV.

(2) The average of all tests equals or exceeds the TV.

402.18 Test Result Validation Procedure. Provide the CO with a duplicate of all samples specified in table 402-4. If the Government decides to run assurance tests on the duplicate samples, the CO will determine the number to be run. Normally, the first three samples submitted will be tested, and 10 percent thereafter.

As testing is completed, the CO will evaluate all the Contractor testing. If Contractor testing is verified by Government testing, the Contractor's test results may be used by the Government to evaluate work for acceptance. If Contractor testing is not verified by Government testing, the Contractor has the option of either retesting or having the Government test the duplicate sample. The Contractor or the CO may witness the testing of the remaining sample portions. If the Contractor retests the sample, the test results will again be evaluated based on Government verification testing. If the test results are not valid, the Government test results will then be used for acceptance.

If it becomes necessary for the Government to test all of the samples for a work item due to the Contractor's tests being declared invalid, a payment deduction equal to the total cost of performing all of the testing for the applicable item will be made.

If the Contractor's test results are shown to be valid, but significant differences or shifts make the test results questionable, the CO will review the Contractor's equipment and test procedures.

If any deficiencies are identified that would account for the significant differences or shifts, the CO will suspend acceptance of all material until the deficiencies have been corrected. If no deficiencies that would account for the significant differences or shifts in test results can be identified, continue testing. In order to identify the deficiencies, the CO will increase testing frequency of sample portions.

Measurement

402.19 Method. Use the method of measurement that is DESIGNATED IN THE SCHEDULE OF ITEMS.

Calculate the tonnage as the weight used in the accepted pavement, and make no deduction for the weight of asphalt cement in the mixture.

Payment

402.20 Basis. The accepted quantities will be paid at the contract unit price for each PAY ITEM listed below that is DESIGNATED IN THE SCHEDULE OF ITEMS, except that payment for sampling and testing will be made as follows:

(a) Twenty-five percent of the lump sum, not to exceed 0.5 percent of the original contract amount, will be paid after all the testing facilities are in place, qualified sampling and testing personnel are identified, and the work being tested has started.

(b) Payment for the remaining portion of the lump sum will be prorated based on the total work completed.

(c) Payment for all or part of this PAY ITEM may be retained if the Government assurance tests invalidate the Contractor's testing.

Payment will be made under:

Pay Item		Pay Unit
402 (01)	Hot asphalt concrete pavement, class _____, grading _____	Ton
402 (02)	State asphalt concrete pavement, class _____	Ton
402 (03)	Hot asphalt concrete pavement, class _____, grading _____, wedge, and level course	Ton
402 (04)	Asphalt cement, grade _____	Ton
402 (05)	Asphalt cement, State department of transportation grade _____	Ton
402 (06)	Sampling and testing	Lump Sum

Section 403—Minor Hot Asphalt Concrete Pavement

Description

403.01 Work. Construct one or more courses of hot asphalt concrete plant mix on a prepared surface as SHOWN ON THE DRAWINGS. Have the surface approved by the CO in writing prior to placing hot asphalt concrete plant mix.

Materials

403.02 Asphalt Cement. Ensure that asphalt cement meets the requirements specified in Subsection 702.01. The exact percent of asphalt cement and the grade to be used will be furnished by the CO after requirements in Subsection 403.05 have been reviewed and evaluated. Ensure that mixing temperatures meet the requirements specified in Subsection 702.04.

403.03 Aggregate. Ensure that aggregate meets the requirements specified in Subsection 703.07, except for aggregate gradation. Maximum gradation size or suggested gradation designations will be SHOWN ON THE DRAWINGS.

403.04 Additives. Additives, such as filler, hydrated lime, and antistrip agents, may be used as necessary to meet specifications. Ensure that filler meets the requirements of AASHTO M 17, hydrated lime meets the requirements of AASHTO M 216, type N, and antistrip materials meet the requirements specified in Subsection 702.07.

403.05 Job-Mix Formula. Submit a job-mix formula and supporting documentation, test results, and calculations for the material to be incorporated into the work. Include copies of laboratory test results and mix design data that demonstrate that the properties of the aggregate, additives, and mixture meet those requirements and criteria of local public agencies or the AI. After reviewing the Contractor's proposed job-mix formula, the CO will determine the final values for the job-mix formula to be used and notify the Contractor in writing.

Construction

403.06 Asphalt Mixing Plant. Ensure that plants used for preparing hot asphalt concrete mixtures are manufactured for that purpose, in good repair, and capable of mixing the material to a uniform consistency.

403.07 Hauling Equipment. Ensure that trucks used for hauling asphalt concrete mixtures have tight, clean, smooth metal beds that have been thinly coated with a material to prevent the mixture from adhering to the beds. Do not use petroleum derivatives or other coating material that contaminates or alters the characteristics of the mixture. Drain truck beds prior to loading, and ensure that each truck has a cover to protect the mixture from weather. When necessary to ensure that the mixture will be delivered at the specified temperature, ensure that truck beds are insulated and covers securely fastened.

403.08 Pavers. Use pavers that are in good working order and have an adjustable vibrating screed or strike-off assembly, heated if necessary, and an auger ahead of the screed to distribute the mixture. Use pavers that are capable of spreading and finishing courses of asphalt concrete plant mix material in lane widths and thickness SHOWN ON THE DRAWINGS. Unless otherwise SHOWN ON THE DRAWINGS, towed-type pavers and Layton-type pavers or graders may be used to place and spread the asphalt concrete plant mix material.

403.09 Rollers. Ensure that all rollers meet the requirements specified in Subsections 203.15(b), (c), and (d). Where it is impractical to operate larger rollers, 3- to 5-t rollers may be used. On walkways, 1-t rollers may be used.

403.10 Weather Limitations. Do not place the asphalt concrete mixture when weather conditions prevent the proper compaction of the mixture, the base course is frozen, or the average temperature of the underlying surface upon which the asphalt concrete mixture is to be placed is less than 7 °C, or when it is raining or snowing.

403.11 Conditioning of Existing Surface. Immediately before placing the asphalt concrete mixture, clean the existing surface of loose or deleterious material.

Before placing the asphalt concrete mixture, paint the contact surfaces of curbing, gutters, manholes, and other structures with a thin, uniform coating of asphalt material.

403.12 Control of Asphalt Concrete Mixture. Supply a certification from the mixing plant stating that the mix conforms to the approved job-mix formula. The CO may reject any batch, load, or section of roadway that appears defective in gradation, asphalt cement content, or moisture content. Do not incorporate material rejected before placement into the pavement. Remove any rejected section of roadway. No payment will be made for the rejected materials or the removal of the materials, unless the Contractor requests that the rejected material be tested, at the Contractor's expense, under the following provisions:

(a) Obtain three representative samples and have them tested at a laboratory approved by the CO.

(b) If test results show that the material conforms to the tolerance shown in table 403-1, payment will be made for the material, and for its removal and testing.

Table 403-1.—Allowable tolerances.

Mixture Characteristic	Tolerances
Asphalt content	Job-mix formula ± 0.5
Sieve size:	
9.5 mm and larger	Job-mix formula ± 5.0
4.75 to 9.49 mm	Job-mix formula ± 7.0
76 µm to 4.74 mm	Job-mix formula ± 5.0
75 µm	Job-mix formula ± 2.0

403.13 Transporting, Spreading, & Finishing. Transport the mixture from the mixing plant to the point of use in vehicles that meet the requirements specified in Subsection 403.07.

Spread the mixture and strike it off to the grade and elevation established. Provide a maximum compacted lift thickness of 100 mm unless otherwise SHOWN ON THE DRAWINGS.

Ensure that the longitudinal joint in any layer offsets that in the layer immediately below by approximately 150 mm. Where laydown requires placement of two adjacent panels to cover the surface of a traveled way, ensure that the longitudinal joint of the top layer is at the centerline. This requirement does not apply to turnouts, extra widening, or parking areas. Offset transverse joints in succeeding layers and in adjacent lanes at least 3 m, where possible.

On areas where irregularities or unavoidable obstacles make the use of mechanical spreading and finishing equipment impracticable, the mixture may be placed and finished using hand tools.

403.14 Compaction. Thoroughly and uniformly compact the surface with rollers that meet the requirements specified in Subsection 403.09, and perform initial compaction while the mixture is above 120 °C. Perform finish rolling with steel-wheel rollers and continue until no roller tracks remain.

Measurement

403.15 Method. Use the method of measurement that is DESIGNATED IN THE SCHEDULE OF ITEMS.

Calculate the quantity of hot asphalt concrete mix that is the tonnage of combined aggregate and asphalt cement used in the accepted work. No separate payment will be made for asphalt cement used in the mixture.

Payment

403.16 Basis. The accepted quantities will be paid for at the contract unit price for each PAY ITEM DESIGNATED IN THE SCHEDULE OF ITEMS.

Payment will be made under:

Pay Item		Pay Unit
403 (01)	Hot asphalt concrete plant mix	Ton
403 (02)	Hot asphalt concrete plant mix	Square Meter

Section 404—Major Cold Asphalt Concrete Pavement

Description

404.01 Work. Construct one or more courses of cold asphalt concrete pavement on a prepared surface that has been approved in writing by the CO.

Cold asphalt concrete pavement grade is designated as shown in table 703-5 or table 703-6.

Cutback asphalt grade is designated as shown in AASHTO M 81, AASHTO M 82, or ASTM D 2026. Emulsified asphalt grade is designated as shown in AASHTO M 140 or M 208.

A prepaving conference will be held at least 10 working days prior to the beginning of paving operations. At that time, the Contractor and the CO will discuss methods of accomplishing all phases of the paving work, including laydown operations, work schedules, work force, quality control systems, spill prevention and contingency plans, and asphalt concrete mix delivery.

Materials

404.02 Requirements. Ensure that the material conforms to specifications in the following subsections:

Material	Subsection
Antistrip Additive	702.07
Cement	701.01
Choker Aggregate	703.11
Cold Asphalt Concrete Pavement Aggregate	703.08
Cutback Asphalt	702.02
Emulsified Asphalt	702.03
Hydrated Lime	725.03
Mineral Filler	725.05
Water	725.01

Ensure that mixing temperature meets the requirements specified in Subsection 702.04.

Construction

404.03 Composition of Mixture (Job-Mix Formula). Ensure that the composition of cold asphalt concrete mixtures conforms to the following:

(a) Furnish a job-mix formula at least 21 days prior to production. Base the formula on a mix design using the type and grade of asphalt material that will be furnished for the project and of aggregate that will be produced for the project. Acceptable job-mix procedures and criteria are found in AI Manual Series number 14 and number 19. After reviewing the Contractor's proposed job-mix formula, the CO will determine a job-mix formula with TV's and will notify the Contractor in writing.

(b) In the proposed job-mix formula, include definite single-value TV's for:

(1) The percentage of aggregate passing each specified sieve, based on the dry weight of aggregate. These percentages shall be within the range shown in Subsection 703.08, table 703-5 or table 703-6, as applicable.

(2) The percentage of bituminous material to be added, based on the total weight of mixture and corresponding residual asphalt content.

(3) The kind and percentages of additives to be used.

(4) The percentage of water, based on the total dry weight of the mixture.

(5) For emulsified asphalt only, the percentage to total fluids at compaction, based on the total dry weight of the mixture.

404.04 Performance. Perform construction in accordance with the following:

(a) Mixing Plant. Use asphalt mixing plants or pugmills that:

(1) Are manufactured for that purpose.

(2) Are in good working order.

(3) Are equipped with weighing or volumetric equipment capable of providing accurate control of the material entering the mixer.

(4) Interlock the aggregate feed controls with the asphalt material and other additives.

(b) Pavers. Use pavers that are:

(1) Self-contained, power-propelled units with adjustable vibratory screeds with full-width screw augers.

(2) Capable of spreading and finishing courses of asphalt mixture in widths at least 300 mm more than the width of one lane.

(3) Equipped with a receiving hopper with sufficient capacity to ensure a uniform spreading operation.

(4) Equipped with automatic feed controls that are properly adjusted to maintain a uniform depth of material ahead of the screed.

(5) Capable of being operated at forward speeds consistent with satisfactory laying of the mixture.

(6) Capable of producing a finished surface of the required smoothness and texture without segregating, tearing, shoving, or gouging the mixture.

(7) Equipped with automatic screed controls with sensors capable of sensing grade from an outside reference line, sensing the transverse slope of the screed, and providing the automatic signals that operate the screed to maintain grade and transverse slope.

(c) Surface Preparation. Prepare the surface in accordance with Sections 304, 306, 307, and 308, as applicable. Apply an asphalt tack coat to contact surfaces of curbing, gutters, manholes, and other structures in accordance with Section 407.

(d) Weather Limitations. Place cold asphalt concrete pavement on unfrozen, reasonably dry surface when the temperature of the road surface, in the shade, is above 15 °C, and it is not raining or snowing, or predicted to rain or snow within 24 hours after placement.

(e) Mixing. Introduce the material into the mixing plant according to the approved job-mix formula. Control the moisture content by adding water in the plant, covering the stockpile, drying the aggregate, or a combination of these methods as necessary to comply with the job-mix formula. When approved in writing by the CO, additives such as lime or cement may be incorporated into the mixture to correct moisture content.

When the aggregate is combined with asphalt materials other than emulsified asphalt, ensure that the aggregate does not contain more than 3 percent moisture and is at a temperature not less than 16 °C and not more than 107 °C. When the aggregate is combined with emulsified asphalt, ensure that the aggregate is at a temperature not less than 16 °C and not more than 79 °C. Determine the mixing time for each phase of the mixing operation from the nature of the aggregates, the job-mix formula, and the size of the batch. If the mixture is stockpiled, do not allow the pile to segregate such that the emulsified asphalt breaks.

(f) Hauling. Use vehicles conforming to Subsection 402.11.

(g) Placing & Finishing. Do not use mixtures produced from different plants unless the mixtures are produced in accordance with the same job-mix formula, use material from the same sources, and are approved.

Place the mixture with a paver that conforms to Subsection 404.04(b). Control horizontal alignment using a reference line. Automatically control the grade and slope from reference lines, a ski and slope control device, or dual skis. Use skis with a minimum length of 6 m.

Offset and locate longitudinal joint according to Subsection 402.12.

For dense-graded mixtures, allow the surface to cure for not less than 10 days, and for open-graded mixtures, not less than 4 days, before covering with the next course. During this period, maintain the surface and keep it free of corrugations. Use an approved material to patch all holes. Remove all excess blotter, dirt, or other objectionable substances before placing the following course or treatment.

(h) Compacting. Perform initial compaction through a minimum of three complete coverages with a steel-wheel roller that meets the requirements of Subsection 203.15(b). If necessary for dense-graded mixtures, aerate the material by periodically moving and exposing it in the stockpile or through manipulation in a windrow to remove excess moisture or cutter. When DESIGNATED IN THE SCHEDULE OF ITEMS, and prior to intermediate rolling, apply choker aggregate to the top layer only using aggregate spreading equipment designed for the controlled spreading of fine material. Uniformly spread the material to a depth that, when compacted, is sufficient to fill the voids of the asphalt concrete mat. Remove excess choker material by brooming.

Perform intermediate compaction through a minimum of two complete coverages of a self-propelled pneumatic-tire roller with a maximum tire pressure of 275 kPa.

Perform final compaction through two complete coverages with a static roller that meets the requirements of Subsection 203.15, and until all roller marks are eliminated. For open-graded mixtures, use a steel-wheel roller. When no choker aggregate is required, perform final compaction while the emulsion is still tacky.

Along forms, curbs, headers, walls, and other places not accessible to the rollers, use other equipment to obtain the minimum compaction of the mixture.

404.05 Acceptance Sampling & Testing. Perform acceptance sampling and testing in accordance with Subsections 402.15 through 402.18, with the following modifications:

(a) Use table 404-1.

(b) Use table 404-2.

Table 404-1.—Sampling and testing.

Type of Acceptance	Material or Product	Property or Characteristics	Test Method or Specification	Frequency	Sampling Point
Production certification (Subsection 105.04)	Asphalt	Contract requirements	AASHTO M 81, AASHTO M 82, AASHTO M 140, AASHTO M 208, ASTM D 2026, as applicable	Daily	–
Tested conformance	Material source	Los Angeles abrasion	AASHTO T 96	Three times for each undeveloped source,[b] or once for all other sources	Material source
		Sodium sulfate soundness loss	AASHTO T 104		
		Durability index (coarse and fine)	AASHTO T 210		
	Aggregate	Fracture faces (coarse)[a]	FLH T 507	Three times for each undeveloped source,[b] or once for all other sources	Cold feed prior to entering mixer
		Sand equivalent value (fine)	AASHTO T 176, alternate method number 2 (referee method)		
	Asphalt	Sample	Subsection 105.04(b)	Once for each 500 t of mix, and not more than three times per day	At point of shipment delivery
	Job-mix formula	Contract requirements	Subsection 404.03	Once for each product or material change	–
Mix evaluation	Cold asphalt concrete pavement	Asphalt content	AASHTO T 164, method B or E	Once for each 500 t, and not more than three times per day	At plant, in hauling units, or behind laydown machine before rolling
		Gradation	AASHTO T 30		

a. Use only for gravel sources.

b. An undeveloped source is a source that has not supplied aggregate for asphalt concrete within 365 days of the start of producing asphalt concrete for this particular project.

Table 404-2—Mix tolerances.

Mixture Characteristic[a]	Tolerances
Residual asphalt content	TV ± 0.5
Total fluids in aggregate at compaction	TV ± 1.5

a. For sieve size tolerance, refer to table 703-5 or 703-6, according to the grading DESIGNATED IN THE SCHEDULE OF ITEMS.

(c) Compaction for the lot will be accepted if the requirements specified in Subsection 404.04(g) have been met and all roller marks are eliminated.

Measurement

404.06 Method. Use the method of measurement that is DESIGNATED IN THE SCHEDULE OF ITEMS.

Calculate tonnage as the weight used in the accepted pavement. Make no deduction for the weight of bituminous material or water. No separate payment will be made for water or additives used in the mixture.

Payment

404.07 Basis. The accepted quantities will be paid for at the contract unit price for each PAY ITEM DESIGNATED IN THE SCHEDULE OF ITEMS, except that payment for sampling and testing will be made as follows:

(a) Twenty-five percent of the lump sum, not to exceed 0.5 percent of the original contract amount, will be paid after all the testing facilities are in place, qualified sampling and testing personnel are identified, and the work being tested has started.

(b) Payment for the remaining portion of the lump sum will be prorated based on the total work completed.

(c) Payment for all or part of this pay item may be withheld if the Government assurance tests invalidate the Contractor's testing.

Payment will be made under:

Pay Item		Pay Unit
404 (01)	Cold bituminous pavement, ________ grading	Ton
404 (02)	Cutback asphalt, grade ______________	Ton
404 (03)	Emulsified asphalt, grade _____________	Ton
404 (04)	Choker aggregate	Ton
404 (05)	Sampling and testing	Lump Sum

Section 405—Minor Cold Asphalt Concrete Pavement

Description

405.01 Work. Construct one or more courses of cold asphalt concrete plant mix on a prepared surface as SHOWN ON THE DRAWINGS. Have the surface approved by the CO in writing prior to placing cold asphalt concrete plant mix.

Cutback asphalt grade is designated as shown in AASHTO M 81 or M 82. Emulsified asphalt grade is designated as shown in AASHTO M 140 or M 208.

Materials

405.02 Asphalt Material. Ensure that asphalt material meets the requirements specified in Subsection 702.02 or 702.03, as applicable. The exact percent of asphalt material and the grade to be used will be furnished by the CO after requirements in Subsection 405.05 have been reviewed and evaluated. Ensure that mixing temperatures shall meet the requirements specified in Subsection 702.04.

405.03 Aggregate. Ensure that aggregate meets the requirements specified in Subsection 703.09, except for aggregate gradation. Maximum gradation size or suggested gradation designations will be SHOWN ON THE DRAWINGS.

405.04 Additives. Additives, such as filler, hydrated lime, and antistrip agents, may be used as necessary to meet specifications. Ensure that filler meets the requirements of AASHTO M 17; hydrated lime meets the requirements of AASHTO M 216, type N; antistrip materials meet the requirements specified in Subsection 702.07; and choker aggregate meets the requirements specified in Subsection 703.11

405.05 Job-Mix Formula. Submit a job-mix formula and supporting documentation, test results, and calculations for the material to be incorporated into the work. Include copies of laboratory test results and mix design data that demonstrate that the properties of the aggregate, additives, and mixture meet those requirements and criteria of local public agencies or the AI. After reviewing the Contractor's proposed job-mix formula, the CO will determine the final values for the job-mix formula to be used and notify the Contractor in writing.

Construction

405.06 Asphalt Concrete Mixing Plant. Ensure that plants used for preparing cold asphalt concrete mixtures are manufactured for that purpose, in good repair, and capable of mixing the material to a uniform consistency.

405.07 Hauling Equipment. Ensure that trucks used for hauling asphalt concrete mixtures have tight, clean, smooth metal beds that have been thinly coated with a material to prevent the mixture from adhering to the beds. Do not use petroleum derivatives or other coating material that contaminates or alters the characteristics of the mixture. Drain truck beds prior to loading, and ensure that each truck has a cover to protect the mixture from weather. When necessary to ensure that the mixture will be delivered at the specified temperature, ensure that truck beds are insulated and covers securely fastened.

405.08 Pavers. Use pavers that are in good working order and have an adjustable vibrating screed or strike-off assembly, and an auger ahead of the screed to distribute the mixture. Use pavers that are capable of spreading and finishing courses of asphalt concrete plant mix material in the lane widths and thickness SHOWN ON THE DRAWINGS. Unless otherwise SHOWN ON THE DRAWINGS, towed-type pavers and Layton-type pavers or graders may be used to place and spread the asphalt concrete plant mix material.

405.09 Rollers. Ensure that all rollers meet the requirements specified in Subsections 203.15(b), (c), and (d). Where it is impractical to operate larger rollers, 3- to 5-t rollers may be used. On walkways, 1-t rollers may be used.

405.10 Weather Limitations. Do not place the asphalt concrete mixture when the base course is frozen, when the average temperature of the underlying surface upon which the asphalt concrete mixture is to be placed is less than 10 °C in the shade, or when it is raining or snowing, or predicted to rain or snow within 24 hours after placement.

405.11 Conditioning of Existing Surface. Immediately before placing the asphalt concrete mixture, clean the existing surface of loose or deleterious material.

Before placing the asphalt concrete mixture, paint the contact surfaces of curbing, gutters, manholes, and other structures with a thin, uniform coating of asphalt material.

405.12 Control of Asphalt Concrete Mixture. Supply a certification from the mixing plant stating that the mix conforms to the approved job-mix formula. The CO may reject any batch, load, or section of roadway that appears defective in gradation, asphalt content, or moisture content. Do not incorporate material rejected before placement into the pavement. Remove any rejected section of roadway. No payment will be made for the rejected materials or the removal of the materials, unless the Contractor requests that the rejected material be tested, at the Contractor's expense, under the following provisions:

(a) Obtain three representative samples and have them tested at a laboratory approved by the CO.

(b) If test results show that the material conforms to the tolerance shown in table 405-1, payment will be made for the material and for its removal and testing.

Table 405-1.—Allowable tolerances.

Mixture Characteristic	Tolerances
Residual asphalt content	Job-mix formula ± 0.5
Sieve size:	
9.5 mm and larger	Job-mix formula ± 5.0
4.75 to 9.49 mm	Job-mix formula ± 7.0
76 µm to 4.74 mm	Job-mix formula ± 5.0
75 µm	Job-mix formula ± 2.0

405.13 Transporting, Spreading, & Finishing. Transport the mixture from the mixing plant to the point of use in vehicles that meet the requirements specified in Subsection 405.07.

Spread the mixture and strike it off to the grade and elevation established. Provide a maximum compacted lift thickness of 100 mm unless otherwise SHOWN ON THE DRAWINGS.

Ensure that the longitudinal joint in any layer offsets that in the layer immediately below by approximately 150 mm. Where laydown requires placement of two adjacent panels to cover the surface of a traveled way, ensure that the longitudinal joint of the top layer is at the centerline. This requirement does not apply to turnouts, extra widening, or parking areas. Offset transverse joints in succeeding layers and in adjacent lanes at least 3 m.

On areas where irregularities or unavoidable obstacles make the use of mechanical spreading and finishing equipment impracticable, the mixture may be placed and finished by using hand tools.

405.14 Compaction. Perform compaction with rollers that meet the requirements specified in Subsection 405.09. Perform initial compaction with steel-wheel rollers for a minimum of three complete coverages. Between initial and final rolling on open-graded mixtures, apply a choker aggregate to the top layer only using aggregate spreading equipment designed for the controlled spreading of fine material. Uniformly spread the material to a depth that, when compacted, will be sufficient to fill the surface voids of the bituminous mat. Remove excessive choke material by brooming. Continue rolling, with a minimum of four complete coverages and until no roller tracks remain, and while the bituminous material is still tacky.

Measurement

405.15 Method. Use the method of measurement that is DESIGNATED IN THE SCHEDULE OF ITEMS.

Calculate the quantity of cold asphalt concrete plant mix as the tonnage of combined aggregate and asphalt material used in the accepted work. No separate payment will be made for asphalt material, water, or additives used in the mixture.

Payment

405.16 Basis. The accepted quantities will be paid for at the contract unit price for each PAY ITEM DESIGNATED IN THE SCHEDULE OF ITEMS.

Payment will be made under:

Pay Item		Pay Unit
405 (01)	Cold asphalt concrete plant mix, grade ____	Ton
405 (02)	Cold asphalt concrete plant mix, grade ____	Square Meter

Section 407—Asphalt Tack Coat

Description

407.01 Work. Apply an emulsified asphalt tack coat. Have the surface to be treated approved by the CO in writing prior to treatment.

Tack coat emulsified asphalt grade is designated as shown in AASHTO M 140 or M 208.

Materials

407.02 Requirements. Ensure that material conforms to the specifications in the following subsections:

Emulsified Asphalt 702.03
Water 725.01

Construction

407.03 Equipment. Use equipment that conforms to specifications in Subsection 410.04.

407.04 Surface Preparation. Immediately before the application of the tack coat, patch the surface to be treated and remove all foreign and loose material.

407.05 Weather Limitations. Apply asphalt tack coat on a dry, unfrozen surface when the surface temperature in the shade is above 5 °C and rising.

407.06 Asphalt Application. Where slow-setting emulsified asphalt is used, dilute it by adding an equal amount of water to the emulsified asphalt.

Apply the asphalt in accordance with Subsection 410.08 at a rate of 0.15 to 0.70 L/m^2. When a tack coat cannot be applied with an asphalt distributor spray bar, apply the tack coat uniformly and completely by fogging with a hand spray attachment or by another approved method. Ensure that the surfaces of adjacent structures and trees are protected from splattering and marring.

If excess asphalt material is applied, squeegee the excess from the surface. Allow the tacked surfaces to completely cure before placing the covering course. Place the covering course within 4 hours of placing the tack coat.

407.07 Acceptance. Emulsified asphalt will be evaluated for acceptance under Subsection 106.05.

Provide the minimum number of samples specified in table 410-4.

Measurement

407.08 Method. Use the method of measurement that is DESIGNATED IN THE SCHEDULE OF ITEMS.

Water used for diluting emulsified asphalt will not be included in the quantities for PAY ITEMS 407(01) or 407(02), and will not be paid for separately.

Payment

407.09 Basis. The accepted quantities will be paid for at the contract unit price for each PAY ITEM DESIGNATED IN THE SCHEDULE OF ITEMS.

Payment will be made under:

Pay Item	Pay Unit
407 (01) Tack coat, grade ___	Ton
407 (02) Tack coat, grade ___	Liter

Section 408—Asphalt Prime Coat

Description

408.01 Work. Apply a cutback or emulsified asphalt prime coat. Have the surface approved in writing by the CO prior to applying the prime coat.

Prime coat asphalt grade is designated as shown in AASHTO M 140 or M 208 for emulsified asphalt, and AASHTO M 81 or M 82 for cutback asphalt.

Materials

408.02 Requirements. Ensure that material conforms to the specifications in the following subsections:

Blotter	703.12
Choker Aggregate	703.11
Cutback Asphalt	702.02
Emulsified Asphalt	702.03
Water	725.01

Construction

408.03 Equipment. Use equipment that conforms to Subsection 410.04.

408.04 Surface Preparation. Immediately before applying the prime coat, lightly blade the surface and roll it with a smooth-wheel roller. Ensure that the moisture content of the top 25 mm of the surface to be treated is slightly damp.

408.05 Weather Limitations. Apply prime coat when the air temperature in the shade and the pavement surface temperature are at least 10 °C and rising, and when the weather is not foggy or rainy.

408.06 Asphalt Application. When required by the CO, lightly spray the surface with water before applying the prime coat. In order to obtain optimum penetration, apply cutback asphalt in accordance with Subsection 410.08 at a rate of 0.45 to 2.25 L/m^2.

Where using an emulsified asphalt that is not formulated as a penetrating prime coat material, dampen the roadway surface and scarify 25 to 50 mm deep. Dilute the emulsified asphalt by adding an equal amount of water. Apply the diluted emulsified asphalt in accordance with the Subsection 410.08 at a rate of 0.45 to 1.35 L/m^2. Immediately process, respread, and compact the material.

Cure surfaces primed with emulsified asphalt for not less than 24 hours, and surfaces primed with cutback asphalt for not less than 5 days before covering with the next course.

Until the next course is placed, maintain the primed surface and keep it free of corrugations.

Where traffic is routed over a primed surface before the asphalt material has been completely absorbed, or to minimize damage by rain, spread blotter to cover the unabsorbed asphalt. If an emulsified asphalt is used, use choker aggregate. When cutback asphalt is used, do not apply the blotter material for at least 4 hours following application of the asphalt. Remove all excess blotter, choke, dirt, or other deleterious material and repair all damaged areas before placing the next course. Dispose of asphalt material in accordance with Subsection 202.04(a).

408.07 Acceptance. Cutback asphalt and emulsified asphalt will be evaluated for acceptance under Subsection 106.05.

Blotter will be evaluated for acceptance under Subsection 105.03.

Provide the minimum number of samples and tests specified in table 410-4.

Measurement

408.08 Method. Use the method of measurement that is DESIGNATED IN THE SCHEDULE OF ITEMS.

Water used for diluting emulsified asphalt will not be included in the quantity for PAY ITEMS 408(01) or 408(02) and will not be paid for separately.

Payment

408.09 Basis. The accepted quantities will be paid for at the contract unit price for each PAY ITEM DESIGNATED IN THE SCHEDULE OF ITEMS.

Payment will made under:

Pay Item		Pay Unit
408 (01)	Prime coat, grade ___	Ton
408 (02)	Prime coat, grade ___	Liter
408 (03)	Blotter	Ton
408 (04)	Blotter	Cubic Meter

Section 409—Slurry Seal

Description

409.01 Work. Apply an asphalt slurry seal mixture. Have the surface approved by the CO in writing prior to placing the slurry seal.

Slurry seal type is designated as shown in table 703-8.

Materials

409.02 Requirements. Ensure that material conforms to the specifications in the following subsections:

Emulsified Asphalt .... 702.03(d)
Mineral Filler 725.05
Slurry Seal Aggregate 703.10
Water 725.01

Construction

409.03 Composition of Mixture (Job-Mix Formula). Furnish a slurry seal mixture of aggregate, water, emulsified asphalt, and additives in accordance with ASTM D 3910 and ISSA T 114. Ensure that the mixture meets the applicable aggregate gradation shown in table 703-8 and has the following residual asphalt contents, based upon weight of dry aggregate:

- Type I—Residual asphalt between 10.0 percent and 16.0 percent.
- Type II—Residual asphalt between 7.5 percent and 13.5 percent.
- Type III—Residual asphalt between 6.5 percent and 12.0 percent.

Submit a written job-mix formula for approval at least 21 days before production that includes the following:

(a) Aggregate Gradation Values. Provide the representative value for each sieve size for the aggregate blend.

(b) Emulsified Asphalt Content. Provide the residual asphalt content, as a percent by weight of dry aggregate.

(c) Samples. Provide samples of the aggregate, emulsified asphalt, and mineral filler when SHOWN ON THE DRAWINGS.

(d) Laboratory Test Reports & Mix Design Data. Provide copies of all laboratory test reports and mix design data verifying that the material meets the requirements of Subsection 409.02 and the job-mix formula.

The job-mix formula will be evaluated and approved in accordance with Subsection 401.03.

409.04 Equipment. Furnish equipment with the following capabilities:

(a) Slurry Seal Mixer. Furnish a slurry seal mixer with the following features and capabilities:

(1) Self-propelled.

(2) Continuous-flow mixing.

(3) Calibrated controls.

(4) Easily readable metering devices that accurately measure all raw material before entering the pugmill.

(5) Automated system for sequencing in all raw material to ensure constant slurry mixture.

(6) Mixing chamber to thoroughly blend all ingredients together.

(7) Fines feeder with an accurate metering device for introducing additive into the mixer, where the aggregate is introduced into the mixer.

(8) A pressurized water system with a fog-type spray bar capable of fogging the surface immediately ahead of the spreading equipment at a rate of 0.13 to 0.27 L/m^2.

(9) Proportioning system that is accurate for measuring all material independent of the engine speed.

(10) Minimum speed of 20 m/min and maximum speed of 55 m/min.

(11) Minimum storage capacity of 6 t.

(12) Capability in accordance with ISSA Performance Guidelines A 105.

(b) Mechanical-Type Single Squeegee Spreader Box. Furnish with the following capabilities:

(1) Attaches to the slurry seal mixer.

(2) Flexible squeegee in contact with the surface to prevent loss of slurry.

(3) Adjustable to assure a uniform spread over varying grades and crowns.

(4) Adjustable in width with a flexible strike-off.

(5) Augers for uniform flow to edges.

(c) Auxiliary Equipment. Furnish hand squeegees, shovels, and other equipment necessary to perform the work. Provide cleaning equipment that includes, but is not limited to, power brooms, air compressors, water-flushing equipment, and hand brooms for surface preparation.

(d) Pneumatic-Tire Roller. When SHOWN ON THE DRAWINGS, provide a pneumatic-tire roller with the following features:

(1) Smooth tread tires of equal size.

(2) Minimum ground pressure of tire greater than 345 kPa.

409.05 Surface Preparation. Immediately before placing the slurry seal, clean the existing surface of loose or deleterious material.

409.06 Weather Limitations. Apply slurry seal when the air temperature in the shade and the surface temperature are at least 15 °C and rising, and when the weather is not foggy, rainy, or overcast.

409.07 Slurry Seal Application. Mix the slurry seal using a slurry seal mixer. Fog the surface with water immediately preceding the spreader.

Blend the additive with the aggregate using the fines feeder. Prewet the aggregate in the pugmill immediately before mixing with the emulsified asphalt. Stockpile aggregate accordingly to Subsection 305.04.

Mix the slurry seal for a maximum of 4 minutes. Ensure that the slurry seal mixture is of the desired consistency as it leaves the mixer, and that it conforms to the approved job-mix formula. If approved by the CO, the mineral filler and the emulsified asphalt content may be adjusted during construction to conform to variations in field conditions.

Carry sufficient slurry seal mixture in the spreader to completely cover the surface. Spread the mixture with a mechanical-type single squeegee spreader box. In areas not accessible to the spreader box, use hand squeegees to work the slurry seal mixture.

Remove or repair ridges or bumps in the slurry surface.

When required, roll the slurry surface, providing a minimum of five coverages, to completely cure it prior to opening to traffic. Cure is complete when clear water can be pressed out of the slurry mixture with a piece of paper without discoloring the paper.

409.08 Acceptance. Emulsified asphalt will be evaluated for acceptance under Subsection 106.05.

Aggregate will be evaluated for acceptance under Subsection 105.03.

Provide the minimum number of samples and tests specified in table 409-1, in accordance with Section 160.

Measurement

409.09 Method. Use the method of measurement that is DESIGNATED IN THE SCHEDULE OF ITEMS.

Payment

409.10 Basis. The accepted quantities will be paid for at the contract unit price for each PAY ITEM DESIGNATED IN THE SCHEDULE OF ITEMS.

Payment will be made under:

Pay Item	Pay Unit
409 (01) Slurry seal, type ___	Square Meter

Table 409-1.—Sampling and testing.

Type of Acceptance	Material or Product	Property or Characteristic	Test Method or Specification	Frequency	Sampling Point
Production certification	Emulsified asphalt[a]	Contract requirements	AASHTO M 140 or M 208, as applicable	Each shipment	–
Tested conformance	Material source[b] (Contractor-located source)	Los Angeles abrasion	AASHTO T 96	Three times for each undeveloped source,[c] or once for all other sources	Material source
	Slurry seal aggregate	Gradation, table 703-8	AASHTO T 27 and T 11	Once for each 1,700 m^2, but not more than three times per day	
		Sand equivalent value	AASHTO T 176, alternate method number 2 (referee method)		
	Emulsified asphalt	Sample	Subsection 106.04(b)	Each tanker, including all trailers	At point of shipment delivery

a. Use in work permitted before sampling and testing for conformance.

b. See Subsection 105.03. Testing not required when using Government-provided material source.

c. An undeveloped source is a source that has not supplied slurry seal aggregate within 365 days of the start of producing slurry seal aggregate for this project.

Section 410—Asphalt Surface Treatment

Description

410.01 Work. Construct a single or multiple asphalt surface treatment course. Have the surface approved by the CO in writing prior to placing the asphalt surface treatment.

Surface treatment aggregate is designated as shown in tables 410-1, 410-2, and 410-3.

Table 410-1.—Approximate quantities of material for single-course surface treatment.

Sequence of Operations	Treatment Designation and Aggregate Gradation				
	B	C	D	E	F
Apply asphalt material (L/m^2):					
Emulsified asphalt	2.5	2.0	1.6	1.1	1.1
Cutback asphalt	1.9	1.6	1.2	0.9	0.9
Asphalt cement	1.7	1.4	1.1	0.8	0.8
Spread aggregate[a,b] (kg/m^2)	24	16	12	8	11

a. See table 703-7 for aggregate gradations.
b. Aggregate weights are for aggregates that have a bulk specific gravity of 2.65, as determined by AASHTO T 84 and T 85. Make proportionate corrections when the aggregate furnished has a bulk specific gravity above 2.75 or below 2.55.

The grade of asphalt is designated as shown in AASHTO M 20 or M 226 for asphalt cement, Subsection 702.03 for emulsified asphalt, and AASHTO M 81 or M 82 for cutback asphalt, or in applicable State department of transportation specifications for the grade specified.

A presurface treatment conference will be held at least 10 working days prior to the beginning of surface treatment operations. At that time, the Contractor and the CO will discuss methods of accomplishing all phases of the work, including operations, work schedules, work force, quality control systems, spill prevention and contingency plans, and material application rates.

Materials

410.02 Requirements. Ensure that material conforms to the specifications in the following subsections:

Table 410-2—Approximate quantities of material for multiple-course surface treatment (using cutback asphalt).

Sequence of Operations	Treatment Designation Aggregate Gradation[a,b]				
	CT–20	CT–24	CT–36	CT–44	CT–48
First course:					
Apply asphalt material (L/m^2)	1.2	1.6	1.9	1.9	1.9
Spread aggregate (kg/m^2)					
Grading D	12	–	–	–	–
Grading C	–	16	–	–	–
Grading B	–	–	24	24	24
Second course:					
Apply asphalt material (L/m^2)	0.9	0.9	1.2	1.2	1.6
Spread aggregate (kg/m^2)					
Grading E	8	8	–	–	–
Grading D	–	–	12	12	–
Grading C	–	–	–	–	16
Third course:					
Apply asphalt material (L/m^2)	–	–	–	0.9	0.9
Spread aggregate (kg/m^2)					
Grading E	–	–	–	8	8
Totals:					
Asphalt material (L/m^2)	3.1	2.5	3.1	4.0	4.4
Aggregate (kg/m^2)	20	24	36	44	48

a. See table 703-7 for aggregate gradations.

b. Aggregate weights are for aggregates that have a bulk specific gravity of 2.65, as determined by AASHTO T 84 and T 85. Make proportionate corrections when the aggregate furnished has a bulk specific gravity above 2.75 or below 2.55.

Asphalt Cement 702.01
Asphalt Surface Treatment Aggregate 703.09
Blotter 703.12
Cutback Asphalt 702.02
Emulsified Asphalt 702.03

Construction

410.03 Material Submittals. For surface treatments, submit the information and samples shown below for approval at least 21 days before production.

(a) Aggregate Samples. Provide 35 kg from each stockpile produced and the gradation range represented by each.

(b) Aggregate Gradation TV's. Submit the proposed percentage of each stockpile to be used and the proposed TV for each sieve size.

Table 410-3—Approximate quantities of material for multiple-course surface treatment (using emulsified asphalt or asphalt cement[a]).

Sequence of Operations	Treatment Designation Aggregate Gradation[b,c]				
	ET–20	ET–24	ET–36	ET–44	ET–48
First course:					
Apply asphalt material (L/m²)	1.6	1.9	2.5	2.1	2.3
Spread aggregate (kg/m²)					
Grading D	12	–	–	–	–
Grading C	–	16	–	–	–
Grading B	–	–	24	24	24
Second course:					
Apply asphalt material (L/m²)	1.1	1.3	1.6	2.1	2.3
Spread aggregate (kg/m²)					
Grading E	8	8	–	–	–
Grading D	–	–	12	12	–
Grading C	–	–	–	–	16
Third course:					
Apply asphalt material (L/m²)	–	–	–	1.0	1.1
Spread aggregate (kg/m²)					
Grading E	–	–	–	8	8
Totals:					
Asphalt material (L/m²)	2.7	3.2	4.1	5.2	5.7
Aggregate (kg/m²)	20	24	36	44	48

a. For asphalt cement spread rates, multiply the asphalt material spread rates shown in the table by 0.68.
b. See table 703-7 for aggregate gradations.
c. Aggregate weights are for aggregates that have a bulk specific gravity of 2.65, as determined by AASHTO T 84 and T 85. Make proportionate corrections when the aggregate furnished has a bulk specific gravity above 2.75 or below 2.55.

(c) Asphalt Temperature. Ensure that asphalt application temperatures conform to table 702-1.

(d) Spread Rates. Furnish proposed spread rates for the asphalt material and aggregate.

410.04 Equipment. Ensure that all equipment is in good working order. Furnish the equipment shown below:

(a) Asphalt Distributor. Furnish an asphalt distributor with the following features and capabilities:

(1) Capable of heating asphalt evenly.

(2) Full-circulation spray bar adjustable to at least 4.6 m wide.

(3) Positive controls, including tachometer, pressure gage, volume-measuring device, or calibrated tank, to uniformly deposit asphalt over the full width within 0.08 L/m^2 of the required rate.

(4) Thermometer for measuring the asphalt temperature in the tank.

(b) Rotary Power Broom. Furnish a rotary power broom equipped to control the vertical broom pressure.

(c) Pneumatic-Tire Rollers. Furnish a minimum of two pneumatic-tire rollers with the following features and capabilities:

(1) Minimum compacting width of 1.5 m.

(2) Minimum ground contact pressure of 550 kPa, with all tires exerting equal contact pressure.

(3) Gross weight adjustable within the range of 35 to 65 kg/mm of compaction width.

(4) Self-propelled.

(d) Aggregate Spreader. Furnish an aggregate spreader with the following features:

(1) Self-propelled.

(2) Minimum of four pneumatic tires on two axles.

(3) Positive controls to uniformly deposit the aggregate over the full width of asphalt within 10 percent by weight of the required rates.

(e) Two-Way Communication. Provide two-way radio communication between the asphalt distributor and aggregate spreader.

(f) Other Equipment. When approved, other equipment of proven performance may be used in addition to, or in lieu of, the equipment specified.

410.05 Surface Preparation. Immediately before placing any layer of the surface treatment, remove loose dirt and other objectionable material from the existing surface.

Apply surface treatments to an existing asphalt surface only when the surface is dry. Prior to application, allow a newly constructed cold or road mix surface to cure for at least 21 days for a cutback asphalt mix, and at least 14 days for an emulsified asphalt mix, unless otherwise approved by the CO.

When applying surface treatments to existing aggregate surfaces, ensure that the surface is dry if the aggregate was primed, and slightly damp if not primed. If the aggregate surface is primed, allow a prime coat curing period of at least 5 days for cutback asphalt and 24 hours for emulsions, unless otherwise approved by the CO.

Fog seal patches SHOWN ON THE DRAWINGS or listed in the SPECIAL PROJECT SPECIFICATIONS using CSS–1 emulsion, diluted with an equal part of water, at 0.65 L/m^2, unless another rate is SHOWN ON THE DRAWINGS.

410.06 Weather Limitations. Apply surface treatments with aggregate only when the ambient air and surface temperatures are above 18 °C and rising, when the weather is not foggy or rainy, and when rain is not forecast for at least 24 hours after application.

Apply fog seals only when the ambient air and surface temperatures are above 10 °C and rising, when the weather is not foggy or rainy, and when rain is not forecast for at least 24 hours after application.

For all work:

(a) Ensure that humidity is less than 75 percent as measured by the sling psychrometer method.

(b) Complete application of the surface treatment 2 hours before sunset.

(c) Unless otherwise approved by the CO, construct fog seals and single-course surface treatments between June 1 and September 1, and multiple-course surface treatments between June 1 and September 15.

410.07 Production Startup Procedures for Surface Treatment. Provide 7 days advance notice before constructing any asphalt surface treatments containing aggregate, and also use these startup procedures when resuming production after termination due to nonconforming work.

Calibrate each asphalt distributor's bar height, nozzle angle, pump pressure, and longitude and transverse spread rates in accordance with ASTM D 2995. If different asphalt distributors are used throughout the project, calibrate each prior to use on the project.

On the first day of production of each surface treatment layer, whenever there is a change in the surface texture or the aggregate TV's, construct a 150-m control strip one lane wide. Locate the control strip on the project as designated.

Construct the control strip using material, laydown, and compaction procedures intended for the remainder of the surface treatment. Cease production after construction of the control strip until the material, the control strip, and the asphalt distributor calibration procedures are evaluated and accepted.

Acceptable control strips may remain in place, and will be accepted as a part of the completed surface treatment.

Repeat this control strip process until an acceptable control strip is produced.

410.08 Asphalt Application. Protect the surfaces of nearby objects to prevent spattering or marring. For transverse construction joints, spread building paper on the surface for a sufficient distance from the beginning and end of each application so that the flow through the distributor nozzles may be started and stopped on the paper.

The CO may make adjustments for variations in field conditions. Apply the asphalt uniformly with an asphalt distributor with the spray bar height set for triple overlap. Move the distributor forward at the proper application speed at the time the spray bar is opened. Use care not to apply excess asphalt at the junction of spreads.

Ensure that the length of spread is no more than what can be covered with aggregate within 1 minute of the asphalt application.

Correct skipped areas or deficiencies. Remove and dispose of paper or other material used, in accordance with Subsection 202.04(a)

410.09 Aggregate Application. When using emulsified asphalt, moisten the aggregate to remove its dust coating.

Stockpile aggregate according to Subsections 305.03 and 305.04.

Apply the aggregate uniformly with an aggregate spreader immediately after the asphalt is applied. Operate the aggregate spreader so the asphalt is covered with the aggregate before wheels pass over it. During part-width construction, leave a strip of the sprayed asphalt approximately 150 mm wide uncovered to permit an overlap of the asphalt material.

Immediately correct excesses and deficiencies by brooming, or by the addition or removal of aggregate, until a uniform texture is achieved. Use hand methods in areas not accessible to power equipment.

Make the first roller pass to seat the aggregate immediately after the aggregate is applied. Operate rollers at a maximum speed of 8 km/h. Do not permit the aggregate to be displaced by pickup or sticking of materials to the tire surface. Ensure that the amount of rolling is sufficient to uniformly and thoroughly bond the aggregate over the full width. Make a minimum of three complete coverages. Ensure that rolling is completed within 1 hour after the asphalt is applied to the surface.

410.10 Fog Seal. To construct a fog seal, apply a slow-setting emulsified asphalt diluted with an equal amount of water onto an existing asphalt surface. Apply the diluted emulsified asphalt in accordance with Subsection 410.08 at a rate of 0.45 to 0.70 L/m^2, depending on the condition of the existing surface. Allow the fog seal to penetrate undisturbed for at least 2 hours, or until the emulsified asphalt breaks and is substantially absorbed into the existing surface. Then lightly cover remaining spots of excess asphalt with blotter before opening the surface to traffic.

410.11 Single-Course Surface Treatment. To construct a single-course surface treatment, apply asphalt onto an existing asphalt surface, immediately followed by a single, uniform application of aggregate. Apply the asphalt and aggregate in accordance with Subsections 410.08 and 410.09 at the approximate rates shown in table 410-1. Determine the exact application rates based on approved control strips.

Unless the road is closed to all traffic for the duration of the placement of the surface treatment, use a pilot car to limit traffic speeds. During the initial 45 minutes after rolling, limit the traffic speeds to 15 km/h. Limit traffic speeds to 30 km/h for 24 hours. At all times, operate hauling equipment in a prudent manner and at speeds that will not damage the new surface treatment or create a hazard to the traveling public.

Lightly broom the aggregate surface on the morning after construction. Do not displace embedded material.

Maintain the surface for 4 days after the application of the last layer of aggregate by distributing blotter to absorb any free asphalt, by repairing areas deficient in aggregate, and by sweeping excess material from the surface using a rotary power broom. Broom when the air temperature is less than 24 °C. Do not displace embedded material when brooming.

410.12 Multiple-Course Surface Treatment. To construct a multiple-course surface treatment, apply multiple layers of asphalt and aggregate. Apply each asphalt and aggregate layer in accordance with Subsections 410.08 and 410.09, and at the approximate rates shown in table 410-2 or 410-3. Determine the exact application rates based on approved control strips. When approved by the CO, a steel-wheeled roller with a minimum weight of 8 t may be substituted for one of the pneumatic tire rollers.

Maintain the surface and limit traffic in accordance with Subsection 410.11.

Wait at least 72 hours between application of the layers when using a cutback asphalt, and 24 hours when using an emulsified asphalt. No wait is necessary when using asphalt cement.

410.13 Acceptance. Asphalt cement, emulsified asphalt, and cutback asphalt will be evaluated for acceptance under Subsection 106.05.

Asphalt treatment aggregate and blotter will be evaluated for acceptance under Subsection 105.03.

Provide the minimum number of samples and tests specified in table 410-4, in accordance with Section 160.

Measurement

410.14 Method. Use the method of measurement that is DESIGNATED IN THE SCHEDULE OF ITEMS.

Water used for diluting emulsified asphalt will not be included in the quantity for PAY ITEM 410(04), and will not be paid for separately.

Payment

410.15 Basis. The accepted quantities will be paid for at the contract unit price for each PAY ITEM DESIGNATED IN THE SCHEDULE OF ITEMS.

Payment will be made under:

Pay Item		Pay Unit
410 (01)	Surface treatment aggregates, designation ___	Ton
410 (02)	Surface treatment aggregates, designation ___	Cubic Meter
410 (03)	Asphalt cement, grade ___	Ton
410 (04)	Emulsified asphalt, grade ___	Ton
410 (05)	Cutback asphalt, grade ___	Ton
410 (06)	Blotter	Ton
410 (07)	Blotter	Cubic Meter

Table 410-4—Sampling and testing.

Type of Acceptance	Material or Product	Property or Characteristic	Test Method or Specification	Frequency	Sampling Point
Production certification	Asphalt cement[a]	Contract requirements	AASHTO M 20 or M 226, as applicable	Each shipment	
	Emulsified asphalt[a]		Subsection 702.03		
	Cutback asphalt[a]		AASHTO M 81 or M 82, as applicable		
Tested conformance	Material source[b] (Contractor-located source)	Los Angeles abrasion	AASHTO T 96	Three times for each undeveloped source,[c] or once for all other sources	Material source
		Sodium sulfate soundness loss	AASHTO T 104		
		Durability index (coarse and fine)	AASHTO T 210		
		Density	AASHTO T 19		
		Coating and stripping of bitumen-aggregate	AASHTO T 182		
	Asphalt surface treatment aggregate	Gradation, table 703-8[d]	AASHTO T 27 and T 11	Once for each 500 t, but not less than once per day of production	At stockpile
		Fractured faces (coarse)[d,e]	FLH T 507		
		Flakiness index[d]	FLH T 508		
	Blotter or choker aggregate	Gradation	AASHTO T 27	Once for each 500 t, but not less than once per day of production	Material source
		Liquid limit	AASHTO T 89		
		Sand equivalent	AASHTO T 176		
		Plasticity index	AASHTO T 90		
	Asphalt cement	Sample	Subsection 105.04(b)	Each tanker, including all trailers	At point of shipment delivery
	Emulsified asphalt				
	Cutback asphalt				

a. Use in work is permitted prior to sampling and testing for conformance.
b. See Subsection 105.06. Testing not required when using Government-provided material source.
c. An undeveloped source is a source that has not supplied surface treatment aggregate within 365 days of the start of producing surface treatment aggregate for this project.
d. Applies to each aggregate grade furnished.
e. Use only for gravel sources.

Section 411—Miscellaneous Asphalt Pavement Seals

Description

411.01 Work. Apply pavement sealing material(s) or rejuvenator to pavement surfaces SHOWN ON THE DRAWINGS. The pavement seal may require blotter or color treatment as SHOWN ON THE DRAWINGS. Have the surface approved by the CO in writing prior to placing the miscellaneous asphalt pavement seal.

Materials

411.02 Requirements. Meet the following requirements for materials:

(a) Seal Material. As the primary component of the pavement seal material, use slow-setting asphalt emulsion that conforms to AASHTO M 140 or M 208. Submit a manufacturer's Certificate of Compliance with each shipment of asphalt emulsion. Mix asphalt emulsions with additives that prevent bleeding. Seal materials may be diluted with water to improve penetration into the pavement surface or to improve consistency.

(b) Rejuvenation Material. Dilute rejuvenation materials by two parts concentrate to one part water. Ensure that the concentrate meets the requirements specified in table 411-1.

(c) Additive Materials. Materials such as clays, slates, fibers, carbon black, and polymers may be added to emulsified asphalt, provided that the manufacturer's product literature is reviewed and approved by the CO prior to application. If the pavement seal material is blended with sand, use sand that has clean, hard, durable, uncoated particles, and is free of clay lumps and organic matter, with 100 percent passing the 600-µm sieve and a maximum of 5 percent passing the 75-µm sieve. The sand must have a gradation that helps control segregation and promotes suspension during product application. Do not add more than 360 g of sand to each liter of applied material.

(d) Color Treatment. When SHOWN ON THE DRAWINGS, use color treatments. Carefully mix and apply color additives according with the manufacturer's instructions, so that a consistent color and durable seal are obtained. Asphacolor is the only known material to meet these requirements. If other products are proposed, provide manufacturer's literature that shows similar color, consistency, and durability characteristics.

Table 411-1—Rejuvenator concentration requirements.

Test	Test Method		Requirements	
	ASTM	AASHTO	Min.	Max.
On emulsion:				
Viscosity at 25 °C, SFS	D 244	T 59	15	40
Residue, percent[a]	D 244 (mod.)	T 59 (mod.)	60	65
Miscibility test[b]	D 244 (mod.)	T 59 (mod.)	No coagulation	
Sieve test, percent[c]	D 244 (mod.)	T 59 (mod.)	–	0.1
Particle charge test	D 244	T 59	Positive	–
On residue from distillation:				
Flash point, COC, °C	D 92	T 48	196	–
Viscosity at 60 °C, cSt	D 445	–	100	200
Asphaltenes, %	D 2006–70	–	–	1.0
Maltene distribution ratio,[d] $\frac{PC + A1}{S + A2}$	D 2006–70	–	0.3	0.6
PC/S ratio[d]	D 2006–70	–	0.5	–
Saturated hydrocarbons, S	D 2006–70	–	21	28

a. For the ASTM D 244 modified evaporation test for percent of residue, heat a 50-g sample to 149 °C until foaming ceases, then cool immediately and calculate results.
b. Use test procedure identical to ASTM D 244–60, but use .02 percent normal calcium chloride solution in place of distilled water.
c. Use test procedure identical to ASTM D 244, but use distilled water in place of 2 percent sodium oleate solution.
d. Chemical composition by ASTM method D 2006–70, where PC = polar compounds, A2 = first acidaffins, A1 = second acidaffins, and S = saturated hydrocarbons.

(e) Blotter. Meet the requirements specified in Subsection 703.12.

(f) Water. Meet the requirements specified in Subsection 725.01.

411.03 Sampling. When directed by the CO, sample materials that are used in the work. Give the CO the opportunity to witness sampling. The CO will be responsible for testing.

Construction

411.04 Weather Limitations. Apply asphalt pavement seals only when the ambient air and surface temperatures are above 10 °C and rising, when the weather is not foggy or rainy, and when rain is not forecast for at least 24 hours after application.

411.05 Equipment. For all materials, use the distribution equipment that is capable of placing a uniform consistency of the material and enabling a uniform application over variable widths of surface. The required application rate may be obtained by making multiple applications. Use equipment for storing and applying of liquid and slurry products that include accurate volume measuring devices and a calibrated tank. The CO may require transport or application equipment to be weighed when full and when empty, if volume-measuring equipment is inadequate.

411.06 Preparation of Surface. Prior to placing the pavement seal, ensure that the surface of the pavement is clean and free from dust, dirt, or other loose foreign matter, grease, oil, or any type of objectionable surface film. Accumulations of oil or grease may be removed by pressure washing, grinding, or burning and scraping. Remove existing painted stripes if SHOWN ON THE DRAWINGS.

Clean all cracks wider than 3 mm by removing accumulated dirt and vegetation, and blow the cracks out with compressed air to a depth of at least three times the crack width. Fill transverse cracks that are more than 3 mm wide according to Subsections 414.04 and 414.05. Other cracks that are more than 3 mm wide may be filled with pavement seal materials prior to application of seal materials over the entire surface.

411.07 Pavement Seal Application With or Without Color Treatment. Application may be by hand-held spray equipment, asphalt distributor, squeegee, or slurry seal spreader box. Spread the seal material in two directions, 180° from each other. The CO may require an additional application on small areas of pavement that have more surface voids, where voids have prevented adequate coverage, or where uneven application exists.

Obtain the minimum residue application rate of 0.55 kg/m^2 over the entire surface to be treated. Ensure that residue consists of the asphalt cement portion of the emulsified asphalt and nonevaporative additive materials (such as clays, slates, fibers, polymers, and sand) that are suspended in the emulsion when applied. Application rates may be adjusted up or down by 0.08 kg residue per square meter by the CO to compensate for pavement surface absorption and roughness.

Readily determine the volume of emulsion used prior to, during, and after application. Final payment will be based on meeting the specified application rate of residue. The CO will determine the weight of residue per liter by drying field samples from the project to constant weight.

411.08 Blotter Application. When blotter is included as a PAY ITEM, apply it at 5 kg/m^2 to areas SHOWN ON THE DRAWINGS. Complete the application within 24 hours of the emulsified asphalt application The CO may require redistribution of blotter during the first 3 days of the pavement seal curing period.

411.09 Rejuvenator Application. Apply diluted rejuvenator at a rate between 0.2 and 0.4 L/m^2. The exact rate will be determined by the CO. Two hours after application, blot the treated surface with 0.5 to 1.5 kg of blotter sand per square meter. Allow the required cure time of 24 to 48 hours before brooming the sand from the surface.

411.10 Surface Maintenance. Prevent asphalt pickup under traffic for 4 days after treatment. Open the treated pavement surface to traffic within 24 hours following treatment. When directed by the CO, apply blotter sand to the sealed pavement to prevent asphalt pickup by vehicles. The CO may require removal of loose blotter by brooming after the maintenance period. After the treatment has been open to traffic for 4 days, repair any areas that are damaged by traffic or that are peeling or cracking. All damage repair is the responsibility of the Contractor.

411.11 Acceptance. Acceptance for pavement seal with or without color will be as follows:

If laboratory-quality assurance tests on samples taken during application do not contain enough residue to meet the specified application rates established by the CO in Subsection 411.07, the CO may require the application of more material or may reduce payment. The following factors will be used to determine whether the specified application rate has been obtained:

(a) The average unit weight of the field samples, determined by weighing 1,000 mL to within ± 0.1 g.

(b) The percent residue per liter on a by-weight basis.

(c) The total volume of product applied.

The rejuvenator will be evaluated for acceptance under Subsection 106.05.

Blotter will be evaluated for acceptance under Subsection 105.03.

Measurement

411.12 Method. Use the method of measurement that is DESIGNATED IN THE SCHEDULE OF ITEMS.

Water used for diluting the pavement seal or rejuvenator will not be included in the quantity for PAY ITEMS 411(01), 411(02), or 411(04), and will not be paid for separately.

Payment

411.13 Basis. The accepted quantities will be paid for at the contract unit price for the PAY ITEM DESIGNATED IN THE SCHEDULE OF ITEMS.

Payment will be made under:

Pay Item		Pay Unit
411 (01)	Pavement seal	Square Meter
411 (02)	Pavement seal with color treatment	Square Meter
411 (03)	Blotter	Square Meter
411 (04)	Asphalt rejuvenator	Square Meter

Section 413—Asphalt Pavement Milling

Description

413.01 Work. Remove asphalt pavement using a cold milling process.

Construction Requirements

413.02 Equipment—Milling Machine. Furnish equipment in good working order and with the following capabilities and features:

(a) Self-propelled.

(b) Sufficient power, traction, and stability to accurately maintain depth of cut.

(c) Capable of removing the pavement thickness to provide profile and cross slope.

(d) Automatic system to control grade elevations by referencing from the existing pavement by means of a ski or matching shoe or from an independent grade control.

(e) Automatic system to maintain cross slope.

(f) System to effectively limit dust and other particulate matter from escaping removal operations.

(g) Loading system or adequate support equipment to completely recover milled material at removal rate.

(h) Cutting width equal to at least one-third of the lane width.

413.03 Milling. Use a longitudinal reference to accurately guide the machine. References may include a curb, the edge of pavement, or a string attached to the pavement surface. Mill in a longitudinal direction to the depth SHOWN ON THE DRAWINGS.

Mill the transverse slope to within 6 mm in 3 m of the required slope. Transition from one transverse slope to another at a uniform rate. Uniformly mill the entire roadway lane width so the cross section of the new surface forms a straight line.

Transition between different depths of cut at a uniform rate of 17 mm of depth per 10 m. At the beginning and end of the milling work, construct a smooth transition to the original surface at this rate. Do not leave an exposed vertical edge perpendicular to the direction of travel.

Mill the surface to a smoothness such that a 3-m metal straightedge, measured at right angle and parallel to the centerline, does not have more than a 7-mm surface deviation between any two contact points.

Use a rotary broom and vacuum immediately behind the milling operations to remove and completely recover all loose material. Minimize the escape of dust into the air. Dispose of recovered milled material in accordance with Subsection 202.04(a) or as SHOWN ON THE DRAWINGS.

413.04 Acceptance. Asphalt pavement milling will be evaluated for acceptance based on visual and measured conformance based upon contract requirements and customary construction tolerances.

Measurement

413.05 Method. Use the method of measurement that is DESIGNATED IN THE SCHEDULE OF ITEMS.

Payment

413.06 Basis. The accepted quantities will be paid for at the contract unit price for each PAY ITEM DESIGNATED IN THE SCHEDULE OF ITEMS.

Payment will be made under:

Pay Item		Pay Unit
413 (01)	Asphalt pavement milling	Square Meter
413 (02)	Asphalt pavement milling	Kilometer

Section 414—Asphalt Pavement Joint & Crack Treatments

Description

414.01 Work. Cut or rout open cracks. Clean and either seal or fill joints and cracks in asphalt pavement. Joint sealant classes are designated in Subsection 414.02.

Materials

414.02 Requirements. Ensure that material conforms to specifications in the following subsections:

Asphalt Cement	702.01
Backer Rod	712.01(g)
Blotter	703.12
Crack Fillers	712.01(a)(3) and (4)
Emulsified Asphalt	702.03
Fine Aggregate for Portland Cement Concrete	703.01
Joint Sealant, Class 1	712.01(a)(1)
Joint Sealant, Class 2	712.01(a)(2)
Joint Sealant, Class 3	712.01(f)
Slurry Seal	409.02 and 409.03

Construction

414.03 Equipment. Furnish equipment with the features and capabilities shown below:

(a) Power Saw & Blades. A saw and blades of such size and configuration that saw cuts can be made with one pass to the required depth and width. Spacers are not allowed.

(b) Router. A power rotary impact router or vertical spindle router capable of cleaning cracks or joints to the required depth and width.

(c) Hot-Compressed Air Lance. A lance capable of providing clean, oil-free compressed air at a volume of 2.8 m^3/min, at a pressure of 830 kPa, and at a temperature of 1,000 °C.

(d) Application Wand. A crack sealant applicator wand attached to a heated hose that is attached to a heated sealant chamber.

(e) Heating Kettle. An indirect-heating-type double boiler with the space between the inner and outer shells filled with oil or other heat transfer medium capable of constant agitation and able to maintain the temperature of the sealant within manufacturer's tolerances. Provide an accurate and calibrated thermometer with a range from 100 °C to 300 °C in 2 °C graduations. Locate the thermometer such that the temperature of the joint sealant may be safely checked.

(f) ***Squeegee.*** Provide a hand-held squeegee for ensuring that the crack is filled to the existing surface.

414.04 Joint Cutting & Cleaning. Saw cut or rout, clean, and seal joints in a continuous operation. Either dry or wet cutting is allowed. The depth and width of joint cutting will be as SHOWN ON THE DRAWINGS.

Clean dry-sawed joints with a stream of air sufficient to remove all dirt, dust, or deleterious matter adhering to the joint walls or remaining in the joint cavity. Blow or brush dry material off the pavement surface.

Immediately after sawing, clean wet-sawed joints with a water blast, 350 kPa minimum, to remove any sawing slurry, dirt, or deleterious matter adhering to the joint walls or remaining in the joint cavity. Immediately flush all sawing slurry from the pavement surface. Blow wet-sawed joints with air to dry joint surfaces.

Do not allow traffic to knead together or damage the sawed joints. If cleaning operations cause interference with traffic, provide protective screening.

414.05 Joint Cleaning & Sealing. If necessary, clean the joint according to Subsection 414.06. Place the sealant when the pavement surface temperature is 4 °C and rising. Discontinue operations when weather conditions detrimentally affect the quality of forming joints and applying sealant.

Submit a copy of, and adhere to, the manufacturer's recommendations for heating and applying the sealant. Heat the sealant in a heating kettle. Do not heat the sealant above the safe heating temperature recommended by the manufacturer. Do not hold the material at the pouring temperature for more than 6 hours, and do not reheat the material.

Place a backer rod in the bottom of the cut or routed joint. Ensure that the size of the backer rod conforms to table 712-2.

Seal the joints with an applicator wand when the sealant material is at the pouring temperature. Heat or insulate the applicator wand to maintain the pouring temperature of the sealant during placing operation. Return the applicator wand to the machine and recirculate the joint sealant material immediately after sealing each joint.

Immediately screed the joint sealant to the elevation of the existing surface. Use a squeegee to ensure that a 75-mm-wide band is centered on the finished sealed crack.

Wait for the sealant to be tack free before opening the joint to traffic. Do not spread blotter on the sealed joints to allow early opening to traffic.

414.06 Crack Cleaning & Filling. Clean the existing surface of all loose material, dirt, or other deleterious substances by brooming, flushing with water, or other approved methods. Dry cracks before sealing.

When using the hot-compressed air lance, keep it moving so as not to burn the surrounding pavement and the crack. Place and finish sealant within 5 minutes after heating with the hot-compressed air lance.

For cracks 6 mm or less, fill with CSS–1, SS–1, or crack filler. Submit a copy of, and adhere to, the manufacturer's recommendations for heating and applying the crack filler. Use a squeegee to ensure that a 75-mm-wide band is centered on the finished sealed crack. Cover the sealed crack with a light application of blotter.

For cracks with a width greater than 6 mm and less than 25 mm, fill with either a slurry seal mixture, fine aggregate-asphalt cement mixture, or fine aggregate–emulsified asphalt mixture. Have the mixture approved by the CO. Use a squeegee or other suitable equipment to force the mixture into the cracks. Immediately screed the sealant or asphalt mixture to the elevation of the existing surface. Cover the sealed crack with a light application of blotter.

For cracks with a width greater than 25 mm, fill flush to the existing surface with either hot or cold asphalt concrete mix. Have the mixture approved by the CO.

414.07 Resealing Defective Joints or Cracks. Reseal areas that exhibit adhesion failure, damage, missed areas, foreign objects in the sealant, or other problems that will accelerate failure.

414.08 Acceptance. Material for joint sealant and crack filler will be evaluated under Subsection 106.05.

Measurement

414.09 Method. Use the method of measurement that is DESIGNATED IN THE SCHEDULE OF ITEMS.

Payment

414.10 Basis. The accepted quantities will be paid for at the contract unit price for each PAY ITEM DESIGNATED IN THE SCHEDULE OF ITEMS.

Payment will be made under:

Pay Item		Pay Unit
414 (01)	Joint cutting and cleaning	Meter
414 (02)	Joint cleaning and sealing	Meter
414 (03)	Joint sealant, class _____	Liter
414 (04)	Joint sealant, class _____	Kilogram
414 (05)	Joint sealant, class _____	Meter
414 (06)	Crack cleaning and filling	Meter
414 (07)	Crack filler	Liter
414 (08)	Crack filler	Kilogram
414 (09)	Crack filler	Meter

Section 415—Paving Geotextiles

Description

415. 01 Work. Furnish and place a paving geotextile and asphalt sealant between pavement layers to form a waterproofing and stress-relieving membrane within the pavement structure. Have the surface approved by the CO in writing prior to placing the asphalt sealant and paving geotextile.

Materials

415.02 Requirements. Provide material that conforms to specifications in the following subsections:

Asphalt Cement 702.01
Choker Aggregate 703.12
Emulsified Asphalt 702.03
Geotextiles, Type VI 714.01

Construction

415.03 Surface Preparation. Clean the surface on which the geotextile is to be placed using a power broom and/or power blower. Fill cracks that exceed 6 mm according to Subsection 414.06. Allow crack filler and patches to cure before placing the geotextile. Remove all foreign and loose material.

415.04 Weather Limitations. Apply asphalt sealant and paving geotextile on a dry surface when the pavement surface temperature is at least 13 °C and rising.

415.05 Asphalt Sealant Application. Use asphalt cements within a temperature range of 140 °C to 165 °C. Use emulsified asphalts within a temperature range of 55 °C to 70 °C.

Apply the asphalt sealant to the pavement surface in accordance with Subsection 410.08 at a rate of 0.90 to 1.35 L/m^2 for asphalt cement and 1.3 to 2.0 L/m^2 for emulsified asphalt.

Spray the asphalt sealant 150 mm wider than the paving geotextile. Do not apply the asphalt sealant any farther in advance of the paving geotextile placement than can be maintained free of traffic.

Where emulsified asphalt is used, allow the emulsion to completely break before placing the paving geotextile.

Where asphalt cement is used, place the paving geotextile immediately after the asphalt cement is applied.

415.06 Paving Geotextile Placement. Place the paving geotextile onto the asphalt sealant with minimal wrinkling. Slit, lay flat, and tack all wrinkles or folds higher than 25 mm. Broom and/or roll the paving geotextile to maximize fabric contact with the pavement surface.

At geotextile joints, overlap the geotextile 150 mm to ensure full closure. Overlap transverse joints in the direction of paving to prevent edge pickup by the paver. Apply additional asphalt sealant to paving geotextile overlaps to ensure proper bonding of the double fabric layer.

If asphalt sealant bleeds through the fabric, treat the affected areas with choker aggregate. Minimize traffic on the geotextile. If circumstances require traffic on the membrane, apply choker aggregate and place signs that read "Slippery When Wet." Broom the excess choke from the geotextile surface before placing the overlay. Repair all damaged fabric before placing the overlay. Apply a light tack coat in accordance with Section 407 before placing the overlay. To avoid damaging the geotextile, do not turn equipment on the geotextile.

Place a hot asphalt concrete overlay within 48 hours after placing the paving geotextile. Limit the laydown temperature of the mix to a maximum of 165 °C, except when the paving geotextile is composed of polypropylene fibers. In this case, limit the laydown temperature of the mix to a maximum of 150 °C.

415.07 Acceptance. Asphalt cement and emulsified asphalt will be evaluated for acceptance under Subsection 106.05.

Blotter will be evaluated for acceptance under Subsection 105.03.

Paving geotextile material will be evaluated for acceptance under Subsection 714.01.

Provide the minimum number of samples specified in table 410-4.

Measurement

415.08 Method. Use the method of measurement that is DESIGNATED IN THE SCHEDULE OF ITEMS.

Payment

415.09 Basis. The accepted quantities will be paid for at the contract unit price for each PAY ITEM DESIGNATED IN THE SCHEDULE OF ITEMS.

Payment will be made under:

Pay Item		Pay Unit
415 (01)	Paving geotextile	Square Meter
415 (02)	Asphalt sealant	Ton
415 (03)	Choker aggregate	Ton

Section 416—Asphalt Pavement Patching

Description

416.01 Work. Perform deep patching, skin patching of asphalt surfaces, and patching of asphalt berms. Prepare the area to be patched, and furnish and place all necessary materials.

Materials

416.02 Requirements. Ensure that asphalt materials are of the type and grade SHOWN ON THE DRAWINGS, and that they meet the requirements specified in the following subsections:

Asphalt Cement	702.01
Cutback Asphalt	702.02
Emulsified Asphalt	702.03

Ensure that mixing temperatures meet the requirements specified in Subsection 702.04; that aggregates meet the requirements specified in Subsection 703.07, except for gradation; and that fabric meets the requirements specified in Subsection 714.01.

416.03 Job-Mix Formula. Prior to producing asphalt concrete mixtures, submit in writing a proposed job-mix formula and supporting documentation for each mixture to the CO for use in setting the job-mix formula to be used with the proposed materials.

After reviewing the proposed job-mix formula, the CO will determine the final values for the job-mix formula to be used and notify the Contractor in writing.

Construction

416.04 Deep Patching. Remove surface course and base course materials above the subgrade to a minimum depth of 50 mm, or as necessary to reach firm support. If firm support for a patch is unavailable, notify the CO prior to placing any material.

Trim or mill the edges of the prepared hole to form a vertical face in unfractured asphalt surfacing. Make the prepared hole rectangular in shape, and clean it of all loose material. When the hole is dry, spray the bottom and faces with an emulsified asphalt.

Immediately patch or barricade prepared sites.

Place the asphalt concrete mixture in layers not exceeding 100 mm. Thoroughly compact each layer with hand or mechanical tampers or rollers.

Compact the finished surface with a steel-wheel roller or vibratory plate compactor. For hot asphalt concrete mixtures, compact the mix while it is above 110 °C. Ensure that the compacted patch, upon completion, is approximately 3 to 6 mm above the level of the adjacent pavement. Seal the edges of the completed patch with emulsified asphalt, and blot with fine sand.

When SHOWN ON THE DRAWINGS and DESIGNATED IN THE SCHEDULE OF ITEMS, use a geotextile saturated with rubberized asphalt to strengthen the pothole area. Ensure that the geotextile has a minimum grab strength of 90 N. Prepare the surface on which the fabric is placed by digging out and patching as described above, or by cleaning the surface, removing vegetation, and filling all cracks more than 6 mm wide with an approved crack-filling material. Remove excess crack-filling material.

Place the fabric membrane over the repaired area. Extend the fabric a minimum of 150 mm beyond the repaired or patched area onto sound adjoining pavement. Use a minimum of 50 mm overlap where adjacent fabric panels are needed to cover the repaired area.

416.05 Skin Patches. Prior to skin patching, patch all potholes.

Treat minor depressions, light raveling, or surface checking at scattered locations SHOWN ON THE DRAWINGS or marked on the ground by applying a skin patch.

Prior to skin patching, clean the surface of loose and deleterious material, and spray it with emulsified asphalt at the rate ordered by the CO. Do not place mixture until authorized by the CO.

Uniformly distribute asphalt concrete mixture in layers not to exceed 50 mm compacted depth. Feather the edges of skin patches. When multiple layers are necessary, offset all joints at least 150 mm between layers.

Compact each layer with a 7- to 9-t steel roller. For hot asphalt concrete mixtures, compact the mix while it is above 110 °C.

Ensure that the completed patch does not have abrupt transitions that could adversely affect the steering of a passenger car traveling across the area. Provide transition tapers for skin patches that are 100 mm per 1 mm thickness of patch in the direction on travel.

416.06 Asphalt Berm. Remove damaged segments of berm and bevel exposed ends at approximately 45° from vertical. Clean and patch the berm foundation as necessary. Coat the foundation and joining surfaces with emulsified asphalt. Place and compact asphalt mix to conform with the shape of the undamaged segment.

416.07 Waste Material. Dispose of all materials removed from potholes, patches, and berms in accordance with Subsection 202.04(a).

Measurement

416.08 Method. Use the method of measurement that is DESIGNATED IN THE SCHEDULE OF ITEMS.

Payment

416.09 Basis. The accepted quantities will be paid for at the contract unit price DESIGNATED IN THE SCHEDULE OF ITEMS.

Payment will be made under:

Pay Item		Pay Unit
416 (01)	Hot asphalt concrete mixture	Ton
416 (02)	Deep patch hot asphalt concrete mixture	Ton
416 (03)	Skin patch hot asphalt concrete mixture	Ton
416 (04)	Cold asphalt concrete mixture	Ton
416 (05)	Rubberized asphalt saturated geotextile	Square Meter

DIVISION 550
Bridge Construction

Section 551—Driven Piles

Description

551.01 Work. Furnish and drive piles. In addition, furnish and place reinforcing steel and concrete in concrete-filled steel shell and concrete-filled pipe piles.

Piles are designated as steel H-piles, steel pipe piles, concrete-filled steel shell piles, concrete-filled pipe piles, precast concrete piles, prestressed concrete piles, or timber piles. Pile load tests are designated as static or dynamic.

Materials

551.02 Requirements. Furnish material that conforms to the specifications in the following sections and subsections:

Concrete Piles	715.03
Paint	708
Pile Shoes	715.08
Reinforcing Steel	554
Sheet Piles	715.07
Splices	715.09
Steel H-Piles	715.06
Steel Pipes	715.05
Steel Shells	715.04
Structural Concrete	552
Treated Timber Piles	715.02
Untreated Timber Piles	715.01

Construction

551.03 Pile-Driving Equipment. Furnish equipment meeting the following requirements:

(a) Pile Hammers. Furnish pile hammers as shown below.

(1) Gravity Hammers. Use gravity hammers to drive timber piles only, and where the ultimate bearing capacity of the timber pile is less than 800 kN. Furnish a hammer with a ram weighing between 900 and 1,600 kg and limit the drop height to 4 m. Ensure that the ram mass is greater than the combined mass of the drive head and pile. Provide hammer guides to ensure concentric impact on the drive head.

(2) Open-End Diesel Hammers. Equip open-end (single-acting) diesel hammers with a device, such as rings on the ram or a scale (jump stick) extending above the ram cylinder, to permit visual determination of hammer stroke. Submit a chart from the hammer manufacturer equating stroke and blows per minute for the hammer to be used. A speed-versus-stroke calibration may be used if approved.

(3) Closed-End Diesel Hammers. Submit a chart, calibrated to actual hammer performance within 90 days of use, equating bounce chamber pressure to either equivalent energy or stroke for the hammer to be used. Equip hammers with a dial gage for measuring pressure in the bounce chamber. Make the gage readable at ground level. Calibrate the dial gage to allow for losses in the gage hose. Verify the accuracy of the calibrated dial gage during driving operations by ensuring that cylinder lift occurs when bounce chamber pressure is consistent with the maximum energy given in the hammer specifications. Do not use closed-end diesel hammers that do not attain cylinder lift at the maximum energy-bounce chamber pressure relationship given in the hammer specification.

(4) Air or Steam Hammers. Furnish plant and equipment for steam and air hammers with sufficient capacity to maintain the volume and pressure specified by the hammer manufacturer. Equip the hammer with accurate pressure gages that are easily accessible. Use a hammer with the mass of the striking parts equal to or greater than one-third the combined mass of the driving head and pile. Ensure that the combined mass is at least 1,250 kg.

When driving test piles, measure inlet pressures for double-acting and differential-acting air or stream hammers with a needle gage at the head of the hammer. If required, also measure inlet pressures during the driving of the production piles. A pressure-versus-speed calibration may be developed for the specific driving conditions at the project as an alternative to periodic measurements with a needle gage.

(5) Nonimpact Hammers. Do not use nonimpact hammers, such as vibratory hammers, unless permitted in writing, SHOWN ON THE DRAWINGS, or provided in the SPECIAL PROJECT SPECIFICATIONS. If permitted, use such equipment for installing production piles only after the pile tip elevation, or embedment length, for safe support of the pile load is established by static or dynamic load testing. Control the installation of production piles when using vibratory hammers by power consumption, rate of penetration, specified tip elevation, or other acceptable methods that will ensure the required pile load capacity is obtained. On 1 out of every 10 piles driven, strike with an impact hammer of suitable energy to verify that the required pile capacity is obtained.

(b) Approval of Pile-Driving Equipment. Furnish pile-driving equipment of such size that the production piles can be driven with reasonable effort to the required lengths without damage.

The Government will evaluate the suitability of the equipment and will accept or reject the driving system within 21 days of receipt of the pile and driving equipment information. Approval of pile-driving equipment will be based on a wave equation analysis under the following conditions:

- When dynamic load testing is required.
- When ultimate pile capacities exceed 2,400 kN.
- When precast or prestressed concrete piles are used.
- When double-acting or differential hammers, air, steam, or diesel are used.

When the wave equation analysis is not used, approval of the pile-driving equipment will be based on minimum hammer energy in table 551-1. Approval of a pile hammer relative to driving stress damage does not relieve the Contractor of responsibility for damaged piles.

Table 551-1.—Minimum pile hammer energy.

Ultimate Pile Capacity (kN)	Minimum Rated Hammer Energy (kJ)
≤ 800	14.0
1,330	21.2
1,600	28.1
1,870	36.0
2,140	44.9
2,400	54.4
> 2,400	Wave equation required

If the wave equation analysis shows an inability to drive the pile(s) to the required ultimate pile-bearing capacity with an acceptable blow count, or that pile damage will occur, change the proposed driving equipment until the wave equation analysis indicates that piles can be driven as specified. Submit proposed changes to the CO for review.

Approval of the pile-driving system is specific to the equipment submitted. If the proposed equipment is modified or replaced, resubmit the revised data for approval before using. The revised driving system will be accepted or rejected within 21 days of receipt of the revised pile, equipment, and wave equation analysis information (if required). Use only the approved equipment during pile-driving operations.

(1) Equipment Submittal. Submit two copies of the following pile-driving equipment information at least 30 days before driving piles. When dynamic load tests are required, submit a wave equation analysis performed by a pile specialty consultant who meets the requirements specified in Subsection 551.12(a). If dynamic load testing is not required, the Government will perform the wave equation analysis.

(a) General. Project and structure identification, pile driving contractor or subcontractor, and auxiliary methods of installation, such as jetting or preboring, and the type and use of the equipment.

(b) Hammer. Manufacturer, model, type, serial number, rated energy (_____ at _____ length of stroke), and modifications.

(c) Capblock (Hammer Cushion). Material, thickness, area, modulus of elasticity (*E*), and coefficient of restitution (*e*).

(d) Pile Cap. Helmet mass, bonnet mass, anvil block mass, and drivehead weight.

(e) Pile Cushion. Cushion material, thickness, area, modulus of elasticity (*E*), and coefficient of restitution (*e*).

(f) Pile. Pile type, length (in leads), mass per meter, wall thickness, taper, cross-sectional area, design pile capacity, description of splice, and tip treatment description.

(2) Wave Equation. The required number of hammer blows indicated by the wave equation at the ultimate pile capacity shall be between 3 and 15 per 25 mm. In addition, ensure that the pile stresses resulting from the wave equation analysis do not exceed the values at which pile damage is impending. The point of impending damage is defined for steel, concrete, and timber piles as follows:

(a) Steel Piles. Limit the compressive driving stress to 90 percent of the yield stress of the pile material.

(b) Concrete Piles. Limit the tensile (*TS*) and compressive (*CS*) driving stresses to:

$$TS \leq 3f_c{}' + EPV$$
$$CS \leq 0.85f_c{}' - EPV$$

where

$f_c{}'$ = 28-day design compressive strength of concrete
EPV = effective prestress value (prestressed piles only)

(c) Timber Piles. Limit the compressive driving stress to 3 times the allowable static design stress.

(3) Minimum Hammer Energy. Ensure that the energy of the driving equipment submitted for approval, as rated by the manufacturer, is at least the energy specified in table 551-1 that corresponds to the required ultimate pile capacity.

(c) Driving Appurtenances. Furnish the driving appurtenances shown below.

(1) Hammer Cushion. Equip all impact pile-driving equipment, except gravity hammers, with a suitable thickness of hammer cushion material to prevent damage to the hammer or pile and to ensure uniform driving behavior. Fabricate hammer cushions from durable, manufactured material in accordance with the hammer manufacturer's recommendations. Do not use wood, wire rope, or asbestos hammer cushions. Place a striker plate, as recommended by the hammer manufacturer, on the hammer cushion to ensure uniform compression of the cushion material. Inspect the hammer cushion in the presence of the CO when beginning pile-driving at each bent or substructure unit or after each 100 hours of pile-driving, whichever is less. Replace the cushion when its thickness is reduced by more than 50 percent of its original thickness or when it begins to burn.

(2) Pile Drive Head. Provide adequate drive heads for impact hammers, and provide appropriate drive heads, mandrels, or other devices for special piles, in accordance with the manufacturer's recommendations. Align the drive head axially with the hammer and pile. Fit the drive head around the pile head so that it will prevent transfer of torsional forces during driving while maintaining proper alignment of hammer and pile.

(3) Leads. Support piles in line and position with leads while driving. Construct pile driver leads to allow freedom of movement of the hammer while maintaining axial alignment of the hammer and the pile. Do not use swinging leads unless permitted in writing, SHOWN ON THE DRAWINGS, or provided in the SPECIAL PROJECT SPECIFICATIONS. When swinging leads are permitted, fit swinging leads with a pile gate at the bottom of the leads and, in the case of battered piles, with a horizontal brace between the crane and the leads. Adequately embed the leads in the ground or constrain the pile in a structural frame (template) to maintain proper alignment. Provide leads of sufficient length that do not require a follower and will permit proper alignment of battered piles.

(4) Followers. Followers are not permitted unless approved in writing. When followers are permitted, drive the first pile in each bent or substructure unit and every tenth pile driven thereafter, full length without a follower, to verify that adequate pile embedment is being attained to develop the required ultimate pile capacity. Provide a follower of such material and dimensions that will permit the piles to be driven to the required penetration. Hold and maintain the follower and pile in proper alignment during driving.

(5) Jetting. Do not use jetting unless approved in writing. Provide jetting equipment with sufficient capacity to deliver a consistent pressure equivalent to at least 700 kPa at two 20-mm jet nozzles. Jet so as not to affect the lateral stability of the final in-place pile. Remove jet pipes when the pile tip is at least 1.5 m above the prescribed tip elevation, and drive the pile to the required ultimate pile capacity with an impact hammer. Control, treat, if necessary, and dispose of all jet water in an approved manner.

551.04 Pile Lengths. Unless otherwise specified, furnish piles with sufficient length to obtain the required penetration and bearing capacity and extend into the pile cap or footing as SHOWN ON THE DRAWINGS. In addition, increase the length to provide fresh heading and to provide for the Contractor's method of operation. When test piles are required, furnish piles in the lengths determined by the test piles, increased to provide for the Contractor's method of operation.

551.05 Test Piles. Construct test piles at locations SHOWN ON THE DRAWINGS. Excavate the ground at the site of each test pile or production pile to the elevation of the bottom of the footing or pile cap before the pile is driven. Furnish test piles that are longer than the estimated length of production piles. Drive test piles with the same equipment as the production piles.

Drive test piles to the required ultimate capacity at the estimated tip elevation. Allow test piles that do not attain the required ultimate capacity at the estimated tip elevation to "set up" for 24 hours before redriving. Warm the hammer before redriving begins by applying at least 20 blows to another pile. If the required ultimate capacity is not attained on redriving, drive a portion or all of the remaining test pile length and repeat the "set up" and redrive procedure as directed. Splice and continue driving until the required ultimate pile capacity is obtained.

Ensure that test piles that are used in the completed structure conform to the requirements for production piles. Remove test piles that are not incorporated into the completed structure to at least 0.5 m below finished grade.

Do not order piling to be used in the completed structure until test pile data have been reviewed and the production pile order lengths are determined. The CO will provide an estimated length list or pile order list within 10 days after completion of all test pile driving.

551.06 Driven-Pile Capacity. Drive piles with approved pile-driving equipment to the specified penetration and to the depth necessary to obtain the required ultimate pile capacity. Splice piles that do not obtain the required ultimate capacity at the ordered length and drive with an impact hammer until the required ultimate pile capacity is achieved.

Use the dynamic formula to determine ultimate pile capacity of the in-place pile, unless the wave equation is required in accordance with Subsection 551.03(b).

(a) Wave Equation. Adequate penetration will be considered to be obtained when the specified wave equation resistance criteria are achieved within 1.5 m of the designated tip elevation as SHOWN ON DRAWINGS. Drive any piles that do not achieve the specified resistance within these limits to a penetration determined by the CO.

(b) Dynamic Formula. Drive the piles to the penetration necessary to obtain the ultimate pile capacity in accordance with the following formula:

$$Ru = \left(7\sqrt{E}\log(10N)\right) - 550$$

where

Ru = ultimate pile capacity in kilonewtons
E = manufacturer's rated hammer energy in joules at the ram stroke observed or measured in the field
$E = W \times H \times 9.81$
W = mass kilograms of striking parts of hammer
H = meter height of fall of the ram measured during pile driving in the field
$\log(10N)$ = logarithm to the base 10 of the quantity 10 multiplied by N
N = number of hammer blows per 25 mm at final penetration

Solving for N:

$$N = 10^x$$

$$x = \left(\frac{Ru + 550}{7\sqrt{E}}\right) - 1$$

Factor of Safety (FS) = 3.0

(1) Jetted Piles. Determine the in-place ultimate capacity of jetted piles based on impact hammer blow counts (dynamic formula) after the jet pipes have been removed. After the pile penetration length necessary to produce the required ultimate pile capacity has been determined by impact hammer blow count, install the remaining piles in each group or in each substructure unit to similar depths with similar methods. Confirm that the required ultimate pile capacity has been achieved by using the dynamic formula.

(2) Vibratory Hammers. The ultimate bearing capacity of piles driven with vibratory hammers will be based on impact driving blow count after the vibratory equipment has been removed. When vibratory installation of the piles is approved by the CO

and the vibrated piles do not attain the required ultimate pile-bearing capacity at the specified length, splice them as required without compensation, and drive with a specified impact pile hammer until the required ultimate pile-bearing capacity is achieved.

(3) Conditions for Dynamic Formula. The dynamic formula is applicable only if all of the following criteria apply:

(a) The hammer is in good condition and operating in a satisfactory manner.

(b) The hammer ram falls freely.

(c) A follower is not used.

(d) The head of the pile is not broomed or crushed.

(c) "Set Period" & Redriving. If piles do not attain the required bearing capacity when driven to the specified length, allow the piles to stand for a "set period" without driving. The "set period" shall be a minimum of 24 hours unless otherwise approved by the CO. After the "set period," perform check driving on either 2 piles in each bent or on 1 pile in 10 piles, whichever is more. The CO will designate the piles on which check driving is to be performed. Do not use a cold hammer for redriving. Warm up the hammer before redriving begins by applying at least 20 blows to another pile. Perform redriving by driving the pile to the required bearing with a maximum of 15 blows. If the specified hammer blow count is not attained on redriving, the CO may require driving all of the remaining pile length and repeating the "set period" and redriving procedure. Splice any piles driven to plan grade that do not attain the hammer blow count required, and drive until the required bearing is obtained. If the required bearing capacity is attained for each pile that is redriven, then the remaining piles in that bent will be considered satisfactory when driven to at least the same penetration and resistance as the redriven piles.

551.07 Preboring. Unless otherwise provided in the SPECIAL PROJECT SPECIFICATIONS, prebore holes to natural ground when piles are driven through compacted embankments more than 1.5 m in depth. Use augering, wet rotary drilling, or other approved methods of preboring. Except for piles end bearing on rock or hardpan, stop preboring at least 1.5 m above the pile tip elevation and drive the pile with an impact hammer to a penetration that achieves the required ultimate pile capacity. Preboring may extend to the surface of the rock or hardpan where piles are to be end bearing on rock or hardpan. Seat installed piles into the end bearing strata.

Prebore holes smaller than the diameter or diagonal of the pile cross section while allowing penetration of the pile to the specified depth. If subsurface obstructions such as boulders or rock layers are encountered, the hole diameter may be increased to the least dimension adequate for pile installation. After driving is completed, fill any void space remaining around the pile with sand or other approved material. Do not use a punch or a spud in lieu of preboring.

Do not impair the carrying capacity of existing piles or the safety of adjacent structures. If preboring disturbs the load carrying capacities of previously installed piles or structures, restore the required ultimate capacity of piles and structures by approved methods.

551.08 Jetting. Jetting will be permitted only when SHOWN ON THE DRAWINGS or approved in writing by the CO. When jetting is not required, but approved at the Contractor's request, determine the number of jets and the volume and pressure of water at the jet nozzles necessary to freely erode the material adjacent to the pile without affecting the lateral stability of the final in-place pile. Control, treat if necessary, and dispose of all jet water in a satisfactory manner. Drive all jetted piles with an approved impact hammer.

551.09 Preparation & Driving. Perform the work specified in Section 206 prior to driving piles. Make the heads of all piles plane and perpendicular to the longitudinal axis of the pile. Coordinate pile driving so as not to damage other parts of the completed work.

Drive piles to within 50 mm of plan location at cutoff elevation for bent caps, and within 150 mm of plan location for piles capped below finished ground. Ensure that the pile is no closer than 100 mm to any cap face and no closer than 225 mm to the face of any footing. Drive piles so that the axial alignment is within 20 mm/m of the required alignment. The CO may stop driving to check the pile alignment. Check the alignment of piles that cannot be internally inspected after installation before the last 1.5 m are driven. Do not pull laterally on piles or splice to correct misalignment. Do not splice a properly aligned section on a misaligned pile.

Unless otherwise SHOWN ON THE DRAWINGS, drive piles at least 5 m below the footing or cap. If the required minimum penetration cannot be obtained, provide a larger hammer, prebore or jet holes, or use other methods approved by the CO and in accordance with Subsection 551.03.

If the specified location and/or alignment tolerances are exceeded, the effect of the pile misalignment on the substructure design will be investigated. If the CO determines that corrective measures are necessary, implement suitable measures to correct the problem without compensation.

Place individual piles in pile groups, either starting from the center of the group and proceeding outward in both directions, or starting at the outside row and proceeding progressively across the group.

In an approved manner, correct all piles that are driven improperly, driven out of proper location, misaligned, or driven below the designated cutoff elevation. Replace piles damaged during handling or driving. Obtain approval for the proposed method(s) of correcting or repairing deficiencies.

Ensure that the method used in driving piles does not produce crushing and spalling of the concrete; injurious splitting, splintering, and brooming of the wood; or deformation of the steel.

(a) Timber Piles. Use piling that meets the minimum diameter requirements SHOWN ON THE DRAWINGS. Do not use piles with checks wider than 15 mm. Drive treated timber piles within 6 months after treatment. Handle and care for pressure-treated piles in accordance with American Wood Preservers Association (AWPA) standard M 4 and applicable portions of Subsection 557.04.

Install pile shoes as SHOWN ON THE DRAWINGS. Carefully shape the pile tip to secure an even, uniform bearing for the pile shoe. Fasten the shoe securely to the pile. Treat all holes, cuts, or daps in treated piles with two brush applications of creosote-coal tar solution or other preservative, as provided in the SPECIAL PROJECT SPECIFICATIONS.

Regulate the drop of the hammer to avoid damage to the pile if driving with a gravity hammer is permitted.

Select piles for any one bent to avoid undue bending or distortion of the sway bracing. Exercise care in the distribution of piles of various sizes to obtain uniform strength and rigidity in the bents of any given structure.

(b) Steel Piles. Furnish full length unspliced piles for lengths up to 18 m. If splices are required in the first pile driven and it is anticipated that subsequent piles will also require splices, place the splices in the lower one-third of the pile. Splice lengths less than 3 m are not permitted, and only two splices per pile are allowed, unless otherwise approved by the CO.

Load, transport, unload, store, and handle steel piles so that the metal is kept clean and free from damage. Do not use piles that exceed the camber and sweep permitted by allowable mill tolerance. Steel piles damaged during installation are considered unsatisfactory unless load tests prove that the bearing capacity is 100 percent of the required ultimate capacity. Load tests performed will be at no cost to the Government.

(c) Precast & Prestressed Concrete Piles. Support concrete piles during lifting or moving at the points SHOWN ON THE DRAWINGS or approved shop drawings. If points are not shown, provide support at the quarter points. Provide slings or other equipment when raising or transporting concrete piles to avoid bending the pile or breaking edges.

Protect the heads of concrete piles with a pile cushion at least 100 mm thick. Cut the pile cushion to match the cross section of the pile top. Replace the pile cushion if it is either compressed more than one-half its original thickness or begins to burn. Provide a new pile cushion for each pile.

A concrete pile is defective if any defect is observed that will affect the strength or long-term performance of the pile.

(d) Concrete-Filled Pipe or Steel Shell Piles. Furnish and handle the steel shells or pipes in accordance with Subsection 551.09(b). Cutting shoes for shells or pipes may be inside or outside the shell. Use high-carbon structural steel with a machined ledge for shell bearing or cast steel with a ledge designed for attachment with a simple weld.

When practicable, drive all pile shells or pipes for a substructure unit prior to placing concrete in any of the shells or pipes. Do not drive pile shells or pipes within 5 m of any concrete-filled pile shell or pipe until the concrete has cured for at least 7 days, or 3 days if using high-early-strength concrete. Do not drive any pile shell or pipe after it is filled with concrete.

Remove and replace shells that are determined to be unacceptable for use due to breaks, bends, or kinks.

551.10 Splices. Submit details for pile field splices for approval. Align and connect pile sections so the axis of the spliced pile is straight.

(a) Steel Piles. Submit a welder certification for each welder. Use welders certified for structural welding.

Make surfaces to be welded smooth, uniform, and free from loose scale, slag, grease, or other material that prevents proper welding. Steel may be oxygen cut. Carbon-arc gouging, chipping, or grinding may be used for joint preparation.

Weld in accordance with AASHTO/American Welding Society (AWS) D 1.5, Bridge Welding Code. Weld the entire pile cross section using prequalified AWS groove weld butt joints. Weld so there is no visual evidence of cracks, lack of fusion, undercutting, excessive piping, porosity, or inadequate size. Manufactured splices may be used in place of full penetration groove butt welds.

(b) Concrete Pile Splices. Submit drawings of proposed splices for approval. Use dowels or other acceptable mechanical means to splice precast concrete or precast prestressed concrete piles. Ensure that the splice develops strengths in compression, tension, and bending equal to or exceeding the strength of the pile being spliced.

(c) Concrete Pile Extensions. Construct precast concrete piles and prestressed piles as shown below.

(1) Precast Concrete Piles. Extend precast concrete piles by removing the concrete at the end of the pile and leaving 40 diameters of reinforcement steel exposed. Remove the concrete to produce a face perpendicular to the axis of the pile. Securely fasten reinforcement of the same size as that used in the pile to the projecting reinforcing steel. Form the extension to prevent leakage along the pile.

Immediately before placing concrete, wet the top of the pile thoroughly and cover with a thin coating of neat cement, retempered mortar, or other approved bonding material. Place concrete of the same mix design and quality as that used in the pile. Keep forms in place for not less than 7 days after the concrete has been placed. Cure and finish in accordance with Section 552.

(2) Prestressed Piles. Extend prestressed precast piles in accordance with Subsection 551.10(c)(1). Include reinforcement bars in the pile head for splicing to the extension bars. Do not drive extended prestressed precast piles.

(d) Timber Piles. Do not splice timber piles.

551.11 Heaved Piles. Check for pile heave during the driving operation. Take level readings immediately after each pile is driven and again after piles within a radius of 5 m are driven. Redrive all piles that heave more than 5 mm. Redrive to the specified resistance or penetration. Continue readings until the CO determines that such checking is no longer required.

551.12 Pile Load Tests. Pile load tests are not required unless SHOWN ON THE DRAWINGS.

(a) Dynamic Load Test. Use a qualified pile specialty consultant with at least 3 years experience in dynamic load testing and analysis, to perform the dynamic load test, the Case Pile Wave Analysis Program (CAPWAP), and the wave equation analysis including the initial wave analysis specified in Subsection 551.03(b)(1). Submit a resume of the specialty consultant for approval by the CO.

Furnish a shelter to protect the dynamic test equipment from the elements. Locate the shelter within 15 m of the test location. Provide a shelter with a minimum floor size of 6 m^2 and minimum ceiling height of 2 m. Maintain the inside temperature between 10 °C and 35 °C.

Furnish equipment and perform dynamic load tests in accordance with ASTM D 4945 under the supervision of the CO.

Place the piles designated as dynamic load test piles in a horizontal position and not in contact with other piles. Drill holes for mounting instruments near the head of the pile. Mount the instruments and take wave speed measurements. Place the designated pile in the leads. Provide at least a 1.2 × 1.2-m rigid platform, with a 1.1-m safety rail, that can be raised to the top of the pile.

Provide a suitable electrical power supply for the test equipment. If field generators are used as the power source, provide functioning meters for monitoring power voltage and frequency.

Drive the pile to the depth at which the dynamic test equipment indicates that the required ultimate pile capacity is achieved. If necessary to maintain stresses in the pile below the values shown in Subsection 551.03(b)(2), reduce the driving energy transmitted to the pile by using additional cushions or reducing the energy output of the hammer. If nonaxial driving is indicated, immediately realign the driving system.

At least 24 hours after the initial driving, redrive each dynamic load test pile with instrumentation attached. Warm the hammer before redriving by applying at least 20 blows to another pile. Redrive the dynamic load test pile for a maximum penetration of 150 mm or a maximum of 50 blows, whichever occurs first. Practical driving refusal is defined as 15 blows per 25 mm for steel piles, 8 blows per 25 mm for concrete piles, and 5 blows per 25 mm for timber piles.

Verify the assumptions used in the initial wave equation analysis submitted in accordance with Subsection 551.03(b) using CAPWAP. Analyze one blow from the original driving and one blow from the redriving for each pile tested.

Perform additional wave equation analyses with adjustments based on the CAPWAP results. Provide a graph showing blow count versus ultimate capacity. For open-end diesel hammers, provide a blow count versus stroke graph for the ultimate capacity. Provide the driving stresses, transferred energy, and pile capacity as a function of depth for each dynamic load test.

Based on the results of the dynamic load testing, CAPWAP analyses, and wave equation analyses, the production driving criteria may be approved by the CO, who will provide the order list and the required cutoff elevations, or additional pile penetration and testing may be specified. This information will be provided within 10 days after receipt of all required test data for the test piles driven.

(b) Static Load Tests. Perform static load tests in accordance with ASTM D 1143 using the quick load test method, except as modified herein. Submit drawings of the proposed loading apparatus for approval by the CO, in accordance with the following:

(1) Have a licensed professional engineer prepare the drawings.

(2) Furnish a loading system capable of applying 150 percent of the ultimate pile capacity or 9,000 kN, whichever is less

(3) Construct the apparatus to allow increments of load to be placed gradually without causing vibration to the test pile.

When tension (anchor) piles are required, drive tension piles at the location of permanent piles when feasible. Do not use timber or tapered piles installed in permanent locations as tension piles. Take the test to plunging failure or the capacity of the loading system.

The safe axial pile load is defined as 50 percent of the failure load. The failure pile load is defined as follows:

- For piles 600 mm or less in diameter or diagonal width:

$$S_f = S + (3.8 + 0.008D)$$

- For piles greater than 600 mm in diameter or diagonal width:

$$S_f = S + \frac{D}{30}$$

where

S_f = settlement at failure in millimeters
D = pile diameter or diagonal width in millimeters
S = elastic deformation of pile in millimeters

Determine top elevation of the test pile immediately after driving and again just before load testing to check for heave. Wait a minimum of 3 days between the driving of any anchor or the load test piles and the commencement of the load test. Prior to testing, redrive or jack to the original elevation any pile that heaves more than 6 mm.

After completion of static testing, remove or cut off any test or anchor piling not a part of the finished structure at least 500 mm below either the bottom of the footing or the finished ground elevation.

Based on the results of the static load testing, the production driving equipment may be approved by the CO, who will provide the order list and the required cutoff elevations, or additional tests may be specified. This information will be provided within 10 days after receipt of all required test data for the test piles driven.

551.13 Pile Cutoffs. Cut off the tops of all production piles and pile casings at the required elevation. Cut off the piles clean and straight parallel to the bottom face of the structural member in which they are embedded.

Ensure full bearing between timber caps and piles by making accurate, square cuts.

Remove all unused pile cutoff lengths and dispose of them in accordance with applicable State and local laws and regulations. Dispose of treated timber pile cut-offs in accordance with the requirements of Subsection 202.04(a) for disposal of treated material.

(a) Steel Piles. Do not paint steel to be embedded in concrete. Before painting the exposed steel pile, thoroughly clean the metal surface of any substance that will inhibit paint adhesion. Paint in accordance with Section 563. Paint portions of completed trestle or other exposed piling to a point not less than 1 m below finished groundline or to the waterline, as SHOWN ON THE DRAWINGS or as provided in the SPECIAL PROJECT SPECIFICATIONS.

(b) Wood Piles. When possible, cut the top of the pile on a bevel. Treat the heads of all treated timber piles that are not embedded in concrete using one of the following methods:

(1) Where possible, reduce the moisture content of the wood to no more than 25 percent and allow no free moisture on the surface. Brush on one application of creosote-coal tar solution as required in AWPA standards, or preservatives as provided in the SPECIAL PROJECT SPECIFICATIONS.

(2) Build up a protective cap by applying alternate layers of loosely woven fabric and hot asphalt or tar similar to membrane waterproofing, using three layers of asphalt or tar and two layers of fabric. Use fabric at least 150 mm wider in each direction than the diameter of the pile. Turn the fabric down over the pile and secure the edges by binding with two turns of 3-mm-diameter galvanized wire. Apply a final layer of asphalt or tar to cover the wire. Neatly trim the fabric below the wires.

(3) Cover the sawed surface with three applications of a hot mixture of 60 percent creosote and 40 percent roofing pitch, or thoroughly brush coat with three applications of hot creosote and cover with hot roofing pitch.

551.14 Unsatisfactory Piles. Correct unsatisfactory piles using an approved method. Methods of correcting unsatisfactory piles may include one or more of the following:

(a) Using the pile at a reduced capacity.

(b) Installing additional piles.

(c) Repairing damaged piles.

(d) Replacing damaged piles.

(e) Splicing on additional length(s) and driving, when necessary.

(f) Building up pile(s).

551.15 Placing Concrete in Steel Shell or Pipe Piles. After driving, clean the inside of shells and pipes by removing all loose material. Keep the shell or pipe substantially watertight. Provide suitable equipment for inspecting the entire inside surface of the driven shell or pipe just before placing concrete.

(a) Reinforcing Steel. When reinforcing steel is required, make the spacing between adjacent cage elements at least 5 times the maximum size of aggregate in the concrete.

Securely tie concrete spacers or other approved spacers at fifth points around the perimeter of the reinforcing steel cage. Install spacers at intervals not to exceed 3 m measured along the length of the cage.

Place the reinforcement cage into the driven shell or pipe when the concrete reaches the lower limits of the reinforcement. Support the reinforcement so it remains within 50 mm of the required vertical location. Support the cage from the top until the concrete reaches the top of the pile.

(b) Concrete. Construct concrete in accordance with Section 552. Place concrete in one continuous operation from the bottom to the top of the pile. Before the initial concrete set, consolidate the top 3 m of the concrete pile using approved vibratory equipment.

Measurement

551.16 Method. Use the method of measurement that is DESIGNATED IN THE SCHEDULE OF ITEMS.

Measure piles by the meter or by the each. When measurement is by the meter, measure the length of pile from the cutoff elevation rounded to the tip.

Measure pile load tests by the each or by the lump sum.

Measure preboring by the meter.

Measure splices by the each for those made as required to drive piling in excess of the estimated plan tip elevation.

Measure test piles and pile shoes by the each.

Payment

551.17 Basis. The accepted quantities will be paid for at the contract unit price for each PAY ITEM DESIGNATED IN THE SCHEDULE OF ITEMS.

Payment will be made under:

Pay Item		Pay Unit
551(01)	__________ piles, furnished	Meter
551(02)	__________ piles, driven	Meter
551(03)	__________ piles, furnished	Each
551(04)	__________ piles, driven	Each
551(05)	_________ pile load test	Each
551(06)	__________ pile load test	Lump Sum
551(07)	Preboring	Meter
551(08)	Splices	Each
551(09)	Test piles	Each
551(10)	Pile shoes	Each

Section 552—Structural Concrete

Description

552.01 Work. Furnish, place, finish, and cure concrete in bridges, culverts, and other structures.

Structural concrete class is designated as shown in table 552-1.

Table 552-1.—Composition of concrete.

Class of Concrete	Minimum Cement Content (kg x m^3)	Maximum W/C Ratio	Slump[a] (mm)	Minimum Air Content[b] (%)	Coarse Aggregate AASHTO M 43
A	360	0.49	50–100	–	No. 57
A(AE)	360	0.44	25–100	5.0	No. 57
C	390	0.49	50–100	–	No. 7
C(AE)	390	0.44	25–75	6.0	No. 7
P	390	0.44	0–100	–	No. 67
Seal	390	0.54	100–200	–	No. 57

a. Maximum slump is 200 mm if approved mix design includes a high-range water reducer.
b. See Subsection 552.03 for maximum air content.

Materials

552.02 Requirements. Furnish material that conforms to specifications in the following subsections:

Air-Entraining Admixtures 711.02
Boiled Linseed Oil 725.14
Chemical Admixtures 711.03
Coarse Aggregate 703.02
Color Coating 725.23
Curing Material 711.01
Elastomeric Bearing Pads 717.10
Elastomeric Compression Joint Seals 717.16
Epoxy Resin Adhesives 725.21
Fine Aggregate 703.01
Fly Ash 725.04
High-Strength Nonshrink Grout 701.02
Latex Modifier 711.04
Low-Strength Grout 701.03
Mortar 701.04
Portland Cement 701.01

Sealants, Fillers, Seals, & Sleeves 712.01
Water 725.01

Construction

552.03 Composition (Concrete Mix Design). Design and produce concrete mixtures that conform to table 552-1 for the class of concrete specified and the minimum strength requirements as SHOWN ON THE DRAWINGS or in Subsection 552.04. Determine design strength values in accordance with ACI 214. Ensure that structural concrete also conforms to the following ACI specifications:

- ACI 211.1 for normal and heavyweight concrete.
- ACI 211.2 for lightweight concrete.
- ACI 211.3 for no-slump concrete.

Submit written concrete mix designs for approval at least 30 days before production. Include the following in each mix design submittal:

(a) Project identification.

(b) Name and address of Contractor and concrete producer.

(c) Mix design designation.

(d) Class of concrete and intended use.

(e) Material proportions.

(f) Name and location of material sources for aggregate, cement, admixtures, and water.

(g) Type of cement and type of cement replacement, if used. Fly ash, ground iron blast-furniture slag, or silica fume may partially replace cement as follows in any mix design except for prestressed concrete:

(1) Fly ash.

(a) Class F. Not more than 20 percent of the minimum weight of Portland cement in table 552-1 may be replaced with class F fly ash at the rate of 1.5 parts fly ash per 1 part cement.

(b) Class C. Not more than 25 percent of the minimum mass of Portland cement in table 552-1 may be replaced with class C fly ash at the rate of 1 part fly ash per 1 part cement.

(2) Ground iron blast-furnace slag. Not more than 50 percent of the minimum mass of Portland cement in table 552-1 may be replaced with ground iron blast-furnace slag at the rate of 1 part slag per 1 part cement.

(3) Silica fume (microsilica). Not more than 10 percent of the minimum mass of Portland cement in table 552-1 may be replaced with silica fume at the rate of 1 part silica fume per 1 part cement.

The water/cement ratio for modified concrete is the ratio of the mass of water to the combined masses of Portland cement and cement substitute.

(h) Cement content in kilograms per cubic meter of concrete.

(i) The saturated surface dry batch weight of the coarse and fine aggregate in kilograms per cubic meter of concrete.

(j) Water content (including free moisture in the aggregate plus water in the drum, exclusive of absorbed moisture in the aggregate) in kilograms per cubic meter of concrete.

(k) Target water/cement ratio.

(l) Dosage of admixtures. Entrained air may be obtained either by the use of an air-entraining Portland cement, or by the use of an air-entraining admixture. Do not use set-accelerating admixtures with class P (prestressed) concrete. Do not mix chemical admixtures from different manufacturers.

(m) Sieve analysis of fine and coarse aggregate.

(n) Absorption of fine and coarse aggregate.

(o) Bulk specific gravity (dry and saturated surface dry) of fine and coarse aggregate.

(p) Dry rodded unit mass of coarse aggregate in kilograms per cubic meter.

(q) Fineness modulus (FM) of fine aggregate.

(r) Deleterious substances (coarse and fine aggregate); clay lumps and friable particles; material finer than the 75-µm sieve; coal and lignite (AASHTO M 80 7.1.6); chert (coarse aggregate only); and organic impurities (fine aggregate only).

(s) Evaluation of potential aggregate reactivity.

(t) Percentage of wear (L.A.R.) for coarse aggregate only.

(u) Sand equivalent (fine aggregate only).

(v) Material certifications for cement, admixtures, and aggregate.

(w) TV's for concrete slump with and without high-range water reducers.

(x) TV's for concrete air content. Include the proposed range of air content for concrete to be incorporated into the work. Describe the methods by which air content will be monitored and controlled. Provide acceptable documentation that the slump and compressive strength of the concrete are within specified limits throughout the full range of proposed air content. In the absence of such acceptable documentation, ensure that the maximum air content is 10 percent.

(y) Concrete unit mass.

(z) Compressive strengths of 7- and 28-day concrete. Pending 28-day strength results, a mix design may be approved on the basis that the 7-day compressive strength results equal or exceed 85 percent of the minimum strength requirements, when no accelerators or early strength cements are used.

(aa) Material samples, if requested.

Use a testing laboratory that is fully equipped and capable of performing the required tests and services. Base the mix design on representative samples of aggregates, cement, water, and admixtures to be used on the project. Take aggregate samples in accordance with AASHTO T 2 and reduce to testing size in accordance with AASHTO T 248. Submit a separate proposed mix design for each class of concrete to the CO for review.

Current mix designs for other projects may be acceptable, provided that all items required herein are covered by certified submittals. Ensure that mix design and aggregate quality tests from other projects have been run within 12 months of the date of submittal, and that the aggregate source is the same.

Begin production only after the mix design is approved.

Furnish a new mix design for approval if there is a change in a source of material, or when the FM of the fine aggregate changes by more than 0.20.

Use type II cement for all classes of concrete, but use type III cement when concrete work is permitted by the CO in air temperatures below 2 °C. Type III cement may be used in class A and seal concrete with the approval of the CO. Type III cement may be used in class P concrete when documented in the approved mix design.

552.04 Concrete Compressive Strength. Use the minimum 28-day compressive strength for the given classes of concrete shown in table 552-2, unless otherwise SHOWN ON THE DRAWINGS.

Table 552-2.—Specified minimum concrete strength (MPa).

Concrete Class	At Time of Transfer of Prestress Force	7-Day	28-Day
A & A(AE)	–	15.9	24.2
C & C(AE)	–	18.2	27.7
P	31.5	–	40
P (AE)	31.5	–	34.5
Seal	–	13.8	20.7

Make two standard test specimens for a strength test. Take enough specimens to make at least one 7-day strength test and one 28-day strength test (a minimum total of four specimens) for each structural element. Use the average of the strengths of the two specimens for test result, but discard any specimen that shows definite evidence, other than low strength, of improper sampling, molding, handling, curing, or testing, and consider the strength of the remaining cylinder to be the test result.

Extend the standard 28-day curing period for compressive strength tests for fly-ash-modified concrete by 1 day (rounded to the nearest whole day) for each 1.5 percent of Portland cement replaced with fly ash at the selected rate. (Example: If the maximum of 20 percent cement is replaced, the curing period for cylinders is 41 days.)

552.05 Storage & Handling of Material. Store and handle all material in a manner that prevents segregation, contamination, or other harmful effects. Do not use cement and fly ash containing evidence of moisture contamination. Store and handle aggregate in a manner that ensures a uniform moisture content at the time of batching.

Obtain the CO's approval before using cement that has been stored on the site for more than 60 days. Provide separate storage of cement that is of different blends, types, or from different mills.

552.06 Measuring Material. Batch the concrete in accordance with the approved mix design and the following tolerances:

Cement .... ± 1%
Water .... ± 1%
Aggregate .... ± 2%
Additive .... ± 3%

Submit to the CO, for approval, a written procedure for adding the specified amount of admixture. Provide separate scales for the admixtures that are to be proportioned by mass and accurate measures for those to be proportioned by volume.

A calibrated volumetric system may be used if the specified tolerances are maintained.

552.07 Batching Plant, Mixers, & Agitators. Use a batching plant, mixer, and agitator conforming to AASHTO M 157. Use continuous volumetric mixing equipment that conforms to AASHTO M 241.

552.08 Mixing. Mix the concrete in a central-mix plant or in truck mixers. Operate all equipment within manufacturer's recommended capacity. Produce concrete of uniform consistency.

(a) Central-Mix Plant. Dispense liquid admixtures through a controlled flowmeter. Use dispensers with sufficient capacity to measure, at one time, the full quantity of admixture required for each batch. If more than one admixture is used, dispense each with separate equipment.

Charge the coarse aggregate, one-third of the water, and all air-entraining admixture into the mixer first, then add remainder of the material.

Mix for at least 50 seconds. Begin mixing time after all cement and aggregate are in the drum. Add the remaining water during the first quarter of the mixing time. Add 4 seconds to the mixing time if timing starts the instant the skip reaches its maximum raised position. Transfer time in multiple-drum mixers is included in mixing time. Mixing time ends when the discharge chute opens.

Remove the contents of an individual mixer before a succeeding batch is charged into the drum.

(b) Truck Mixer. Do not use mixers with any section of the blades worn 25 mm or more below the original manufactured height. Do not use mixers and agitators with accumulated hard concrete or mortar in the mixing drum.

Add admixtures to the mix water before or during mixing.

Charge the batch into the drum so a portion of the mixing water enters in advance of the cement.

Mix each batch of concrete not less than 70 or more than 100 revolutions of the drum or blades at mixing speed. Begin the count of mixing revolutions as soon as all material, including water, is in the mixer drum.

Do not allow the sum of all drum revolutions at both mixing and agitating speeds to exceed 300 before all concrete has been discharged from the drum; but ensure that the sum of all drum revolutions does not exceed 200 if the outside air temperature is over 30 °C. If mixing is done before arrival of the truck at the point of delivery, rotate the drum at mixing speed for 10 to 15 revolutions to reblend possible stagnant spots.

If set-retarding admixture is used, do not allow the sum of all drum revolutions at both mixing and agitating speeds to exceed 550 before all concrete has been discharged from the drum; but ensure that the sum of all drum revolutions does not exceed 450 if the outside air temperature is over 30 °C.

Do not handmix except in case of emergency and with the written approval of the CO. When permitted, perform only on watertight platforms. Do not exceed 0.1 m^3 volume for handmixed batches. Do not permit handmixing for concrete that is to be placed under water.

552.09 Delivery. Submit a written schedule of concreting operations, including scheduling, personnel, and equipment, when requested by the CO. Provide the CO 24-hour notice prior to placing any concrete.

Produce and deliver concrete to permit a continuous placement. Do not permit concrete to achieve initial set before the remaining concrete is placed adjacent to it. Never allow the time interval between placement to exceed 30 minutes. Use methods of delivering, handling, and placing that will minimize rehandling of the concrete and prevent any damage to the structure.

Do not place concrete that has developed an initial set. Never retemper concrete by adding water.

*(a) **Truck Mixer/Agitator.*** Use the agitating speed for all rotation after mixing. When a truck mixer or truck agitator is used to transport concrete that is completely mixed in a stationary central construction mixer, mix during transportation at manufacturer's recommended agitating speed.

Water and admixtures (if in the approved mix design) may be added at the project to obtain the required slump or air content, provided that the total of all water in the mix does not exceed the maximum water/cement ratio. If additional water is necessary, add only once and remix with 30 revolutions at mixing speed. Complete the remixing within 45 minutes (75 minutes for type I, IA, II, or IIA cements with water-reducing or -retarding admixture) after the initial introduction of the mixing water to the cement or the cement to the aggregates.

After the beginning of the addition of the cement, complete the discharge of the concrete within the time specified in table 552-3, unless otherwise approved by the CO or as allowed by the SPECIAL PROJECT SPECIFICATIONS.

Table 552-3.—Concrete discharge time limits.

Cement Type With and Without Admixtures	Time Limit (h) ≤ 30 °C	 > 30 °C
Type I, IA, II, or IIA	1.50	1.00
Type I, IA, II, or IIA with water-reducing or -retarding admixture	2.00	1.50
Type III	1.25	0.75
Type III with water-reducing or -retarding admixture	1.75	1.25

Note: Temperatures are ambient air measured on formwork.

(b) Nonagitating Equipment. Nonagitating equipment may be used to deliver concrete if the concrete discharge is completed within 20 minutes from the beginning of the addition of the cement to the mixing drum. Use equipment with smooth, mortar-tight, metal containers capable of discharging the concrete at a controlled rate without segregation. Provide covers when needed for protection.

552.10 Quality Control of Mix. Submit and follow a quality control plan for the following:

(a) Mixing. Designate a competent and experienced concrete technician to be at the mixing plant in charge of the mixing operations and to be responsible for the overall quality control, including:

(1) The proper storage and handling of all components of the mix.

(2) The proper maintenance and cleanliness of plant, trucks, and other equipment.

(3) The gradation testing of fine and coarse aggregates.

(4) The determination of the FM of fine aggregate.

(5) The measurement of moisture content of the aggregates and adjustment of the mix proportions, as required before each day's production, or more often if necessary, to maintain the required water/cement ratio.

(6) The computation of the batch weights for each day's production and the checking of the plant's calibration as necessary.

(7) The completion of batch tickets. Include the following information:

(a) Concrete supplier.

(b) Ticket serial number.

(c) Date and truck number.

(d) Contractor.

(e) Structure or location of placement.

(f) Mix design and concrete class.

(g) Component quantities and concrete total volume.

(h) Moisture corrections for aggregate moisture.

(i) Total water in mix at plant.

(j) Time of batching and time at which discharge must be completed.

(k) Maximum water that may be added to the mix at the jobsite.

Provide equipment necessary for the above tests and controls. Furnish copies of work sheets for items (3), (4), (5), and (6) as they are completed.

(b) Delivery & Sampling. Designate at least one competent and experienced concrete technician to be at the project and be responsible for concrete delivery, discharge operations, and sampling, including:

(1) The verification that adjustments to the mix before discharge comply with the specifications.

(2) The completion of the batch ticket, the recording of the apparent water/cement ratio, and the time discharge is completed. Furnish a copy of each batch ticket at the time of placement.

(3) The furnishing of all equipment and the performance of temperature, unit weight, air content, slump, and other tests to verify specification compliance before and during each placement operation.

Sample every batch after at least 0.2 m^3 are discharged and before placing any of the batch in the forms. When continuous mixing is used, sample approximately every 10 m^3. Test the air content in accordance with AASHTO T 152 or T 196, and evaluate the result based on a single test or the average of two tests.

Test slump and temperature of each batch in accordance with AASHTO T 119 and AASHTO T 152 or T 196.

If three successive samples are tested and compliance with the specifications is indicated, screening tests may be reduced to a frequency approved by the CO. Resume initial testing frequency if a test shows a failing temperature, air content, or slump, or when directed.

If there is no prior experience with the approved mix design or if special handling procedures, such as pumping, change one or more of the characteristics between discharge of the load and placement in the forms, correlate the discharge tests with the placement tests to define these changes. Provide documentation. Repeat the correlations as often as necessary or as directed.

(4) The taking of samples for strength tests in accordance with AASHTO T 141 and T 23 from batches specified by the CO. Composite samples are not required. The point of sampling is from the discharge stream at the point of placement. Provide cylinder molds. Make compressive strength test cylinders

as directed by the CO, provide the appropriate initial curing, and carefully transport the cylinders to the jobsite curing facility. Cylinders will be used for 28-day breaks, verification, projected strengths, or other purposes specified. Assist in the performing of other tests as requested.

(c) Testing. Determine compressive strength of concrete test cylinders in accordance with AASHTO T 22, and of drilled concrete cores in accordance with AASHTO T 24.

Ensure that the average of all the strength tests representing the concrete in each structural element meets the following requirements:

(1) For concrete in structures designed by the service load method, when seven or more strength tests are available, not more than 20 percent of the strength tests shall have values less than the specified strength, and the average of any six consecutive strength tests shall be equal to or greater than the specified strength. This paragraph does not apply to designs by the strength method where the service load method was used to check fatigue and crack control.

(2) For concrete in structures designed by the strength method and in all prestressed members, when seven or more strength tests are available, not more than 10 percent of the strength tests shall have values less than the specified strength, and the average of any three consecutive strength tests shall be equal to or greater than the specified strength. In applying this requirement to prestressed members, the strength tests performed on all similar members, such as all beams, shall be grouped together for purposes of counting the number of tests available. This paragraph also applies to designs by the strength method where the service load method was used to check fatigue and crack control.

If six or fewer strength tests are available, the average of all the tests shall be equal to or greater than the strengths shown in the following:

	Required Average Strength (% of specified strength)	
Number of Strength Tests	Class A, C, and Seal	Class P
1	79	86
2	90	97
3	94	102
4	97	105
5	99	107
6	100	108

If the concrete strength tests fail to meet the requirements of this specification, the CO may order the Contractor to have a testing laboratory that is acceptable to the Forest Service take and test core samples of questionable concrete. The CO may order all low-strength concrete removed and replaced if core strengths are below specified strengths. All costs connected with concrete coring and removal and replacement of concrete that fails to meet these requirements shall be borne by the Contractor.

552.11 Field Adjustment of Concrete Mix. Field adjustment of the concrete mix designs will be necessary to compensate for the free-water content in the aggregates.

After initial mixing, if the consistency (slump) is outside the specification limits (table 552-l) by less than 25 mm, the CO may approve the addition of water or cement, provided all that the following conditions are met:

(a) Addition of Water. Water may be added, provided that:

(1) The maximum allowable water content in kilograms per cubic meter of concrete (table 552-1) is not exceeded.

(2) The maximum allowable mixing time (or number of drum revolutions) is not exceeded.

(3) Concrete is remixed for at least half of the minimum mixing time (or number of drum revolutions).

(b) Addition of Cement. Cement may be added, except to class P concrete, provided that:

(1) The amount of cement added does not exceed 55.7 kg/m^3 more than the mix design or a total of 418 kg/m^3, unless otherwise DESIGNATED IN THE SPECIAL PROJECT SPECIFICATIONS.

(2) The maximum allowable mixing time (or number of drum revolutions) is not exceeded.

(3) Concrete is remixed for at least half of the minimum mixing time (or number of drum revolutions).

(c) Adjustment for Percent Entrained Air. Vary the amount of air-entraining admixture used in each batch as necessary from that given in the approved mix design to produce concrete with the percent entrained air specified in table 552-1.

552.12 Temperature and Weather Conditions. Maintain the temperature of the concrete mixture just before placement between 10 °C and 32 °C; except maintain the concrete for bridge decks between 10 °C and 25 °C.

(a) Cold Weather. Cold weather is defined as any time during the concrete placement or curing period that the ambient temperature at the worksite drops below 2 °C or the ambient temperature at the site drops below 10 °C for a period of 12 hours or more.

When cold weather is reasonably expected or has occurred within 7 days of anticipated concrete placement, submit a detailed plan for producing, transporting, placing, protecting, curing, and temperature monitoring of concrete during cold weather. Include procedures for accommodating abrupt changes in weather conditions. Do not commence placement until plan is approved. Approval of an acceptable plan will take at least 1 day.

Before commencing cold weather concreting, have all material and equipment required for protection available at or near the project and subject to the approval of the CO.

Remove all snow, ice, and frost from the surfaces, including reinforcement and subgrade, against which the concrete is to be placed. Ensure that the temperature of any surface that will come into contact with fresh concrete is at least 2 °C and is maintained at a temperature of 2 °C or above during the placement of the concrete.

Place heaters and direct ducts so as not to cause concrete drying or fire hazards. Vent exhaust flue gases from combustion heating units to the outside of any enclosures. Heat the concrete components in a manner that is not detrimental to the mix. Do not heat cement or permit the cement to come into contact with aggregates that are hotter than 40 °C. Ensure that concrete at the time of placement is of uniform temperature and free of frost lumps. Do not heat aggregates with a direct flame or on sheet metal over fire. Do not heat fine aggregate by direct steam. Do not add salts to prevent freezing.

Provide heat within the housing by steam or hot air. Maintain a humid condition within the housing during the heating period. Do not use stoves or open-burning salamanders within the housing.

Do not use any heating method that will endanger forms, falsework, or any part of the structure, or that will subject the concrete to drying out or other injury due to excessive temperatures. Do not allow the concrete deck surface temperature to exceed 32 °C throughout the curing period.

Maintain a reasonably uniform temperature within the enclosure throughout the curing period.

Provide adequate fire protection when heating is in progress, and maintain watchmen or other attendants to keep heating units in continuous operation.

Furnish and place continuously recording surface temperature measuring devices that are accurate within ± 1 °C.

Make outside air temperature recordings at the same time that recordings are made within the enclosure. Provide a copy of temperature records to the CO.

During cold weather, protect the concrete for at least 7 days at or above the minimum temperatures shown in table 552-4.

Table 552-4.—Cold weather concrete surface temperatures.

Concrete Surface Temperatures	Minimum Section Size Dimension (mm)			
	< 300	300–900	900–1,800	> 1,800
Minimum temperature during protection period	13 °C	10 °C	7 °C	4 °C
Maximum allowable temperature drop in any 24-hour period after end of protection	28 °C	22 °C	17 °C	11 °C

When pozzolan or fly ash cement is used, adjust the required period of controlled temperature and moisture as follows:

Percentage of Cement Replaced by Weight	Required Period of Controlled Temperature and Moisture
10%	9 days
10–15%	10 days
16–20%	11 days

The above requirement for an extended period of controlled temperature and/or moisture may be waived if a compressive strength of 65 percent of the specified 28-day strength is achieved in 7 days.

At the end of the protection period, allow the concrete to cool gradually over 24 hours at a rate not to exceed the maximum values shown in table 552-4. All protection may be removed when the concrete surface temperature is within 15 °C of the ambient air temperature.

If the concrete temperatures cannot be maintained within the limits specified in table 552-4 through insulated forms or blankets, enclose each section of the structure with adequate housing before placing the concrete in the section.

Make the protective housing of sufficient size to allow all concrete placing and finishing operations for any one placement to proceed under cover without hindrance. However, to facilitate the placement of concrete, install the covering material immediately after depositing the concrete. Construct the housing to be weathertight and in a manner that will ensure that specified temperatures will be maintained uniformly throughout the enclosure during the protection period.

When housing of the structure is not initially installed, but may be subsequently required in accordance with the specifications, protect structural concrete in bridge decks or similar thin sections with insulating blankets or other methods approved by the CO. Ensure that the curing method prevents moisture loss on all exposed surfaces, including those protected by insulating blankets.

Provide insulation that consists of bats or blankets of fiberglass, rock wool, balsam wool, insulation boards, or other approved material.

Completely encase the bats or blankets in suitable wind- and water-resistant covers that are be fastened securely to wood forms between the studs and walls, with edges and ends sealed to the framing to minimize heat loss. Attach insulation to steel forms by adhesive or other approved methods. Cover ribs and flanges of steel forms with insulating blankets or separate strips of insulation. Ensure that the edges and corners of concrete are well insulated. Protect horizontal surfaces of concrete with a layer of the insulating material securely fastened in place. Protect the tops of placements, such as bridge decks and similar flat slab sections, with tarpaulins over the insulation. Cover large insulating blankets around and securely fasten in place for curing concrete columns cast in prefabricated forms and similar concrete items. Seal all joints in the blankets with tape.

Use electric heating blankets and other suitable materials instead of insulated blankets or bats only when specifically approved by the CO for each application.

Assume entire responsibility for the proper protection and final satisfactory condition of all concrete placed during cold weather or exposed to cold weather within the required protection period. Remove and replace any concrete that has been frozen or damaged due to other causes.

(b) Hot Weather. Hot weather is defined as any time during the concrete placement that the ambient temperature at the work site is above 35 °C.

In hot weather, cool all surfaces that will come in contact with the mix to below 35 °C. Cool by covering with wet burlap or cotton mats, fog spraying with water, covering with protective housing, or using other approved methods.

Immediately prior to and during placement, maintain concrete at a temperature not to exceed 35 °C; but ensure that bridge superstructure (deck) concrete does not exceed a temperature of 25 °C.

When placing concrete deck slabs, if the air temperature near the slab's surface is expected to rise above 25 °C, schedule operations so that finishing of the top of the slab is completed before this occurs, or use hot-weather concreting practices to maintain the deck surface temperature at 25 °C or less until finishing is completed.

Maintain concrete temperature by using any combination of the following methods:

(1) Shade the material storage areas or production equipment.

(2) Cool aggregate by sprinkling.

(3) Cool aggregate and/or water by refrigeration or replace a portion or all of the mix water with flaked or crushed ice to the extent that the ice will completely melt during mixing of the concrete.

(c) Evaporation. When placing concrete in bridge decks or other exposed slabs, limit the expected evaporation rate to less than 0.5 kg/m²/h, as determined by figure 552-1 or the following:

$$EVAP = \frac{1+0.2374WV}{2{,}906} \times \left[CT^2 - 4.762CT + 220.8 - RH \times \left[\frac{AT^3 + 127.8AT^2 + 665.6AT + 34{,}283}{20{,}415} \right] \right]$$

where

$EVAP$ = evaporation rate (kg/m²/h)
WV = wind velocity (km/h)
RH = relative humidity (%)
AT = air temperature (°C)
CT = concrete temperature (°C)

When necessary, take one or more of the following actions:

(1) Construct windbreaks or enclosures to effectively reduce the wind velocity throughout the area of placement and for a period of 12 hours following completion of deck placement or until the evaporation rate is less than 0.5 kg/m²/h.

(2) Use fog sprayers upwind of the placement operation to effectively increase the relative humidity.

(3) Reduce the temperature of the concrete in accordance with Subsection 552.12(b).

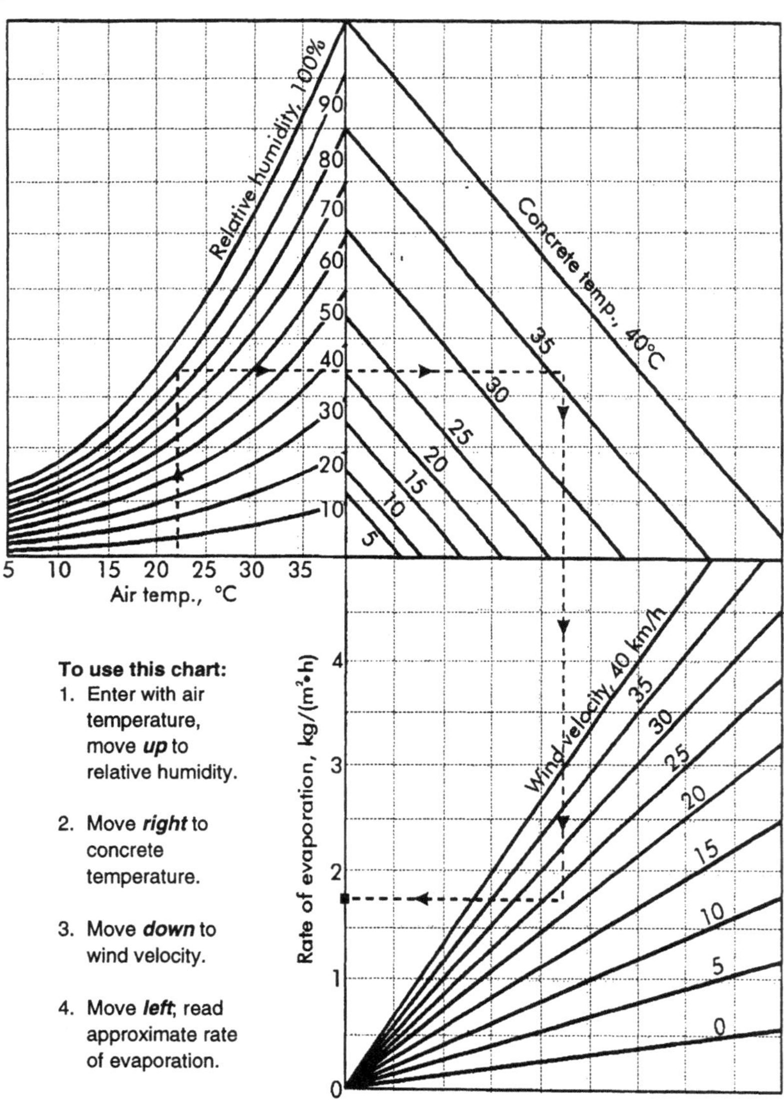

Note: Example shown by dashed lines is for an air temperature of 22.5 °C, relative humidity of 90 percent, concrete temperature of 36 °C, and wind velocity of 22.5 km/h. This results in a rate of evaporation of 1.75 kg/m²/h.

Figure 552-1.—Evaporation rate of surface moisture.

(d) Rain. At all times during and immediately after placement, protect the concrete from rain.

552.13 Handling & Placing Concrete. Perform the work specified in Section 206. Construct reinforcing steel, structural steel, bearing devices, joint material, and miscellaneous items in accordance with the appropriate sections.

(a) General. Use falsework and forms in accordance with Section 562. Handle, place, and consolidate concrete using methods that will not cause segregation and will result in dense, homogeneous concrete that is free of voids and rock pockets. Use placement methods that do not cause displacement of reinforcing steel or other material that is embedded in the concrete. Place and consolidate concrete before initial set. Do not retemper concrete by adding water to the mix except as provided for in Subsection 552.11.

Do not place concrete until the forms, all embedded material, and the adequacy of the foundation material have been inspected and approved by the CO.

Remove all mortar, debris, and foreign material from the forms and reinforcing steel before commencing placement. Thoroughly moisten the forms and subgrade immediately before concrete is placed against them. Temporary form spreader devices may be left in place until concrete placement precludes their need. Remove them when no longer needed.

Place concrete continuously without interruption between planned construction or expansion joints. Ensure that the delivery rate and placing sequence and methods are such that fresh concrete is always placed and consolidated against previously placed concrete before initial set has occurred in the previously placed concrete. Do not allow the time between the placement of successive batches to exceed 30 minutes (20 minutes under hot weather conditions).

During and after placement of concrete, do not damage previously placed concrete or break the bond between the concrete and reinforcing steel. Keep workers off fresh concrete. Do not support platforms for workers and equipment directly on reinforcing steel. After the concrete is set, do not disturb the forms or reinforcing bars that project from the concrete until it is of sufficient strength to resist damage.

Five to 10 working days before placing concrete in a cast-in-place bridge deck, hold a preplacement conference to discuss the construction procedures, personnel, and equipment to be used. At this time, provide full details on plans for the placement operation, including finishing machine data, workforce, contingency plans, concrete delivery, and other information requested by the CO.

(b) Sequence of Placement. Observe the following sequence of placement:

(1) Substructures. Do not place loads on finished bents, piers, or abutments until concrete strength cylinder tests from the same concrete cured under the same conditions as the substructure element indicate that all concrete has at least 80 percent of its required 28-day compressive strength.

(2) Vertical Members. For vertical members more than 5 m in height, allow the concrete to set for at least 4 hours before placing concrete for integral horizontal members. For vertical members less than 5 m in height, allow the concrete to set for at least 30 minutes. Do not apply loads from horizontal members until the vertical member has attained its required strength.

(3) Superstructures. Do not place concrete in the superstructure until substructure forms have been stripped sufficiently to determine the acceptability of the supporting substructure concrete. Do not place concrete in the superstructure until the substructure has attained the required strength.

Place concrete for T-beams in two separate operations. Wait at least 5 days after stem placement before placing the top deck slab concrete.

Concrete for box girders may be placed in two or three separate operations consisting of bottom slab, girder webs, and top slab, or as SHOWN ON THE DRAWINGS. However, place the bottom slab first, and do not place the top slab until the girder webs have been in place for at least 5 days.

(4) Arches. Place concrete in arch rings so that the centering is loaded uniformly and symmetrically.

Place centering upon approved jacks to provide means of correcting any slight settlement that may occur after concrete placement has begun. Make any adjustments made necessary by settlement before the concrete has taken its initial set.

(5) Box Culverts. Place the base slab of box culverts and allow to set 24 hours before the remainder of the culvert is constructed. For sidewall heights of 1.5 m or less, the sidewalls and top slab may be placed in one continuous operation. For sidewalls greater than 1.5 m but less than 5 m in height, allow sidewall concrete to set at least 30 minutes before placing concrete in the top slab. For sidewalls 5 m or higher, allow sidewall concrete to set at least 12 hours before placing concrete in the top slab.

(6) Precast Elements. Place and consolidate concrete so that shrinkage cracks are not produced in the member.

(c) Placing Methods. Use equipment of sufficient capacity, and ensure that it is designed and operated to prevent mix segregation and mortar loss. Do not use equipment that causes vibrations that could damage the freshly placed concrete. Do not use equipment with aluminum parts that come in contact with the concrete. Remove set or dried mortar from inside surfaces of placing equipment.

Place concrete as near as possible to its final position. Do not place concrete in horizontal layers greater than 0.5 m thick. Do not exceed the vibrator capacity to consolidate and merge the new layer with the previous layer. Do not place concrete at a rate that, when corrected for temperature, exceeds the design loading of the forms.

Do not drop unconfined concrete more than 2 m. Concrete may be confined by using a tube fitted with a hopper head or other approved device that prevents mix segregation and mortar spattering. This does not apply to cast-in-place piling when concrete placement is completed before initial set occurs in the concrete placed first.

In thin sections where there is not sufficient space inside the form to place by chute, place concrete through form windows.

Operate concrete pumps so that a continuous stream of concrete without air pockets is delivered at the tube discharge. Do not use conveyor belt systems longer than 170 m when measured from end to end of the total belt assembly. Arrange the belt assembly so that each section discharges into a vertical hopper to the next section without mortar adhering to the belt. Use a hopper, chute, and deflectors at the discharge end of the conveyor belt system to cause the concrete to drop vertically.

Arrange the equipment so that no vibrations result that might damage freshly placed concrete.

(d) Consolidation. Provide sufficient hand-held internal concrete vibrators suitable for the conditions of concrete placement. Ensure that the vibrators meet requirements shown in table 552-5. Provide rubber-coated vibrators when epoxy-coated reinforcement is used.

Table 552-5.—Hand-held vibratory requirements.

Head Diameter (mm)	Frequency (vibrations/min)	Radius of Action (mm)
19–38	10,000–15,000	75–125
32–64	9,000–13,500	125–255
50–89	8,000–12,000	180–485

Provide a sufficient number of vibrators to consolidate each batch as it is placed. Provide a spare vibrator at the site in case of breakdown. Use external form vibrators only when the forms have been designed for external vibration and when internal vibration is not possible.

Consolidate all concrete by mechanical vibration immediately after placement. Manipulate vibrators to thoroughly work the concrete around reinforcement, embedded fixtures, corners, and angles in the forms. Do not cause segregation. Do not consolidate concrete placed underwater. Supplement vibration with spading, as necessary, to provide smooth surfaces and dense concrete along form surfaces, in corners, and at locations impossible to reach with the vibrators.

Vibrate the concrete at the point of deposit and at uniformly spaced points not farther apart than one and one-half times the radius over which the vibration is visibly effective. Insert vibrators so that the affected vibrated areas overlap. Do not use vibrators to move concrete. Insert vibrators vertically, and slowly withdraw them from the concrete. Vibrate long and intensely enough to thoroughly consolidate the concrete, but not to cause segregation. Do not vibrate at any one point long enough to cause localized areas of grout to form. Do not vibrate reinforcement.

(e) ***Underwater Placement.*** Underwater placement of concrete is permitted only for seal concrete and drilled shafts. Perform underwater placement only in the presence of the CO. If other than seal concrete is used, increase the minimum cement content by 10 percent. Use tremies, concrete pumps, or other approved methods for placement. Do not place concrete in running water.

(1) Tremies. Use watertight tremies with a diameter of 250 mm or more. Fit the top with a hopper. Use multiple tremies as required. Make tremies capable of being rapidly lowered to retard or stop the flow of concrete and to permit free movement of the discharge end over the entire surface of the cement.

At the start of concrete placement, seal the discharge end and fill the tremie tube with concrete. Keep the tremie tube full of concrete to the bottom during placement, and keep the discharge end completely submerged in the concrete at all times. If water enters the tube, withdraw the tremie and reseal the discharge end. Maintain continuous concrete flow until the placement is completed.

(2) Concrete Pumps. Use pumps with a device at the end of the discharge tube to seal out water while the tube is first being filled with concrete. When concrete flow is started, keep the end of the discharge tube full of concrete and below the surface of the deposited concrete until placement has been completed.

Place underwater concrete continuously from start to finish in a compact mass. Place each succeeding layer of concrete before the preceding layer has taken initial set. Use more than one tremie or pump as necessary to ensure compliance with this requirement. Keep the concrete surface as horizontal as practicable. Do not disturb after placement. Maintain still water at the point of deposit.

Dewater after test specimens cured under similar conditions indicate that the concrete has sufficient strength to resist the expected loads. Remove all laitance or other unsatisfactory material from the exposed concrete.

(f) Concrete Railings & Parapets. Use smooth, tight-fitting, rigid forms. Neatly miter corners. Place concrete railings and parapets after the centering or falsework for the supporting span is released. Remove forms without damaging the concrete. Finish all corners to be true, clean-cut, and free from cracks, spalls, or other defects.

Cast precast railing members in mortar-tight forms. Remove precast members from molds as soon as the concrete has sufficient strength to be self-supporting. Protect edges and corners from chipping, cracking, and other damage. Cure in accordance with Subsection 552.17(b). The curing period may be shortened, as approved, by using moist heat and/or type III cement or water-reducing agents.

552.14 Construction Joints. Provide construction joints at locations as SHOWN ON THE DRAWINGS. Written approval is required for any additional construction joints.

At horizontal construction joints, place gage strips inside the forms along all exposed faces to produce straight joint lines. Clean and saturate construction joints before placing fresh concrete. Keep joints saturated until adjacent fresh concrete is placed. Immediately before placing new concrete, draw forms tight against previously placed concrete. Where accessible, thoroughly coat the existing surface with a very thin coating of cement mortar. Extend reinforcing bars across construction joints.

552.15 Expansion & Contraction Joints. Form expansion and contraction joints as follows:

(a) Open Joints. Form open joints with a wooden strip, metal plate, or other approved material. Remove the joint-forming material without chipping or breaking the corners of the concrete. Do not extend reinforcement across an open joint.

(b) Filled Joints. Cut premolded expansion joint filler to the shape and size of the surface being jointed. Secure the joint filler on one surface of the joint using galvanized nails or other acceptable means so that it will not be displaced by the concrete. Splice in accordance with the manufacturer's recommendations. After form removal, neatly cut and remove all concrete or mortar that has sealed across the joint. Fill all joint gaps 3 mm or wider with hot asphalt or other approved filler. Place all necessary dowels, load transfer devices, and other devices as SHOWN ON THE DRAWINGS or as directed.

(c) Steel Joints. Fabricate plates, angles, or other structural shapes accurately to conform to the concrete surface. Set joint opening to conform to the ambient temperature at the time of concrete placement and as SHOWN ON THE DRAWINGS. Securely fasten the joints to keep them in correct position. Maintain an unobstructed joint opening during concrete placement.

(d) Water Stops. Construct water stops in accordance with Section 712 and as SHOWN ON THE DRAWINGS.

(e) Compression Joint Seals. Use one-piece compression joint seals for transverse joints and the longest practicable length for longitudinal joints. Clean and dry joints and remove spalls and irregularities. Apply a lubricant-adhesive as a covering film to both sides of the seal immediately before installation. Compress the seal and place it in the joint as recommended by the manufacturer. Make sure the seal is in full contact with the joint walls throughout its length.

Remove and discard all seals that are twisted, curled, nicked, or improperly formed. Remove and reinstall joint seals that elongate more than 5 percent of their original length when compressed. Remove all excess lubricant-adhesive before it dries.

(f) Elastomeric Expansion Joint Seal. Install the joint in accordance with the manufacturer's recommendations and the SPECIAL PROJECT SPECIFICATIONS, or as SHOWN ON THE DRAWINGS.

552.16 Finishing Plastic Concrete. Strike off concrete surfaces that are not placed against forms. Float finish the concrete surface. Remove any laitance or thin grout. Carefully tool all nonchamfered edges with an edger. Leave edges of joint filler exposed.

Protect the surface from rain damage.

Finish all concrete surfaces used by traffic to a skid-resistant surface.

(a) Striking Off & Floating. For bridge decks or top slabs of structures serving as finished pavements, use an approved power-driven finishing machine equipped with oscillating screed. If approved, use hand-finishing methods for irregular areas where the use of a machine is impractical.

Limit placement of concrete to that which can be properly finished before the beginning of initial set. Never allow concrete to be placed more than 2.5 m ahead of the finishing machine.

Strike off all surfaces using equipment supported by and traveling on screed rails or headers. Do not support rails within the limits of the concrete placement without approval.

Set rails or headers that can be readily adjusted for elevation on nonyielding supports so the finishing equipment operates without interruption over the entire surface being finished. Extend rails beyond both ends of the scheduled concrete placement a sufficient distance to enable the finishing machine to finish the concrete being placed.

Set rails the entire length of steel girder superstructures.

Use rails or headers that are of a type that can be installed so that no springing or deflection will occur under the weight of the finishing equipment.

Adjust rails, headers, and strike-off equipment to the required profile and cross section, allowing for anticipated settlement, camber, and deflection of falsework.

Before beginning delivery and placement of concrete, operate the finishing machine over the entire area to be finished to check for excessive rail deflections, proper deck thickness, and reinforcing steel cover, and to verify proper operation of equipment. Make necessary corrections before concrete placement begins. Obtain approval to begin deck concrete placement.

Schedule delivery and placement of concrete so as to permit all placement and finishing operations to be completed during daylight hours, unless otherwise approved in advance by the CO.

Place concrete bridge decks continuously along the full length of the structure or superstructure unit, unless otherwise SHOWN ON THE DRAWINGS or approved in writing by the CO. Provide sufficient material, equipment, and manpower to complete bridge deck placement at a minimum rate of 6 m per hour, unless otherwise SHOWN ON THE DRAWINGS.

Should settlement or other unanticipated events occur that would prevent obtaining a bridge deck meeting the requirements of this specification, discontinue placing deck concrete until corrective measures are taken. If satisfactory measures are not taken prior to initial set of the concrete in the affected area, discontinue all placing of concrete and install a bulkhead at a location approved by the CO. Remove all concrete in place beyond the bulkhead.

After placing the concrete, operate the finishing machine over the concrete as needed to obtain the required profile and cross section. Keep a slight roll of excess concrete in front of the cutting edge of the screed at all times. Maintain this excess of concrete to the end of the pour or form and then remove and waste it. Adjust rails or headers as necessary to correct for unanticipated settlement or deflection.

Remove rail supports embedded in the concrete to at least 50 mm below the finished surface, and fill and finish any voids with fresh concrete. Finish the surface with a float, roller, or other approved device as necessary to remove all local irregularities.

Remove all excess water, laitance, or foreign material brought to the surface using a squeegee or straightedge drawn from the center of the slab towards either edge. Do not apply water to the surface of the concrete during finishing operations.

Following the completion of the strike-off, float the roadway slab surface to a smooth, uniform surface by means of floats 3 m or more in length. Use floats to remove roughness and minor irregularities left by the strike board or finishing machine and to seal the concrete surface. Do not permit excessive working of the concrete surface. Ensure that each transverse pass of the float overlaps the previous pass by a distance equal to at least one-half the length of the float.

Operate hand-operated float boards from transverse finishing bridges. Provide finishing bridges that completely span the roadway area being floated. Provide a sufficient number of finishing bridges to permit operation of the floats without undue delay and to permit inspection of the work. Use at least two transverse finishing brid-ges when hand-operated float boards are used, unless otherwise approved bythe CO.

Provide finishing bridges that are of rigid construction, free of wobble and spring when used by the operators of longitudinal floats, and easily moved.

(b) Straightedging. Check all slab and sidewalk surfaces in the presence of the CO. Check the entire surface parallel to the centerline of the bridge with a 3-m metal straightedge. Overlap the straightedge at least one-half the length of the previous straightedge placement.

Correct deviations in excess of 3 mm from the testing edge of the straightedge. For deck surfaces that are to receive an overlay, correct deviations in excess of 6 mm.

(c) Texturing. Produce a skid-resistant surface texture on all driving surfaces by grooving. Use grooved, sidewalk, and troweled and brushed finishes, or a combination thereof, for other surfaces as required.

(1) Grooved Finish. Use a float with a single row of fins or an approved machine designed specifically for sawing grooves in concrete pavements. Space fins 13 to 20 mm on centers. Make the grooves 2 to 5 mm wide and 3 to 5 mm deep. Groove perpendicular to the centerline without tearing the concrete surface or loosening surface aggregate.

If grooves are sawn, cut the grooves 5 mm wide at a spacing of 15 to 25 mm.

On bridge decks, discontinue grooving 300 mm from the curb face and provide a longitudinal troweled finish on the surface of gutters.

(2) Sidewalk Finish. Strike off the surface using a strike board, and then float the surface. Use an edging tool on edges and expansion joints. Broom the surface using a broom with stiff bristles. Broom perpendicular to the centerline from edge to edge, with adjacent strokes slightly overlapped. Produce regular corrugations not more than

3 mm in depth without tearing the concrete. While the concrete is plastic, correct porous spots, irregularities, depressions, small pockets, and rough spots. Groove contraction joints at the required interval using an approved grooving tool.

(3) Troweled & Brushed Finish. Use a steel trowel to produce a slick, smooth surface free of bleedwater. Brush the surface with a fine brush using parallel strokes.

(4) Exposed Aggregate Finish. Strike off the surface using a strike board and then float the surface. Use an edging tool on all transverse and longitudinal joints that are against forms or existing pavement. Do not edge transverse joints in a continuous pour or longitudinal joints in a continuous dual-lane pour.

As soon as the concrete hardens sufficiently to prevent particles of gravel from being dislodged, broom the surface. Use stiff brushes approved by the CO. Exercise care to prevent marring of the surface and cracking or chipping of slab edges or joints. If approved by the CO, apply a light spray of retardant to the unfinished surface to facilitate this work.

First, broom transversely across the pavement. Pull the loosened semistiff mortar entirely off the pavment. Remove the mortar from all adjacent pavements. Then broom parallel to the pavement centerline. Continue this operation until a sufficient amount of coarse aggregate is exposed. Other methods of aggregate exposure, such as using a water spray attachment on a special exposed aggregate broom, will be permitted if satisfactory results are demonstrated.

After curing according to Subsection 501.10, wash the surface with brush and water to remove all laitance and cement from the exposed coarse aggregate.

(d) Surface Underneath Bearings. Finish all bearing surfaces to within 5 mm of plan elevation. When a masonry plate is to be placed directly on the concrete or on filler material less than 5 mm thick, finish the surface with a float to an elevation slightly above plan elevation. After the concrete is set, grind the surface as necessary to provide a full and even bearing.

When a masonry plate is to be set on filler material between 5 and 15 mm thick, finish the surface with a steel trowel. Finish or grind the surface so that it does not vary from a straightedge in any direction by more than 2 mm.

When a masonry plate is to be set on filler material greater than 15 mm thick or when an elastomeric bearing pad is to be used, finish the surface to a plane surface free of ridges.

When required under a masonry plate or elastomeric bearing pad, use mortar in the proportions of 1 part Portland cement and 1-1/2 parts clean sand. Thoroughly mix sand and cement before adding water. Mix only enough mortar for immediate use. Discard mortar that is more than 45 minutes old. Do not retemper mortar. Cure mortar at least 3 days, and do not apply loads to mortar for at least 48 hours. Do not mix and use mortar during freezing conditions. Ensure that mortar sand conforms to AASHTO M 45. Proprietary products may be used with approval.

*(e) **Surface Underneath Waterproofing Membrane Deck Seal.*** Ensure that surfaces that are to be covered with a waterproofing membrane deck seal are not coarse textured, but rather finished to a smooth surface that is free of ridges and other projections.

552.17 Curing Concrete. Begin curing immediately after the free surface water has evaporated and the finishing is complete. If the surface of the concrete begins to dry before the selected cure method can be implemented, keep concrete surface moist using a fog spray without damaging the surface. Unless otherwise approved by the CO, provide fogging equipment for all deck placement operations.

Use fogging equipment capable of applying water to the concrete in the form of a fine mist in sufficient quantity to curb the effects of rapid evaporation of mixing water from the concrete on the deck. Obtain approval by the CO in advance for fogging noz-zles and water supply methods. Produce a true mist that will not harm the surface finish of fresh concrete. Apply the mist at the times and in the manner approved by the CO.

Keep surfaces to be rubbed moist after forms are removed. Cure immediately following the first rub.

Cure the top surfaces of bridge decks using the liquid membrane curing compound method, combined with either the water method or the waterproof cover method. Apply liquid membrane curing compound immediately after finishing. Apply a water method or the waterproof cover method within 4 hours after finishing.

Cure all concrete for at least 7 consecutive days. When pozzolan or fly-ash-modified cement is used, extend the required period of controlled moisture, as called for in Subsection 552.12.

*(a) **Forms-in-Place Method.*** For formed surfaces, leave the forms in place without loosening. If forms are removed during the curing period to facilitate rubbing, strip forms only from areas able to be rubbed during the same shift. During rubbing, keep the surface of the exposed concrete moist. After the rubbing is complete, continue curing process using the water method or by applying a clear curing compound (type 1 or type 1–D) for the remainder of the curing period.

*(b) **Water Method.*** Keep the concrete surface continuously wet by ponding, spraying, or covering with material that is kept continuously and thoroughly wet. Covering material may consist of cotton mats, multiple layers of burlap, or other approved material that does not discolor or otherwise damage the concrete. Do not cure concrete underwater if the temperature of the water is less than 2 °C.

Cover the wet concrete surface with a waterproof sheet material that prevents moisture loss from the concrete.

Use widest sheets practical. Lap adjacent sheets at least 150 mm and tightly seal all seams with pressure-sensitive tape, mastic, glue, or other approved methods. Secure all material so that wind will not displace it. Immediately repair sheets that are broken or damaged.

(c) Liquid Membrane Curing Compound Method. Do not use the liquid membrane method on surfaces to receive a rubbed finish. Use on construction joint surfaces is permitted only if the compound is removed from the concrete and the reinforcing steel by sandblasting before placement of concrete against the joint.

Use type 2 white-pigmented liquid membrane only on the top surfaces of bridge decks or on surfaces not exposed to view in the completed work. Use type 1 or 1–D clear curing compounds on other surfaces.

Mix membrane curing solutions containing pigments before use. Continue to agitate during application. Use equipment capable of producing a fine spray. Apply the curing compound at a minimum rate of 0.25 L/m^2 in one or two uniform applications. If the solution is applied in two applications, follow the first application with the second application within 30 minutes and apply at right angles to the first application.

If the membrane is damaged by rain or other means during the curing period, immediately apply a new coat over the damaged areas.

552.18 Finishing Formed Concrete Surfaces. Remove and replace or repair, as approved, all rock pockets or honeycombed concrete.

Provide a class 1 finish to all formed concrete surfaces, unless another finish is SHOWN ON THE DRAWINGS or called for in the SPECIAL PROJECT SPECIFICATIONS.

Finish sound formed concrete surfaces as shown below.

(a) Class 1—Ordinary Surface Finish. Finish the following surfaces with a class 1, ordinary surface finish:

(1) Undersurfaces of slab spans, box girders, filled spandrel arch spans, and the roadway deck slab between superstructure girders.

(2) Inside vertical surfaces of exterior superstructure girders and all vertical surfaces of interior girders.

(3) Surfaces to be buried and culvert surfaces above finished ground that are not visible from the traveled way or a walkway.

Begin finishing as soon as the forms are removed. Remove fins and irregular projections from all surfaces that are exposed or will be waterproofed. Remove bulges and offsets with carborundum stones or discs. Remove localized poorly bonded rock pockets or honeycombed concrete and replace with sound concrete or packed mortar in an approved manner. Cut back at least 25 mm beneath the concrete surface all projecting wire or other devices used to hold forms in place.

Clean and point all form tie cavities, holes, depressions, voids, broken corners and edges, and other defects. Saturate the area with water. Finish the area with mortar that is less than 1 hour old. After the mortar is set, rub it (if required) and continue curing. Match exposed surfaces to surrounding concrete.

Carefully tool and remove free mortar and concrete from construction and expansion joints. Leave joint filler exposed for its full length with clean, true edges.

Rub or grind bearing surfaces on piers and abutments to the specified elevation and slope.

For patching large or deep areas, add coarse aggregate to the patching material and take special precautions to ensure a dense, well-bonded, and properly cured patch. Areas of honeycomb that exceed 2 percent of the surface area of a structural element may be considered sufficient cause for rejection of the structural element.

Cure mortar patches in accordance with Subsection 552.17.

If the final finished surface is not true and uniform, rub it in accordance with Subsection 552.18(b).

(b) Class 2—Rubbed Finish. Finish the following surfaces with a class 2 rubbed finish:

(1) All surfaces of bridge superstructures, except those surfaces designated to receive a class 1 or other finish.

(2) All surfaces of bridge piers, piles, columns and abutments, and retaining walls above finished ground and to at least 300 mm below finished ground.

(3) All surfaces of open spandrel arch rings, spandrel columns, and abutment towers.

(4) All surfaces of pedestrian undercrossings, except floors and surfaces to be covered with earth.

(5) Surfaces above finished ground of culvert headwalls and endwalls when visible from the traveled way or walkway.

(6) Inside surfaces of culvert barrels higher than 1 m that are visible from the traveled way. Finish for a distance inside the barrel at least equal to the height of the culvert.

(7) All surfaces of railings.

Complete a class 1 finish in accordance with Subsection 552.18(a). Saturate the concrete surface with water. Rub the surface with a medium-coarse carborundum stone using a small amount of mortar on its face. Use mortar composed of cement and fine sand mixed in the same proportions as the concrete being finished. Continue rubbing until form marks, projections, and irregularities are removed and a uniform surface is obtained. Leave the paste produced by this rubbing in place.

After other work that could affect the surface is completed, rub with a fine carborundum stone and water until the entire surface has a smooth texture and uniform color. After the surface has dried, rub it with burlap to remove loose powder. Leave it free from all unsound patches, paste, powder, and objectionable marks.

(c) Class 3—Tooled Finish. Let the concrete set for at least 14 days or longer if necessary to prevent the aggregate particles from being "picked" out of the surface. Use air tools such as a bush hammer, pick, or crandall. Chip away the surface mortar and break the aggregate particles to expose a grouping of broken aggregate particles in a matrix of mortar.

(d) Class 4—Sandblasted Finish. Let the concrete set for at least 14 days. Protect adjacent surfaces that are not to be sandblasted. Sandblast the surface with hard, sharp sand to produce an even fine-grained surface in which the mortar is cut away, leaving the aggregate exposed.

(e) Class 5—Wire Brushed or Scrubbed Finish. Begin as soon as the forms are removed. Scrub the surface with stiff wire or fiber brushes using a solution of muriatic acid. Mix the solution in the proportion of 1 part acid to 4 parts water. Scrub until the cement film or surface is completely removed and the aggregate particles are exposed. Leave an even pebbled texture with the appearance of fine granite to coarse conglomerate, depending upon the size and grading of aggregate. Wash the entire surface with water containing a small amount of ammonia.

(f) Class 6—Color Finish. Build a sufficient number of 0.5 × 1-m concrete color sample panels to obtain a color acceptable to the CO. Protect the approved color sample panel at all times during the work. Color all designated surfaces to match the color of the approved sample.

Complete a class 1 finish in accordance with Subsection 552.18(a). Do not apply the color finish until all concrete placement for the structure is complete. Remove all dust, foreign matter, form oil, grease, and curing compound with a 5-percent solution of trisodium phosphate, and then rinse the concrete surface with clean water.

Use paper, cloth, or other means to protect surfaces not to be color finished. Apply the finish to a dry concrete surface when the surface temperature is 4 °C or higher and the air temperature in the shade is anticipated to be 4 °C or higher during the 24 hours following application.

Apply the color finish in accordance with the manufacturer's recommendations. Spray, brush, or roll on the first coat of penetrating sealer and color base. Spray, brush, or roll on the finish coat after the first coat has thoroughly dried. Apply finish to provide a uniform, permanent color, free from runs and sags to the surfaces.

Clean concrete areas not intended to be covered by the finish using an approved method.

552.19 Concrete Anchorage Devices. Use chemical, grouted, or cast-in-place concrete anchorage devices for attaching equipment or fixtures to concrete.

Furnish the following for approval:

(a) Concrete anchorage device sample.

(b) Manufacturer's installation instructions.

(c) Material data and certifications.

Fabricate all metal parts of the anchorage devices from stainless steel or from steel protected with a corrosion-resistant metallic coating that does not react chemically with concrete. Supply anchorage devices complete with all hardware.

For chemical or grouted anchors, conduct a system approval test on one anchor at the jobsite, not to be incorporated in the work. Conduct a static load test in accordance with ASTM E 488. Demonstrate that the anchorage device will withstand a sustained direct tension test load not less than the values shown in table 552-5 for a period of at least 48 hours with movement not to exceed 1 mm. Also demonstrate that, when loaded to failure, the anchorage device demonstrates a ductile failure of the anchor steel, not a failure of the chemical, grout, or concrete.

Install concrete anchorage devices as recommended by the device manufacturer and so that the attached equipment or fixtures will bear firmly against the concrete. Torque installed nuts to the values specified in table 552-6, unless otherwise

Table 552-5.—Sustained load test values.

Anchorage Device Stud Size	Tension Test Load (kN)
M20	24.0
M16	18.3
M12	12.7
M8	7.1

Table 552-6.—Torque for anchorage devices.

Anchorage Device Stud Size	Torque (N•m)
M20	180
M16	130
M12	80
M8	30

specified in the manufacturer's instructions. Set bearing anchor bolts in accordance with the requirements specified in Section 564.

In the presence of the CO, proof load a random sample of at least 10 percent of the anchors to 90 percent of the yield stress of the steel. If any anchor fails, reset the failed anchor and proof torque the reset anchor and 100 percent of all remaining anchors. The proof load may be applied by torquing against load indicator washers, applying direct tension load to the anchor, or some other method approved by the CO. After proof loading, release the load on the anchor and retighten to the load specified in table 552-6, or in accordance with the manufacturer's instructions.

552.20 Loads on New Concrete Structures. Do not allow vehicles or construction equipment on any span until concrete in the entire superstructure has attained its design compressive strength and has been in place 21 days.

Do not place any loads on finished piers, bents, or abutments until tests on concrete cylinders cast from the same concrete and cured under the same conditions indicate that the concrete has obtained at least 80 percent of the specified minimum 28-day concrete comprehensive strength. This restriction does not apply to placement of upper lifts for substructure elements cast in stages.

For posttensioned concrete structures, do not allow vehicles weighing more than 2,000 kg on any span until the prestressing steel for that span is tensioned, grouted, and cured. Vehicles weighing less than 2,000 kg may be permitted on a span, provided the weight of the vehicle was included in the falsework design.

Permit no public traffic on the bridge until approaches, curbs, bridge rail, and object markers are completed and in place.

Erect barricades at each end of the bridge span upon completion of the deck concreting if road approaches allow vehicles to drive directly onto the structure. Locate barricades so as to physically prevent vehicular access to the bridge. Do not remove barricades until the structure is open to public traffic as approved by the CO.

Measurement

552.21 Method. Use the method of measurement that is DESIGNATED IN THE SCHEDULE OF ITEMS.

Measure structural concrete by the cubic meter or lump sum. Measure in accordance with the neat lines of the structure as SHOWN ON THE DRAWINGS, except as altered by the CO to fit field conditions. Make no deduction for the volume occupied by reinforcing steel, anchors, weep holes, piling, or pipes less than 200 mm in diameter. Do not include the volume of fillet less than 150 mm on a side or the varying thickness haunches between prefabricated girder and bridge decks.

Payment

552.22 Basis. The accepted quantities will be paid for at the contract unit price for each PAY ITEM DESIGNATED IN THE SCHEDULE OF ITEMS.

Payment will be made under:

Pay Item		Pay Unit
552 (01)	Structural concrete, class ___	Cubic Meter
552 (02)	Structural concrete, class ___	Lump Sum
552 (03)	Structural concrete, class ___, for _________ (*Description*)	Cubic Meter
552 (04)	Structural concrete, class ___, for _________ (*Description*)	Lump Sum

Section 553—Prestressed Concrete

Description

553.01 Work. Prestress precast or cast-in-place concrete by furnishing, placing, and tensioning prestressing steel. Manufacture, transport, store, and install all precast prestressed members except piling.

Furnish prestressed members complete, including all concrete, prestressing steel, bar reinforcing steel, and incidentals in connection therewith.

Materials

553.02 Requirements. Furnish material that conforms to specifications in the following sections and subsections:

Elastomeric Bearing Pads	717.10
High-Strength Nonshrink Grout	701.02
Low-Strength Grout	701.03
Mortar	701.04
Prestressing Steel	709.03
Reinforcing Steel	554
Sealants, Fillers, Seals, & Sleeves	712.01
Structural Concrete	552
Structural Steel	717.01

Use class P concrete in prestressed members unless otherwise SHOWN ON THE DRAWINGS. Design the concrete mix in accordance with Subsection 552.03 with a 28-day design compressive strength as SHOWN ON THE DRAWINGS. Do not permit lightweight concrete unless otherwise SHOWN ON THE DRAWINGS.

Construction

553.03 Method Approval. Notify the CO a minimum of 10 days prior to fabrication of any prestressed members.

Inspect all prestressed concrete members by one of the following methods, unless otherwise SHOWN ON THE DRAWINGS or in the SPECIAL PROJECT SPECIFICATIONS:

- Use the quality control engineer of a plant certified by the Prestressed Concrete Institute (PCI). Submit a copy of the transmittal letter of the latest PCI inspection with the shop drawings. Furnish a copy of all testing and inspection reports to the CO upon delivery of the members to the jobsite.

- Use an independent licensed professional engineer experienced in prestressed concrete girder inspection to certify that the prestressed members were built in accordance with the drawings and specifications. Furnish, along with the certification, a copy of all testing and inspection reports to the CO upon delivery of the members to the jobsite.

Ensure that dimensional tolerances for prestressed girders are as given in division 5, section 5 of PCI manual 116–77 ("Manual for Quality Control: Precast Prestressed Concrete Products," Prestressed Concrete Institute, Chicago, Illinois).

Perform prestressing by pretensioning methods. Submit four copies of shop drawings for prestressed members and one copy of reproducible detailed drawings of the method, material, and equipment proposed for approval at least 21 days before starting prestressing.

Show the following:

(a) Method and sequence of stressing.

(b) Complete specifications, details, and test results for the prestressing steel and anchoring devices.

(c) Anchoring stresses.

(d) Arrangement of the prestressing steel in the members to include the strand pattern at midspan and at centerspan of bearing; and the location of total strand center of gravity at midspan, at hold-down points, at quarter points, and at centerline of bearing.

(e) Tendon elongation calculations for jacking procedures to be used, to include the calibration curve for the gauge and jacking system, the stress-strain curve for the prestressed strands, and the pressure gauge readings.

(f) Number, spacing, and method of draping pretensioned strands.

(g) The prestressing bed layout and overall length between grips at fixed and jacking ends, and the type of equipment to be used.

(h) Other substantiating calculations for the prestressing method.

(i) Certification of wire or strand taken in accordance with Subsection 709.03.

(j) Concrete mix design.

Provide the signature and seal of a licensed professional engineer on all shop drawings and calculations prepared for fabrication of prestressed members.

The CO will review mix designs and approve shop drawings prior to fabrication of prestressed members.

553.04 Prestressing Steel. Use prestressing steel that is bright and free of corrosion, dirt, grease, wax, scale, rust, oil, or other foreign material that may prevent bond between the steel and the concrete. Do not use prestressing steel that has sustained physical damage or is pitted.

One approved splice per pretensioning strand is permitted if the splice is between members in the casting bed. Splice so the strands have the same "twist" or "lay."

Do not weld or ground welding equipment on forms or other steel in the member after the prestressing steel is installed.

Allow a seven-wire strand with one broken wire to remain in the member, provided it is within the following limits:

For members with:

(a) Less than 20 strands, no wire breaks permitted.

(b) 20 to 39 strands, one wire break permitted.

(c) 40 to 59 strands, two wire breaks permitted.

(d) 60 or more strands, three wire breaks permitted.

Remove and replace all strands that exceed the permissible number of wire breaks. Remove any strand that has one or more than one broken wire. Securely wrap the broken ends of any wire breaks that are permitted to remain in the member with tie wire to prevent raveling.

553.05 Concrete. Construct prestressed concrete in accordance with Section 552. Construct reinforcing steel in accordance with Section 554.

Ensure that threaded inserts develop the full tensile strength of bars or bolts they secure. Unless otherwise SHOWN ON THE DRAWINGS, provide lifting devices of adequate strength to safely lift the girders within 500 mm of the girder ends.

Straighten wires, wire groups, parallel-lay cables, and any other prestressing elements to ensure proper position in the enclosures. Provide suitable horizontal and vertical spacers, if required, to hold the wires in position.

Do not place concrete in the forms until the placement of reinforcing steel, prestressing steel, ducts, bearing plates, and other embedded material is approved. Place and vibrate concrete with care to avoid displacing the embedded material.

Rough cast the top surface of members against which concrete will be cast.

Determine the strength of precast concrete required prior to release of pretensioned strands by tests on cylinders cast and cured under conditions in which the time-temperature relationship of the cylinder will simulate as nearly as possible that obtained during the curing of the structural member. When the forms are heated by steam or hot air, place the cylinder in the lowest heat zone during the curing period. When forms are heated by some other means, provide a recording of the time-temperature relationship of the test cylinder for comparison with that of the prestressed unit.

Mold, cure, and test the cylinders in accordance with AASHTO T 126 and T 22 for 28-day test cylinders, and AASHTO T 23 for test cylinders cured with the members. When accelerated curing methods are used, allow the cylinders to cool for at least one-half hour prior to capping, and allow caps of sulfur compound to cure one-half hour before testing.

Table 553-1.—Prestressed members (from the same placements).

Number of Members/Day	Release Test Cylinders Taken[a]	Minimum Cylinders Broken (Release Test)	28-Day Strength Test Cylinders Taken and Broken[a]
1	3	2	3
2	3	1 per beam	4[b]
3	3[c]	1 per beam	6[b]
4	4[c]	1 per beam	8[b]
5	5[c]	1 per beam	10[b]
6	6[c]	1 per beam	12[b]
7	7[c]	1 per beam	14[b]
8	8[c]	1 per beam	16[b]

a. Assumes all concrete is air-entrained or nonair-entrained. If both types of concrete are used in the same member, the number of test cylinders listed shall be taken from the air-entrained concrete, and the same number of test cylinders shall be taken from the nonair-entrained concrete.
b. Two test cylinders taken from each member.
c. One test cylinder taken from each member.

As a minimum, take the numbers of test cylinders shown in table 553-1. Take more cylinders if the CO judges it necessary.

Cure the girder in a saturated atmosphere of at least 90 percent relative humidity. Cure time may be shortened by heating the outside of impervious forms with radiant heat, convection heat, conducted steam, or hot air.

Apply radiant heat by means of pipes circulating steam, hot oil, hot water, or electric heating elements. Inspect casting beds to ensure uniform heat application. Use a suitable enclosure to contain the heat. Minimize moisture loss by covering all exposed concrete surfaces with plastic sheeting or liquid membrane curing compound in accordance with Subsection 552.17. Sandblast curing compound from all surfaces to which concrete will be bonded.

When using steam, envelop the entire surface with saturated steam. Completely enclose the casting bed with a suitable type of housing, tightly constructed to prevent the escape of steam and exclude outside air. Use steam at 100 percent relative humidity. Do not apply the steam directly to the concrete.

When using hot air, the CO will approve the method to envelop and maintain the girder in a saturated atmosphere. Never allow dry heat to touch the girder surface.

For all heat curing methods:

(a) Keep all unformed girder surfaces in a saturated atmosphere throughout the curing time.

(b) Embed a thermocouple linked with a thermometer accurate to ± 3 °C 150 to 200 mm from the top or bottom of the girder on its centerline and near its midpoint.

(c) Monitor with a recording sensor (accurate to ± 3 °C) arranged and calibrated to continuously record, date, and identify concrete temperature throughout the heating cycle.

(d) Make this temperature record available to the CO.

(e) Heat concrete to no more the 38 °C during the first 2 hours after placing concrete, and then increase no more than 14 °C per hour to a maximum of 80 °C.

(f) After curing is complete, cool concrete no more than 14 °C per hour to 38 °C.

(g) Keep the temperature of the concrete above 15 °C until the girder reaches release strength.

(h) Do not expose the girders to temperatures below freezing until the specified 28-day strength has been achieved.

(i) To prevent cracking of members, detension strands and transfer their stress to the concrete immediately upon attainment of required release strengths and before the members have been allowed to dry and cool. Should this be impractical, keep the members covered and moist, and hold at a minimum temperature of 15 °C until strands are detensioned.

Cure precast pretensioned members until the concrete has attained the required release compressive strength. The average strength of two test cylinders shall be greater than the minimum required strength. Ensure that the individual strength of any one cylinder is not more than 5 percent below the required strength.

Steam-cure curbs and diaphragms cast after the prestress member has been cured for a minimum of 12 hours at 38 °C to 71 °C or moist-cured for a minimum of 3 days in accordance with Subsection 552.17.

Provide a class 2 rubbed finish to the exterior surface of the exterior girders and the bottom flanges of all girders, as specified in Subsection 552.18(b), unless otherwise SHOWN ON THE DRAWINGS. Provide a class 1 ordinary surface finish to the rest of the girders, as specified in Subsection 552.18(a).

Finish portions of prestressed members that will serve as bridge decks, as provided in Subsection 552.16(c)(1) or (2), as appropriate, or as SHOWN ON THE DRAWINGS.

With the approval of the CO, repair rock pockets and other minor deficiencies of a nonstructural nature in the girders. Reject any girders that are repaired without the approval of the CO, regardless of the extent of the repair work.

553.06 Tensioning. Stress strands only when an inspector (see Subsection 553.03) is present. Record the pretensioning gauge pressures and measured strand elongations, and provide a copy to the CO.

Use hydraulic jacks to tension prestressing steel. Use a pressure gage or load cell for measuring jacking force.

Calibrate measuring devices at least once every 6 months, or if they appear to be giving erratic results. Calibrate the jack and gage as a unit, with the cylinder extension in the approximate position to be at final jacking force. Keep a certified calibration chart with each gage.

If a pressure gage is used, do not gage loads less than one-quarter or more than three-quarters of the total graduated capacity of the gage, unless calibration data clearly establishes consistent accuracy over a wider range. Use a pressure gage with an accurate reading dial at least 150 mm in diameter.

Measure the force induced in the prestressing steel using jacking gages, and take elongation measurements of the prestressing steel. If there is a discrepancy of more than 7 percent between the jacking force and the expected elongation, check the entire operation, determine the reasons for the discrepancy, and correct before proceeding. Recalibrate jacking gages if their readings do not agree within 5 percent of each other. If the jacking system is equipped with an automatic release valve that closes when the required prestressing force is reached, strand elongation measurements are only required for the first and last tendon tensioned and for at least 10 percent of the remaining tendons.

If a load cell is used, do not use the lower 10 percent of the manufacturer's rated capacity of the load cell to determine the jacking force.

Do not exceed a temporary tensile stress in prestressing steel of 80 percent of the specified minimum ultimate tensile strength of the prestressing steel. Anchor prestressing steel at an initial stress that will result in the retention of a working stress after all losses of not less than those required.

For pretensioned members, do not allow the initial release stress after seating and before other losses to exceed 70 percent of the specified minimum ultimate tensile strength of the prestressing steel for stress-relieved strands, and 75 percent for low-relaxation strands. For posttensioned members, do not allow the initial release stress after seating to exceed 70 percent of the specified minimum ultimate tensile strength of the prestressing steel.

553.07 Pretensioned Members. Cast pretensioned members to the tolerances shown in table 553-2.

Cast pretensioned members in commercial prestressing plants that are PCI-Certified Plants in Product Group B—Bridges, category B3 (Prestressed Straight Strand Bridge Members) or category B4 (Prestressed Draped Strand Bridge Members), as applicable to the members to be manufactured.

(a) Prestressing Steel. Protect prestressing steel placed in the stressing bed from contamination and corrosion if the stressing bed will be exposed to weather for more than 36 hours before encasement in concrete.

Table 553-2.—Prestressed concrete member tolerances.

Description	Tolerance
Precast girders with cast-in-place decks:[a]	
Length	± 10 mm/10 m, ± 25 mm max.
Width (overall)	+ 10 mm, – 5 mm
Depth (overall)	+ 15 mm, – 5 mm
Depth (flanges)	– 5 mm
Width (web)	+ 10 mm, – 5 mm
Sweep[b]	3 mm/3 m
Variation from end squareness or skew	± 15 mm/m, ± 25 mm max.
Camber variation from design camber	± 3 mm/3 m ± 15 mm, max. > 25 m length ± 25 mm, max. > 25 m length
Position of strands:	
Individual	± 5 mm—bundled
Bundled	± 15 mm
Position from design location of deflection points for deflected strands	± 500 mm
Position of plates other than bearing plates	± 25 mm
Position of bearing plates	± 15 mm
Tipping and flushness of plates	± 5 mm
Tipping and flushness of bearing plates	± 5 mm
Position of inserts for structural connections	± 15 mm
Position of handling devices:	
Parallel to length	± 150 mm
Transverse to length	± 25 mm
Position of stirrups:	
Longitudinal spacing	± 50 mm
Projection above top	± 20 mm
Local smoothness[c]	± 6 mm in 3 m, any surface

a. AASHTO I Beams and Bulb Tee Girders.
b. Variation from straight line parallel to centerline of member.
c. Does not apply to top surface left rough to receive a topping or to visually concealed surfaces.

Free all strands of kinks or twists. Accurately hold prestressing steel in position, and tension in accordance with Subsection 553.06. Do not allow strands to unwind more than one turn. Keep a record of the jacking force and elongation measurements after the strands are tensioned to 20 percent of final jacking force.

Tension prestressing steel to the required stress. Include in elongation computations strand anchorage slippage, splice slippage, horizontal movement of abutments, and prestressing steel temperature changes between the time of tensioning and the time when the concrete takes its initial set.

Table 553-2.—Prestressed concrete member tolerances (cont.).

Description	Tolerance
Precast girders used in multibeam decks:[d]	
Length	± 20 mm
Width (overall)	± 5 mm
Depth (overall)	± 5 mm
Depth (top flange)	± 15 mm
Depth (bottom flange)	+ 15 mm, – 5 mm
Width (web)	± 10 mm
Sweep[e]–	
Up to 12-m member length	± 5 mm
12- to 18-m member length	± 10 mm
Greater than 18-m member length	± 15 mm
Variation from end squareness or skew:	± 10 mm/m
Horizontal	± 15 mm max.
Vertical	± 15 mm
Camber variation from design camber	± 3 mm/3 m, ± 15 mm max.
Differential camber between adjacent members of the same design	6 mm/3 m, 20 mm max.
Position of strands:	
Individual	± 5 mm
Bundled	± 5 mm
Position from design location of deflection points for deflected strands	500 mm
Position of plates other than bearing plates	± 25 mm
Tipping and flushness of plates	± 5 mm
Position of inserts for structural connections	± 15 mm
Position of handling devices:	
Parallel to length	± 150 mm
Transverse to length	± 25 mm

d. Box beams, slabs, decked bulb tee, and multistem girders.
e. Variation from straight line parallel to centerline of member.

Maintain the prestress bed forms, strands, and reinforcement bar temperature within 14 °C of the temperature of the concrete to be placed in the forms. Support strands with rollers at points of direction change when strands are tensioned in a draped position. Use free-running rollers with minimal friction. Initially, when strands are tensioned and then pulled into the draped position, tension to no more than the required tension minus the increased tension due to forcing the strand to a draped profile. If the load in a draped strand at the dead end, as determined by elongation measurements, is less than 95 percent of the jack load, tension the strand from both ends of the bed. Make the load, as computed from the sum of elongations produced by jacking at both ends, agree within 5 percent of the jack load.

Table 553-2.—Prestressed concrete member tolerances (cont.).

Description	Tolerance
Precast girders used in multibeam decks, position of stirrups:	
Longitudinal spacing	± 25 mm
Projection above top	+ 5 mm, − 20 mm
Tipping of beam seat bearing area	± 5 mm
Position of dowel tubes	± 15 mm
Position of tie rod tubes:	
Parallel to length	± 15 mm
Vertical	± 10 mm
Position of slab void:	
End of void to center of tie hole	± 15 mm
Adjacent to end block	± 25 mm
Local smoothness[f]	± 5 mm in 3 m, any surface
Posttension members:	
Position of posttensioning ducts	± 5 mm
Position of tendon anchorage bearing plates	± 5 mm

f. Does not apply to top surface left rough to receive a topping or to visually concealed surfaces.

Within 3 hours before placing concrete, check the tension on all prestressing steel strands. The method and equipment for checking the loss of prestress shall be subject to approval by the CO. If strands are tensioned individually, check each strand for loss of prestress. Retension to the original computed jacking stress all strands that show a loss of prestress in excess of 3 percent. If strands are tensioned in a group, check the entire group for loss of prestress. Release and retension the entire group if the total prestress shows a loss in excess of 3 percent, or if any individual strand appears significantly different from the rest of the strands in the group.

(b) Releasing Steel. Release the prestress load to the concrete after the concrete has attained its required release compressive strength. Do not expose the concrete to temperatures below freezing for at least 7 days after casting. Cut or release strands such that lateral eccentricity of the prestress force will be minimized. Cut prestress steel off flush with the end of the member unless otherwise SHOWN ON THE DRAWINGS.

553.08 Storing, Transporting, & Erecting. Do not ship prestressed concrete members until concrete cylinder tests manufactured of the same concrete and cured under the same conditions as the members indicate that the concrete in each member has attained the minimum required design strength and is at least 14 days old.

Store, transport, and erect precast and prestressed girders, slab units, and box units in the upright position with the points of support and directions of the reactions, with respect to the member, approximately the same as when the member is in its final position, unless otherwise shown on approved shop drawings. Prevent cracking or damage during storage, hoisting, and handling of the precast units. Replace units damaged by improper storage or handling.

Store, transfer, and erect precast prestressed concrete piling in accordance with the requirements for precast concrete piling specified in Section 551. Place other precast prestressed structural members in the structure as SHOWN ON THE DRAWINGS and in accordance with the SPECIAL PROJECT SPECIFICATIONS.

553.09 Erecting & Placement of Multibeam Members. Advise the CO a minimum of 48 hours before prestressed girders for multibeam bridges are to be field welded, and before any field grout or mortar is to be placed.

Adjust, if necessary, multibeam girders by using galvanized steel shims the same length and width as the bearing pad or plate. Allow no more than 5 mm vertical difference between top of adjacent beam edges at each end of the span. When an asphalt wearing surface or cast-in-place deck is to be placed on top of the prestressed beams, allow a vertical tolerance of only 15 mm. Do not load beams to make them assume the same camber as an adjacent beam.

Perform abrasive blasting on the keyway surfaces of all multibeam prestressed concrete members to provide a new and clean concrete surface that is free of carbonated concrete and other contaminants, and to expose parts of the large aggregate beneath the concrete paste.

Use high-pressure water blasting (20.7 kPa or more) to remove all debris and loosened paste in the keyways immediately prior to placing mortar. Remove all free-standing water and allow keyways to completely surface dry. Test for the presence of carbonated concrete when directed by the CO or called for by the SPECIAL PROJECT SPECIFICATIONS. Repeat abrasive blasting and water washing as needed if tests indicate the presence of carbonated concrete.

Use mortar in keyways between multibeam members and to patch defects, blockouts, or other areas on the concrete roadway portion of the structure 25 mm or more in depth and over 25 mm in width. Patch smaller areas on the concrete roadway with grout.

Maintain air and concrete keyway temperatures between 7 °C and 30 °C before placing mortar. Maintain the temperature within these limits until mortar placement and application of curing method is completed.

Use grout on all anchor bolts and dowels to make all repairs.

Require air and concrete temperatures for grout placement to be the same as required for mortar. Thoroughly saturate the areas to be grouted with water and remove all free-standing water just prior to grout placement.

Strike off exposed grout surfaces flush with the same surface texture finish as the surrounding concrete as soon as the grout has set sufficiently. Cure the exposed surface as specified in Subsection 552.17. When artificial means are used to control the curing temperature of the mortar or grout, as during hot or cold weather, the CO will approve the method in advance. Use combustion heaters only if fully vented outside their enclosure. Store all dry mortar materials and mixing and placing equipment such that their temperature is above freezing. Warm mixing water to provide mortar or grout at desired temperature, but ensure that it is at 30 °C or less when mixed with the dry materials. Use ice as part of the mixing water provided it is completely melted prior to the introduction of the water to the dry materials.

Ensure that patching mortar and grout are the same color as the parent concrete.

Ensure that all field welding meets the requirements specified in Section 555. When welding or burning on precast members, attach the ground lead directly to the base metal; reject any precast prestressed member used as a conductor for the ground, and replace the member without compensation.

553.10 Painting Steel. Use a wire brush or abrasive blast to remove all dirt and residue not firmly bonded to the metal or concrete surfaces. Clean and paint the exposed ends of the prestress steel, posttension anchor head assemblies, and a 25-mm strip of adjoining concrete.

Mix zinc-rich paint conforming to Federal Specifications and Standards (FSS) TT–P–641. Work the paint into all voids in the prestressing tendons. Apply one thick coat to surfaces that will be covered with concrete. Apply two coats to surfaces not covered with concrete.

Measurement

553.11 Method. Use the method of measurement that is DESIGNATED IN THE SCHEDULE OF ITEMS.

Measure precast prestressed structural concrete members by the each or by the meter.

Measure prestressed piling under Section 551.

Payment

553.12 Basis. The accepted quantities will be paid for at the contract unit price for each PAY ITEM DESIGNATED IN THE SCHEDULE OF ITEMS.

Payment will be made under:

Pay Item		Pay Unit
553 (01)	Precast prestressed concrete structural members, ________________ *Description*	Each
553 (02)	Precast prestressed concrete structural members, ________________ *Description*	Meter

Section 553A—Precast Concrete Structures

Description

553A.01 Work. Construct precast concrete members. In addition, manufacture, test materials for, transport, store, and install all precast concrete portions except piling, and perform all necessary grouting, welding, or other connections. Furnish precast concrete members complete and in place, including all concrete reinforcing steel and incidentals connected therewith.

Materials

553A.02 Requirements. Provide materials that meet the requirements specified in the following subsections:

Elastomeric Bearing Pads	717.10
High-Strength Nonshrink Grout	701.02
Low-Strength Grout	701.03
Mortar	701.04
Reinforcing Steel	709.01
Sealants, Fillers, Seals, & Sleeves	712.01
Structural Concrete	552.02
Structural Steel	717.01

Provide precast concrete members of the size, shape, strength, air content, and finish that are SHOWN ON THE DRAWINGS.

Perform all sampling, testing, and inspection necessary to ensure quality control of the component materials and the concrete. Sample and test for quality control and acceptance testing in accordance with the AASHTO or ASTM test methods prescribed in Section 552.

Maintain adequate records of all inspections and tests. Keep records that indicate the nature and number of observations made, the number and type of deficiencies found, the quantities approved and rejected, and the nature of any corrective action taken.

Sample and test every batch (100 percent sampling and testing) for air content and slump at the start of concrete production. Random sampling and testing for air content and slump at the rate of one for every five successive batches may be substituted for 100 percent sampling and testing if the test results for three successive batches are within the specification limitations for air content or slump; but reinstate 100 percent sampling and testing if a test result for any random sample is outside the specification limitations for either air content or slump.

Make compression tests to determine the minimum strength requirements on cylinders. Make a minimum of four cylinders from each day's production, and cure them in the same manner as the precast units. Use testing methods in accordance with AASHTO T 22.

Furnish, or have the supplier furnish, a Certificate of Compliance to the CO certifying that the above materials comply with the applicable specifications. In addition, furnish to the CO a copy of all test results performed by the Contractor or supplier that are necessary to ensure compliance.

Construction

553A.03 Performance. Construct precast concrete structural members in accordance with the following sections and subsections, as applicable:

Erecting and Placement of Multibeam Members 553.09
Reinforcing Steel 554
Storing, Transporting, & Erecting 553.08
Structural Concrete 552

Submit four sets of shop drawings to the CO for approval, including the concrete mix design for each class of concrete proposed for use, a minimum of 21 days before fabrication of the precast member(s).

553A.04 Casting Yard. The precasting of concrete structural members may be done at a casting yard location selected by the Contractor.

553A.05 Handling, Transporting, & Erecting. Provide additional reinforcement, as needed, to meet the requirements of handling, transporting, and erecting precast members.

Measurement

553A.06 Method. Use the method of measurement that is DESIGNATED IN THE SCHEDULE OF ITEMS.

Each member will include the concrete, reinforcement steel, anchorages, plates, nuts, and other material contained within or attached to the unit.

Payment

553A.07 Basis. The accepted quantities will be paid for at the contract unit price for each PAY ITEM DESIGNATED IN THE SCHEDULE OF ITEMS.

Pay Item	Pay Unit
553A (01) Precast concrete member, _______________ (*Description*)	Each
553A (02) Precast concrete structure, _______________ (*Description*)	Lump Sum

Section 554—Reinforcing Steel

Description

554.01 Work. Furnish and place reinforcing steel.

Materials

554.02 Requirements. Furnish material that conforms to specifications in the following subsections:

Reinforcing Steel 709.01

Construction

554.03 Order Lists. When SHOWN ON THE DRAWINGS or in the SPECIAL PROJECT SPECIFICATIONS, submit all order lists and bending diagrams to the CO for approval. Approval does not relieve the Contractor of responsibility for the accuracy of the lists and diagrams. Do not order material until the lists and diagrams are approved.

Do not fabricate vertical reinforcement in columns, walls, piers, and shafts until footing elevations are established in the field, unless otherwise SHOWN ON THE DRAWINGS or in the SPECIAL PROJECT SPECIFICATIONS.

554.04 Identification. Ship bar reinforcement in standard bundles tagged and marked in accordance with the "Manual of Standard Practice" by the Concrete Reinforcing Steel Institute (CRSI).

554.05 Bending. Fabricate reinforcing bars in accordance with ACI SP 66. Cold bend all reinforcing bars that require bending. Limit the overall height or drop bending tolerance of deck truss bars to + 0 mm or – 6 mm. Do not bend bars partially embedded in concrete except as SHOWN ON THE DRAWINGS or otherwise permitted.

When the dimensions of hooks or the diameter of bends is not SHOWN ON THE DRAWINGS, provide standard hooks conforming to ACI SP 66.

554.06 Protection of Material. Store reinforcing steel above the ground on platforms, skids, or other supports. Protect from physical damage, rust, and other surface deterioration.

Use reinforcing steel only when the surface is clean and the minimum dimensions, cross-sectional area, and tensile properties conform to the physical requirements for the size and grade of steel specified.

Do not use reinforcing steel that is cracked, laminated, or covered with dirt, rust, loose scale, paint, grease, oil, or other deleterious material.

554.07 Epoxy-Coated Reinforcing Steel. Support coated bars on padded contact areas. Pad all bundled bands. Lift with a strong back, multiple supports, or a platform bridge. Prevent bar-to-bar abrasion. Do not drop or drag bundles.

Before placement, inspect coated bars for damage to the coating. Patch all defects in the coating that are discernible to the unaided eye with a prequalified patching/repair material, in accordance with AASHTO M 284M. Clean areas to be patched by removing all surface contaminants and damaged coating. Roughen the area to be patched before applying the patching material. Where rust is present, remove the rust by blast cleaning or power-tool cleaning immediately before applying the patching material.

Promptly treat the bar in accordance with the resin manufacturer's recommendations and before detrimental oxidation occurs. Overlap the patching material onto the original coating for 50 mm or as recommended by the manufacturer. Provide a minimum 200-µm dry film thickness on the patched areas.

Take necessary steps to minimize damage to the epoxy coating of installed bars. Clean and patch any damage to the coating noted after installation, as described above.

Field repairs will not be allowed on bars that have severely damaged coatings. Replace bars with severely damaged coatings. A severely damaged coating is defined as a coating with a total damaged area in any 0.5-m length of bar that exceeds 5 percent of the surface area of that portion of the bar. Coat mechanical splices after splice installation in accordance with AASHTO M 284M for patching damaged epoxy coatings.

554.08 Placing & Fastening. Support the bars on precast concrete blocks or metal supports in accordance with the CRSI "Manual of Standard Practice of the Concrete Reinforcing Steel Institute." Attach concrete block supports to the supported bar with 2-mm wire cast in the center of each block. Use class 1 (plastic-protected) or class 2, type B (stainless-steel-protected) metal supports in contact with exposed concrete surfaces. Use stainless steel conforming to ASTM A 493, type 430. Coat chairs, tie wires, and other devices used to support, position, or fasten epoxy-coated reinforcement with a dielectric material. Do not use plastic, wood, aluminum, brick, or rock supports.

Space slab bar supports no more than 1.2 m apart transversely or longitudinally. Do not use bar supports either directly or indirectly to support runways for concrete buggies or other similar construction loads.

Space parallel bars within 38 mm of the required location. Do not accumulate spacing variations. Ensure that the average of any two adjacent spaces does not exceed the required spacing.

Provide 50 mm clear cover for all reinforcement except as otherwise SHOWN ON THE DRAWINGS.

Place reinforcing steel in deck slabs within 6 mm of the vertical plan location. Tie bridge deck reinforcing bars together at all intersections, except where spacing is less than 300 mm in both directions, in which case alternate intersections may be tied. Check the clear cover over deck-reinforcing steel using a template before placing deck concrete. Replace damaged supports.

Tie every reinforcing-steel intersection at the outside edges of decks, in top mats of footings, and in all precast and/or prestressed concrete units.

Tie bundle bars together at intervals not exceeding 2 m. Do not bundle bars unless the location and splice details are specified.

Do not place concrete in any member until the placement of the reinforcement is approved by the CO. Concrete placed without approval may be rejected, and the Contractor may be required to remove it without compensation.

554.09 Splices. Splicing, except as SHOWN ON THE DRAWINGS, is not permitted without approval. Provide lap lengths as SHOWN ON THE DRAWINGS or in accordance with the latest edition of "Standard Specifications for Highway Bridges," published by AASHTO.

Splice reinforcing bars only where SHOWN ON THE DRAWINGS. Do not place slab bar mechanical splices adjacent to each other.

Make lapped splices by placing the reinforcing bars in contact and wiring them together with at least three ties to maintain the alignment and position of the bars.

If welding of reinforcing steel is permitted, ensure that the welds conform to AWS D 1.4. Do not weld reinforcing steel if the chemical composition of the steel exceeds the percentages shown in table 554-1.

Use welders that are currently certified. When required, test each weld using magnetic particle, radiography, or other nondestructive inspection techniques.

Table 554-1.—Reinforcing steel components.

Chemical Composition	Percent
Carbon	0.30
Manganese	1.50
Carbon equivalent	0.55

Do not tack-weld reinforcing steel.

Mechanical couplers may be used in lieu of welding, if approved. Use couplers with a strength that is at least 125 percent of the required yield strength of the reinforcing steel.

If welded wire fabric is shipped in rolls, straighten into flat sheets before placing. Splice sheets of mesh or bar mat reinforcement by overlapping not less than 1 mesh width plus 50 mm. Securely fasten at the ends and edges.

Measurement

554.10 Method. Use the method of measurement that is DESIGNATED IN THE SCHEDULE OF ITEMS.

Measure reinforcing steel by the kilogram or by the lump sum, excluding laps added for the Contractor's convenience.

Payment

554.11 Basis. The accepted quantities will be paid for at the contract unit price for each PAY ITEM DESIGNATED IN THE SCHEDULE OF ITEMS.

Payment will be made under:

Pay Item		Pay Unit
554 (01)	Reinforcing steel	Kilogram
554 (02)	Epoxy-coated reinforcing steel	Kilogram
554 (03)	Reinforcing steel	Lump Sum

Section 555—Steel Structures

Description

555.01 Work. Construct steel structures and the steel structure portions of composite structures. Furnish, fabricate, and erect structural steels, and perform incidental metal construction.

Materials

555.02 Requirements. Furnish material that conforms to specifications in the following sections and subsections:

Bearing Devices	564
Bolts & Nuts	717.01(d)
Castings	717.04
Elastomeric Compression Joint Seals	717.16
Falsework	562
Galvanized Coatings	717.07
High-Strength Bolts, Nuts, & Washers	717.01(e)
Painting	563
Pins & Rollers	717.03
Sheet Lead	717.08
Steel Forgings	717.02
Steel Grid Floors	717.09
Steel Pipe	717.06
Structural Steel	717.01
Welded Stud Shear Connectors	717.05

Construction

555.03 General. Fabricate the structural steel in a fabricating plant that is certified under the American Institute of Steel Construction (AISC) Quality Certification Program. Fabricate "fracture-critical" elements in accordance with the AASHTO "Guide Specifications for Fracture Critical Non-Redundant Steel Bridge Members."

Perform welding and weld qualification tests in accordance with the provisions of American National Standards Institute (ANSI)/AASHTO/AWS Bridge Welding Code D 1.5.

555.04 Notice of Beginning of Work. Give written notice 30 days before beginning work at the shop. Do not manufacture any material or perform any work in the shop before notification.

555.05 Inspection. Structural steel may be inspected at the fabrication site.

Ultrasonically inspect all girder flanges before fabrication, in accordance with ASTM A 578, except as follows:

(a) Inspect after the flanges are stripped from the master plate.

(b) Section 6 and 7 acceptance standards do not apply. Use supplementary requirement S2.1 for acceptance standards.

(c) Flanges may be inspected in the plant or warehouse where the flanges are stripped.

Furnish a copy of all mill orders and certified mill test reports. Show on the mill test reports the chemical analyses and physical test results for each heat of steel used in the work.

If approved, furnish production certificates in lieu of mill test reports for material that normally is not supplied with mill test reports and for items such as fills, minor gusset plates, and similar material when quantities are small and the material is taken from stock.

Include in the certified mill test reports for steels with specified impact values, in addition to other test results, the results of Charpy V-notch impact tests. When fine-grain practice is specified, confirm on the test report that the material was so produced. Furnish copies of mill orders at the time orders are placed with the manufacturer. Furnish certified mill test reports and production certificates before the start of fabrication using material covered by these reports. Furnish, from the manufacturer, a Certificate of Compliance in accordance with Subsection 106.03.

555.06 Drawings (Shop Drawings, Erection Drawings, & Transportation Drawings). Prepare and submit drawings at the times indicated herein. Approval of the drawings covers the requirements for strength and detail only. No responsibility is assumed for errors in dimensions.

(a) Shop Drawings. Submit four copies of shop drawings at least 21 days in advance of the start of fabrication to allow time for review without delaying the work. Show full detailed dimensions and sizes of component parts of the structure and details of all miscellaneous parts (such as pins, nuts, bolts, drains, weld symbols, and so forth) on shop drawings for steel structures.

Where specific orientation of plates is required show the direction of rolling of plates. Cut flanges and webs of plate girders from plates so the long dimension of the girder parallels the rolling direction.

Show the sequence of shop and field assembly and erection, and all welding sequences and procedures.

Identify on the shop drawings the type and grade of each piece.

Show on the shop drawings assembly marks that are cross-referenced to the original pieces of mill steel and their certified mill test reports.

The location of all shop-welded splices shown on the shop drawings is subject to approval. Locate all shop-welded splices to avoid points of maximum tensile or fatigue stress. Locate splices in webs at least 300 mm from shop splices, butt joints in flanges, or stiffeners. Additional nondestructive tests may be required on shop-welded splices.

(b) Erection Drawings. Submit drawings fully illustrating the proposed method of erection a minimum of 21 days before field assembly and erection. Show details of all falsework bents, bracing, guys, dead-men, lifting devices, and attachments to the bridge members. Show the sequence of erection, location of cranes and barges, crane capacities, location of lifting points, and weights of bridge members. Show complete details for all anticipated phases and conditions of erection. Calculations may be required to demonstrate that allowable stresses are not exceeded and that member capacities and final geometry will be correct. See Subsection 562.03 for additional requirements.

(c) Camber Diagram. Along with the shop drawings, furnish a camber diagram complete with substantiating calculations that show the camber at each panel point of trusses or arch ribs and at the location of field splices and fractions of span length (one-quarter points minimum) of continuous beams and girders or rigid frames. On the camber diagram, show calculated cambers to be used in preassembly of the structure, as required in Subsection 555.15.

(d) Transportation Drawings. If required, furnish transportation drawings for approval a minimum of 10 days prior to shipment.

Show all support points, tie-downs, temporary stiffening trusses or beams, and any other details needed to support and brace the member. Provide calculation sheets showing the dead load plus impact stresses induced by the loading and transportation procedure. Use impact stresses of at least 200 percent of the dead load stress. Use a total load, including impact, of not less than 300 percent of the dead load.

Ship and store all members, both straight and curved, with their webs vertical.

555.07 Storage of Material. Store structural material above the ground on platforms, skids, or other supports. Keep material free from dirt, grease, and other foreign matter, and provide appropriate protection from corrosion.

555.08 Fabrication. Provide a workmanship and finish in accordance with the best general practice in modern bridge shops. Finish neatly all portions of the work exposed to view. Perform shearing, flame cutting, and chipping carefully and accurately.

Rolled material must be straight before being laid off or worked. If straightening is necessary, use methods that will not injure the metal. Sharp kinks and bends will be cause for rejection of the material.

Heat curving of steel girders is not allowed.

(a) Identification of Steels. Use a system of assembly-marking of individual pieces and cutting instructions to the shop (generally by cross-referencing of the assembly marks shown on the shop drawings with the corresponding item covered on the mill purchase order) that maintains the identity of the original piece.

Material may be furnished from stock that can be identified by heat number and mill test report.

During fabrication, up to the point of assembling members, show clearly and legibly the specification of each piece of steel by writing the material specification on the piece or using the identification color code shown in table 555-1.

Table 555-1.—Identification color codes.

Grade	Color
345	Green and yellow
345W	Blue and yellow
485W	Blue and orange
690	Red
690W	Red and orange

For other steels not shown in table 555-1 or included in AASHTO M 160M, provide information on the color code used.

Mark for grade by steel-die stamping, or by firmly attaching a substantial tag, pieces of steel that, before assembling into members, will be subject to fabrication operations (such as blast cleaning, galvanizing, heating for forming, or painting) that might obliterate paint color code marking. Where the steel-stamping method is used, place the impressions on the thicker tension-joint member in transition joints.

The maximum allowed depth of the impression is 0.25 mm. Use a tool that will make character sizes with corresponding face radii as shown in table 555-2. Avoid impressions near edges of tensile-stressed plate members.

Table 555-2.—Size of steel die stamp markings.

Character Size	Minimum Face Radii
3 mm	0.2 mm
5 mm	0.1 mm
6 mm	0.3 mm

Use low-stress-type steel die stamps. Do not use die stamps on fracture-critical members.

If requested, furnish an affidavit certifying that the identification of steel has been maintained throughout the fabrication operation.

(b) ***Plates.*** Conform to the following:

(1) Direction of Rolling. Unless otherwise SHOWN ON THE DRAWINGS, cut and fabricate steel plates for main members and splice plates for flanges and main tension members, not secondary members, so that the primary direction of rolling is parallel to the direction of the principal tensile and/or compressive stresses.

(2) Plate Cut Edges. Conform to the following:

(a) Edge Planing. Remove sheared edges on plates thicker than 15 mm to a depth of 5 mm beyond the original sheared edge, or beyond any re-entrant cut produced by shearing. Fillet re-entrant cuts before cutting.

(1) Oxygen Cutting. Oxygen cut structural steel in accordance with ANSI/ AASHTO/AWS Bridge Welding Code D 1.5.

(2) Visual Inspection & Repair of Plate Cut Edges. Visually inspect and repair plate cut edges. Ensure that cut edges conform to ANSI/AASHTO/AWS Bridge Welding Code D 1.5.

(b) Flange Plates. Furnish flange plates with oxygen-cut edges that have the corners chamfered at least 2 mm by grinding, or furnish universal mill plates unless oxygen-cut edges are required.

(c) Web Plates. Oxygen cut to the prescribed camber web plates of built-up beams and girders, box girders, and box arches. Cut sufficient extra camber into the webs to provide for all camber losses due to welding, cutting, and so forth.

(d) Truss Members. Use oxygen cutting to prepare all longitudinal edges of all plates in welded sections of truss web and chord members. Chamfer at least 2 mm by grinding the edges of the corners of plates not joined by welding.

(e) Stiffeners & Connection Plates. Stiffeners and connection plates welded transverse to girder webs and flanges may be furnished with sheared edges, provided that the plate thickness does not exceed 20 mm. Universal mill plate may be used, provided that its thickness does not exceed 25 mm. Furnish other stiffeners and connection plates with oxygen-cut edges.

(f) Lateral Gusset Plates. Oxygen cut, parallel to lines of stress, gusset plates and other connections that are welded parallel to lines of stress in tension members where the plate thickness exceeds 10 mm. Bolted lateral gusset plates may be furnished with sheared edges, provided the thickness is less than or equal to 20 mm.

(g) Splice Plates & Gusset Plates. Furnish girder and stringer splice plates and truss gusset plates with oxygen-cut edges.

(h) Bent Plates. Furnish unwelded, load-carrying, rolled-steel plates to be bent as shown in table 555-3.

Table 555-3.—Minimum bending radii.[a]

Plate Thickness (mm)	Bending Radius[b]
$t \leq 13$	$2(t)$
$13 < t \leq 25$	$2.5(t)$
$25 < t \leq 38$	$3(t)$
$38 < t \leq 64$	$3.5(t)$
$64 < t \leq 102$	$4(t)$

a. t = plate thickness.
b. For all grades of structural steel.

Take material from the stock plates such that the bend line will be at right angles to the direction of rolling, except that cold-bent ribs for orthotropic deck bridges may be bent with bend lines in the direction of rolling.

Before bending, round the corners of the plates to a radius of 2 mm throughout the portion of the plate where the bending occurs.

(1) Cold Bending. Cold bend so that no cracking of the plate occurs. Use the minimum bend radii shown in table 555-3 measured to the concave face of the metal.

Allow for springback of grade 690 and grade 690W steels equal to about 3 times that for grade 250 steel. Use a lower die span of at least 16 times the plate thickness for break press forming.

(2) Hot Bending. If a radius shorter than the minimum specified for cold bending is essential, hot bend the plates at a temperature not greater than 650 °C, except for grades 690 and 690W. When grade 690 and grade 690W steel plates are heated to temperatures greater than 610 °C, requench and temper in accordance with the producing mill's standard practice.

(c) Fit of Stiffeners. Fabricate (mill, grind, or weld as SHOWN ON THE DRAWINGS or as specified) end-bearing stiffeners for girders and stiffeners intended as supports for concentrated loads to provide full bearing on the flanges to which they transmit load or from which they receive load. Fabricate intermediate stiffeners not intended to support concentrated loads to provide a tight fit against the compression flange.

(d) Abutting Joints. Mill or saw-cut abutting joints in compression members of trusses and columns to give a square joint and uniform bearing. The maximum allowed opening at other joints, not required to be faced, is 10 mm.

(e) Facing of Bearing Surfaces. Finish bearing and base plates and other bearing surfaces that will come in contact with each other or with concrete to the ANSI surface roughness defined in ANSI B46.1, "Surface Roughness, Waviness and Lay, Part I," as shown in table 555-4.

Table 555-4.—ANSI surface roughness values.

Bearing Surface	Surface Roughness Value (μm)
Steel slabs	50
Heavy plates in contact in shoes to be welded	25
Milled ends of compression members, milled or ground ends of stiffeners and fillers	13
Bridge rollers and rockers	6
Pins and pin holes	3
Sliding bearings	3

Machine sliding bearings that have a surface roughness greater than 2 μm according to ANSI, so the lay of the cut is parallel to the direction of movement.

Fabricate parts in bearing to provide a uniform, even contact with the adjacent bearing surface when assembled. Limit the maximum gap between bearing surfaces to 1 mm. Base and sole plates that are plane and true and have a surface roughness not exceeding the above-tabulated values need not be machined, except machine sliding surfaces of base plates.

Do not machine surfaces of fabricated members until all fabrication on that particular assembly or subassembly is complete. Machine metal components that are to be heat-treated after heat treatment.

*(f) **Straightening Material.*** If approved, straighten plates, angles, other shapes, and built-up members by methods that will not produce fracture or other damage to the metal. Straighten distorted members by mechanical means or, if approved, by carefully planned procedures and supervised application of a limited amount of localized heat. Use rigidly controlled procedures and do not exceed the temperatures specified in table 555-5 when heat straightening grades 485W, 690, and 690W steel members.

Table 555-5.—Heat-straightening temperatures.

Material To Be Straightened	Maximum Temperature
Grade 485W > 150 mm from weld	580 °C
Grade 485W < 150 mm from weld	480 °C
Grade 690 or 690W > 150 mm from weld	605 °C
Grade 690 or 690W < 150 mm from weld	510 °C

In all other steels, do not exceed 650 °C in the heated area. Control the application by temperature-indicating crayons, liquids, or bimetal thermometers.

Keep parts to be heat-straightened substantially free of external forces and stress, except stresses resulting from mechanical means used in conjunction with the application of heat.

Evidence of fracture following straightening of a bend or buckle will be cause for rejection of the damaged piece.

555.09 Annealing & Stress Relieving. Machine, finish bore, and straighten annealed or normalized structural members subsequent to heat treatment. Normalize and anneal (full annealing) in accordance with ASTM A 919. Maintain uniform temperatures throughout the furnace during the heating and cooling so that the temperature at any two points on the member does not differ by more than 60 °C at any one time.

Do not anneal or normalize members of grades 690/690W or 485W steels. Stress relieve these grades only with approval.

Record each furnace charge, identify the pieces in the charge, and show the temperatures and schedule actually used. Provide proper instruments, including recording pyrometers, for determining at any time the temperatures of members in the furnace. Make records of the treatment operation available for approval. The maximum allowed holding temperature for stress relieving grades 690/690W and 485W steels is 605 °C and 580 °C, respectively.

Stress relieve members (such as bridge shoes, pedestals, or other parts that are built up by welding sections of plate together) in accordance with subsection 4.4 of ANSI/ AASHTO/AWS Bridge Welding Code D 1.5.

555.10 Bolt Holes. Punch or drill all bolt holes. Material forming the parts of a member that is composed of not more than five thicknesses of metal may be punched 2 mm larger than the nominal diameter of the bolts where the thickness of the material is not greater than 20 mm for structural steel, 15 mm for high-strength steel, or 15 mm for quenched and tempered alloy steel, unless subpunching and reaming is required under Subsection 555.10(h), Preparation of Field Connections.

Where there are more than five thicknesses or where any of the main material is thicker than 20 mm for structural steel, 15 mm for high-strength steel, or 15 mm for quenched and tempered alloy steel, either subdrill and ream or drill all holes full size.

If required, either subpunch or subdrill (subdrill if thickness limitation governs) 5 mm smaller and, after assembling, ream 2 mm larger or drill full size to 2 mm larger than the nominal diameter of the bolts.

*(a) **Punched Holes.*** Use a die diameter that is not more than 2 mm larger than the punch diameter. Ream holes that require enlarging to admit bolts. Cut the holes clean without leaving torn or ragged edges.

*(b) **Reamed or Drilled Holes.*** Ream or drill holes so they are cylindrical and perpendicular to the member. Where practical, direct reamers by mechanical means. Remove burrs on the outside surfaces. Ream and drill with twist drills, twist reamers, or roto-broach cutters. Assemble and securely hold together connecting parts that are being reamed or drilled and match-mark before disassembling.

*(c) **Accuracy of Holes.*** Holes not more than 1 mm larger in diameter than the true decimal equivalent of the nominal diameter of the drill or reamer are acceptable. The slightly conical hole resulting from punching operations is acceptable. Ensure that the width of slotted holes produced by flame cutting or a combination of drilling or

punching and flame cutting is no more than 1 mm greater than the nominal width. Grind flame-cut surfaces smooth.

(d) Accuracy of Hole Group Before Reaming. Accurately punch full-size, subpunched, or subdrilled holes so that after assembling (before any reaming is done) a cylindrical pin 3 mm smaller in diameter than the nominal size of the punched hole may be entered perpendicular to the face of the member, without drifting, in at least 75 percent of the contiguous holes in the same plane. Punched pieces not meeting this requirement will be rejected. Holes through which a pin 5 mm smaller in diameter than the nominal size of the punched hole cannot be inserted will be rejected.

(e) Accuracy of Hole Group After Reaming. After reaming, the maximum allowed offset of 85 percent of any contiguous group of holes through adjacent thicknesses of metal is 1 mm.

Use steel templates with hardened-steel bushings in holes accurately dimensioned from the centerlines of the connection, as inscribed on the template. Use connection centerlines when locating templates from the milled or scribed ends of members.

(f) Numerically Controlled (N/C) Drilled Field Connections. In lieu of drilling undersized holes and reaming while assembled, or drilling holes full-size while assembled, drilling or punching bolt holes full-size is allowed in unassembled pieces and/or connections, including templates for use with matching undersized and reamed holes by means of suitable N/C drilling or punching equipment.

(g) Holes for Ribbed Bolts, Turned Bolts, or Other Approved Bearing-Type Bolts. Provide finished holes with a driving fit.

(h) Preparation of Field Connections. Subpunch or subdrill and ream while assembled, or drill full-size to a steel template, holes in all field connections and field splices of main members of trusses, arches, continuous beam spans, bents, towers (each face), plate girders, and rigid frames.

Holes for field splices of rolled beam stringers continuous over floor beams or cross frames may be drilled full-size unassembled to a steel template. Holes for floor beams or cross frames may be drilled full-size unassembled to a steel template. Subpunch and ream while assembled, or drill full-size to a steel template, all holes for floor beam and stringer field end connections.

When reaming or drilling full-size field connection holes through a steel template, carefully locate and position the template and firmly bolt in place before drilling. Use exact duplicates of templates used for reaming matching members, or the opposite faces of a single member. Accurately locate templates used for connections on like parts or members so that the parts or members are duplicates and require no match-marking.

For any connection, in lieu of subpunching and reaming or subdrilling and reaming, holes drilled full-size through all thicknesses or material assembled in proper position may be used.

555.11 Pins & Rollers. Accurately fabricate pins and rollers that are straight, smooth, and free from flaws. Forge and anneal pins and rollers more than 225 mm in diameter. Pins and rollers 225 mm or less in diameter may be either forged and annealed or cold-finished carbon-steel shafting.

In pins larger than 225 mm in diameter, bore a hole not less than 50 mm in diameter full length along the pin axis after the forging has been allowed to cool to a temperature below the critical range (under suitable conditions to prevent damage by too-rapid cooling and before being annealed).

(a) Boring Pin Holes. Bore pin holes true to the specified diameter, smooth and straight, at right angles with the axis of the member and parallel with each other. Produce the final surface using a finishing cut.

Produce a pin hole diameter that does not exceed that of the pin by more than 0.5 mm for pins 125 mm or less in diameter, or by 1 mm for larger pins.

The maximum allowed variation of the outside-to-outside distance of end holes in tension members and the inside-to-inside distance of end holes in compression members is 1 mm from that specified. Bore pin holes in built-up members after the member has been assembled.

(b) Threads for Bolts and Pins. Provide threads on all bolts and pins for structural steel construction that conform to the Unified Standard Series UNC ANSI B1.1, class 2A for external threads and class 2B for internal threads; but when pin ends have a diameter of 35 mm or more, provide six threads per 25 mm.

555.12 Eyebars. Pin holes may be flame cut at least 50 mm smaller in diameter than the finished pin diameter. Securely fasten together (in the order to be placed on the pin) all eyebars that are to be placed side by side in the structure and bore at both ends while clamped. Pack and match-mark eyebars for shipment and erection. Stamp with steel stencils, so as to be visible when the bars are nested in place on the structure, all identifying marks on the edge of one head of each member after fabrication is completed. Use low-stress-type steel die stamps.

Provide eyebars, straight and free from twists, with pin holes accurately located on the centerline of the bar. Do not allow the inclination of any bar to the plane of the truss to exceed 5.25 mm/m.

Simultaneously cut the edges of eyebars that lie between the transverse centerline of their pin holes with two mechanically operated torches abreast of each other, guided by a substantial template to prevent distortion of the plates.

555.13 Assembly—Bolting. Clean surfaces of metal in contact before assembling. Assemble parts of a member. Securely pin and firmly draw together before beginning drilling, reaming, or bolting. Take assembled pieces apart, if necessary, for the removal of burrs and shavings produced by the operation. Assemble members so that they are free from twists, bends, and other deformation.

Drift during assembling only enough to bring the parts into position without enlarging holes or distorting the metal.

555.14 Welded Connections. Fabricate surfaces and edges to be welded smooth, uniform, clean, and free of defects that would adversely affect the quality of the weld. Prepare edge in accordance with ANSI/AASHTO/AWS Bridge Welding Code D 1.5.

555.15 Preassembly of Field Connections. Preassemble field connections of main members of trusses, arches, continuous beams, plate girders, bents, towers, and rigid frames before erection to verify the geometry of the completed structure or unit and to verify or prepare field splices. Present the method and details of preassembly for approval.

Use methods and details of preassembly that are consistent with the procedure shown on the approved erection camber diagrams. Assemble all girders and beams in their cambered (no-load) condition.

When members are assembled with their webs vertical, support them at intervals of 6 m, or two-tenths of the span length, whichever is less. When the webs are horizontal, the above intervals of support may be increased, provided there is no noticeable deflection between points of support.

Assemble trusses in full dead-load position, unless the design of the structure provides for the secondary stresses created by assembling the truss in the fully cambered (no-load) position. Support trusses during assembly at each panel point. Preassemble at least three contiguous panels that are accurately adjusted for line and camber. For successive assemblies, include at least one section or panel of the previous assembly (repositioned if necessary and adequately pinned to assure accurate alignment) plus two or more sections or panels added at the advancing end. For structures longer than 50 m, make each assembly not less than 50 m long, regardless of the length of individual continuous panels or sections. Assembly may start from any location in the structure and proceed in one or both directions, as long as the preceding requirements are satisfied.

(a) Bolted Connections. Where applicable, assemble major components with milled ends of compression members in full bearing and then ream subsized holes to the specified size while the connections are assembled.

(b) Check Assembly—N/C Drilling. When using N/C drilling or punching, make a check assembly for each major structural type of each project. Fabricate the check assembly of at least three contiguous shop sections or, for a truss, all members in at least three contiguous panels, but not less than the number of panels associated with three contiguous chord lengths (such as the length between field splices). Base check assemblies on the proposed order of erection, joints in bearings, special complex points, and similar considerations. Shop assemblies other than the check assemblies are not required.

If the check assembly fails in some specific manner to demonstrate that the required accuracy is being obtained, further check assemblies may be required.

Receive approval of each assembly (including camber, alignment, accuracy of holes, and fit of milled joints) before reaming is commenced or before any N/C-drilled check assembly is dismantled.

(c) Field-Welded Connections. Field-welded connections are prohibited unless specifically SHOWN ON THE DRAWINGS. Verify the fit of members (including the proper space between abutting flanges) with the preassembled segment.

(d) Match-Marking. Match-mark connecting parts preassembled in the shop to assure proper fit in the field. Provide a diagram showing such match-marks.

555.16 Connections Using Unfinished, Turned, or Ribbed Bolts. Use unfinished, turned, or ribbed bolts, where specified, that conform to ASTM A 307 for grade-A bolts. Use bolts with approved single self-locking nuts or double nuts. Use beveled washers where bearing faces have a slope of more than 1:20 with respect to a plane normal to the bolt axis.

(a) Turned Bolts. Furnish turned bolts with a body-surface ANSI roughness not exceeding 3 µm. Furnish hex-headed bolts and nuts of the nominal size specified. Carefully ream holes for turned bolts, and furnish bolts to provide for a light driving fit. Keep bolt threads entirely outside of the holes. Provide a washer under the nut.

(b) Ribbed Bolts. Use approved form of ribbed body with continuous longitudinal ribs. Provide a body diameter measured on a circle through the points of the ribs 2 mm greater than the nominal diameter specified for the bolts.

Furnish ribbed bolts with round heads conforming to ANSI B18.5. Furnish hexagonal nuts that are either recessed or have a washer of suitable thickness. Furnish ribbed bolts that have a driving fit when installed in holes. Provide sufficiently hard ribs such that the ribs do not compress, deform, or allow the bolts to turn in the holes during tightening. If the bolt twists before drawing tight, ream the hole and provide an oversized replacement bolt.

555.17 Connections Using High-Strength Bolts. Assemble structural joints using AASHTO M 164M or M 253M high-strength bolts, or equivalent fasteners, as SHOWN ON THE DRAWINGS, tightened to a high tension.

(a) Bolted Parts. Use steel material within the grip of the bolt with no compressible material such as gaskets or insulation. Fabricate bolted steel parts to fit solidly together after the bolts are tightened. Limit the maximum slope of the surfaces of parts in contact with the bolt head or nut to 1:20 with respect to a plane normal to the bolt axis.

(b) Surface Conditions. At the time of assembly clean all joint surfaces (including surfaces adjacent to the bolt head and nut) of dirt or foreign material and scale, except tight mill scale. Remove burrs that would prevent solid seating of the connected parts in the snug-tight condition.

Paint or other coatings are not permitted on the faying surfaces of slip-critical connections. All connections are considered to be slip-critical, unless otherwise SHOWN ON THE DRAWINGS. Exclude paint (including any inadvertent overspray) from areas closer than one bolt diameter, but not less than 25 mm, from the edge of any bolt hole and all areas within the bolt pattern.

(c) Installation. Install fasteners of the same lot number together. Protect fasteners from dirt and moisture. Take from protected storage only as many fasteners as are anticipated to be installed and tightened during a work shift. Return to protected storage fasteners not used at the end of the shift. Do not clean lubricant from fasteners where the lubricant is required to be present in the as-delivered condition. Clean and relubricate, before installation, fasteners for slip-critical connections that accumulate rust or dirt.

Provide a tension-measuring device (a Skidmore-Wilhelm calibrator or other acceptable bolt-tension-indicating device) at all job-sites where high-strength fasteners are being installed and tightened. Use the tension-measuring device to perform the rotational-capacity test and to confirm all of the following:

- The requirements of table 555-6 of the complete fastener assembly.
- The calibration of the wrenches, if applicable.
- The understanding and proper use of the tightening method.

For short grip bolts, direct tension indicators (DTI's) with solid plates may be used to perform this test. First check the DTI with a longer grip bolt in the Skidmore-Wilhelm calibrator. The frequency of confirmation testing, number of tests to be performed, and test procedure shall conform to Subsection 555.17(c)(3) through (5), as applicable. Confirm the accuracy of the tension-measuring device through an approved testing agency at least once per year.

Install fasteners together with washers of the size and quality specified, located as required below, in properly aligned holes and tightened using any of the methods described in Subsection 555.17(c)(3) through (6) to at least the minimum tension specified in table 555-6 after all the fasteners are tight.

Table 555-6.—Minimum fastener tension.[a]

Nominal Bolt Diameter and Tread Pitch	AASHTO M 164M (kN)	AASHTO M 253M (kN)
M16 x 2	91	114
M20 x 2.5	142	179
M22 x 2.5	176	221
M24 x 3	205	258
M27 x 3	267	334
M30 x 3.5	326	408
M36 x 4	475	595

a. Equal to 70 percent of the specified minimum tensile strength of bolts (as specified for tests of full-size ASTM A 325M and ASTM A 490M bolts), rounded to the nearest kilonewton.

If approved, tightening may be performed by turning the bolt while the nut is prevented from rotating when it is impractical to turn the nut. If impact wrenches are used, provide adequate capacity and sufficient air to tighten each bolt in approximately 10 seconds.

Do not reuse AASHTO M 253M fasteners and galvanized AASHTO M 164M fasteners. If approved, other AASHTO M 164M bolts may be reused once. Touching up or retightening previously tightened bolts that may have been loosened by the tightening of adjacent bolts will not be considered to be reuse, provided the snugging up continues from the initial position and does not require greater rotation, including the tolerance, than that specified in table 555-7.

(1) Rotational-Capacity Tests. Subject high-strength fasteners, black and galvanized, to jobsite rotational-capacity tests performed in accordance with AASHTO M 164M, subsection 8.5, and the following:

(a) After tightening to a snug-tight condition, as defined in Subsection 555.17(c)(3), tighten the fastener twice the required number of turns indicated in table 555-7, in a Skidmore-Wilhelm calibrator or equivalent tension-measuring device, without stripping or failure.

Table 555-7.[a]—Nut rotation from the snug-tight condition.[b]

Bolt Length Measured From Underside of Head to End of Bolt	Geometry of Outer Faces of Bolted Parts		
	Both Faces Normal to Bolt Axis	One Face Normal to Bolt Axis and Other Face Sloped Not More Than 1:20 (Bevel Washers Not Used)	Both Faces Sloped Not More Than 1:20 From Normal to Bolt Axis (Bevel Washers Not Used)
Up to and including 4 diameters	1/3 turn	1/2 turn	2/3 turn
Over 4 diameters, but not exceeding 8 diameters	1/2 turn	2/3 turn	5/6 turn
Over 8 diameters, but not exceeding 12 diameters[c]	2/3 turn	5/6 turn	1 turn

a. Applicable only to connections where all material within the grip of the bolt is steel.
b. Nut rotation is relative to bolt, regardless of the element (nut or bolt) being turned. The tolerance is ± 30° for bolts installed by 1/2 turn or less. The tolerance is ± 45° for bolts installed by 2/3 turn or more.
c. Determine the required rotation by actual tests in a suitable tension device simulating the actual conditions.

(b) During this test, the maximum recorded tension must be equal to or greater than the turn test tension, which is 1.15 times the required minimum fastener tension indicated in table 555-6.

(c) Ensure that the measured torque at a tension P, after exceeding the turn test tension required above, does not exceed the value obtained by the following equation:

$$Torque = PD/4{,}000$$

where

$Torque$ = measured torque in newton meters (N•m)
P = measured bolt tension in newtons (N)
D = nominal bolt diameter in millimeters (mm)

For rotational-capacity tests, use washers even though their use may not be required in the actual installation.

(2) Washers. Where the outer face of the bolted parts has a slope greater than 1:20 with respect to a plane normal to the bolt axis, use a hardened beveled washer to compensate for the lack of parallelism.

Use hardened square or rectangular beveled washers for American Standard Beams and Channels conforming to AASHTO M 293.

Where necessary, washers may be clipped on one side not closer than seven-eighths of the bolt diameter from the center of the washer.

Hardened washers are not required for connections using AASHTO M 164M and M 253M bolts except under the following conditions:

(a) Use hardened washers under the element turned in tightening when the tightening is done by the calibrated wrench method.

(b) Use hardened washers under both the head and the nut when AASHTO M 253M bolts are installed in material with a specified yield point less than 275 MPa, regardless of the tightening method.

(c) Use a hardened washer conforming to ASTM F 436M where AASHTO M 164M bolts of any diameter or AASHTO M 253M bolts equal to or less than M 24 are to be installed in oversize or short-slotted holes in an outer ply.

(d) Use hardened washers conforming to ASTM F 436M, except with 8 mm minimum thickness, under both the head and the nut in lieu of standard-thickness hardened washers where AASHTO M 253 bolts over M 24 are to be installed in an oversize or short-slotted hole in an outer ply. Multiple hardened washers with combined thickness equal to or greater than 8 mm do not satisfy this requirement.

(e) Where AASHTO M 164M bolts of any diameter or AASHTO M 253M bolts equal to or less than M24 are installed in a long-slotted hole in an outer ply, provide a plate washer or continuous bar that has a thickness of at least 8 mm, with standard holes of sufficient size to cover the slot after installation, and is of structural-grade material that need not be hardened.

When AASHTO M 253M bolts over M24 are used in long-slotted holes in external plies, use a single hardened washer conforming to ASTM F 436M with an 8-mm minimum thickness in lieu of washers or bars of structural steel. Multiple hardened washers with combined thickness equal to or greater than 8 mm do not satisfy this requirement.

Alternate design fasteners conforming to Subsection 717.01, with a geometry that provides a bearing circle on the head or nut with a diameter equal to or greater than the diameter of hardened washers conforming to ASTM F 436M, satisfy the requirements for washers specified herein and may be used without washers.

(3) Turn-of-Nut Tightening. At the start of work, test nut tightening using a device capable of indicating bolt tension. Test not less than three bolt-and-nut assemblies of each diameter, length, and grade to be used in the work. Demonstrate with the test that the method to be used for estimating the snug-tight condition and controlling the turns from snug tight develops a tension not less than 5 percent greater than the tension specified in table 555-6. Perform periodic retesting when required.

Install bolts in all holes of the connection and initially tighten to a snug-tight condition. Snug tight is defined as the tightness that exists when the plies of the joint are in firm contact. This may be attained by a few impacts of an impact wrench or the full effort of a worker using an ordinary spud wrench.

Systematically snug-tighten bolt groups from the most rigid part of the connection to the free edges. Then retighten the bolts of the connection in a similar systematic manner as necessary until all bolts are snug tight and the connection is fully compacted. Following the snug-tightening operation, tighten all bolts in the connection by the applicable amount of rotation specified in table 555-7.

During all tightening operations, do not allow rotation of the fastener part not turned by the wrench. Tighten systematically from the most rigid part of the joint to its free edges.

(4) Calibrated Wrench Tightening. Calibrated wrench tightening may be used only when installation procedures are calibrated on a daily basis and when a hardened washer is used under the element turned in tightening. Standard torques taken from tables or from formulas that assume to relate torque to tension are not acceptable.

If calibrated wrenches are used for installation, set them to provide a tension not less than 5 percent in excess of the minimum tension specified in table 555-6. Calibrate the installation procedure at least once each working day for each bolt diameter, length, and grade using fastener assemblies that are being installed in the work.

Perform the calibration with a device capable of indicating actual bolt tension by tightening three typical bolts of each diameter, length, and grade from the bolts and washers being installed using a job-supplied washer under the element turned in tightening. Recalibrate wrenches when significant difference is noted in the surface condition of the bolts, threads, nuts, or washers. Verify during use that the wrench adjustment selected by the calibration does not produce a nut or bolt head rotation from snug tight greater than permitted in table 555-7. Turn nuts in the tightening direction when measuring the torque of manual torque wrenches.

If calibrated wrenches are used to install bolts in a connection, install bolts with hardened washers under the turned element. When tightening bolts in all holes of the connection, tighten to a snug-tight condition. Following this initial tightening operation, tighten all bolts in the connection using a calibrated wrench. Tighten systematically from the most rigid part of the joint to its free edges. "Touch up" previously tightened bolts that may have been relaxed during the subsequent tightening of adjacent bolts until all bolts are properly tightened.

(5) DTI Tightening. When tightening bolts using DTI devices, assemble a representative sample of not less than three devices for each diameter and grade of fastener to be used in the work in a calibration device capable of indicating bolt tension. Include in the test assembly flat-hardened washers, if required in the actual connection, arranged like those in the actual connections to be tensioned. The calibration test must demonstrate that the device indicates a tension not less than 5 percent greater than that specified in table 555-6.

Follow the manufacturer's installation procedures for installation of bolts in the calibration device and in all connections. Give special attention to proper installation of flat-hardened washers when DTI devices are used with bolts installed in oversize or slotted holes, and where the load-indicating devices are used under the turned element.

When bolts are installed using DTI's conforming to ASTM F 959, install bolts in all holes of the connection and bring to a snug-tight condition. Snug tight is indicated by partial compression of the DTI protrusions. Then tighten all fasteners systematically from the most rigid part of the connection to the free edges in a manner that will minimize relaxation of previously tightened fasteners. Comply with the installation instructions portion of section 11.5.6.4.7, division II, of AASHTO's "Standard Specifications for Highway Bridges." Proper tensioning of the bolts may require more than a single cycle of systematic partial tightening before final tightening to deform the protrusion to the specified gap.

(6) Installation of Alternate Design Bolts. When fasteners that incorporate a design feature intended to indirectly indicate the bolt tension or to automatically provide the tension specified in table 555-6 and that conform to Subsection 717.01 are to be installed, test a representative sample of not less than three bolts of each diameter, length, and grade at the jobsite with a device capable of indicating bolt tension.

Include in the test assembly flat-hardened washers, if required in the actual connection, arranged as in the actual connections to be tensioned. The calibration test must demonstrate that each bolt develops a tension not less than 5 percent greater than the tension specified in table 555-6. Follow manufacturer's installation procedure. Perform periodic retesting when required.

When alternate design fasteners that are intended to control or indicate bolt tension of the fasteners are used, install bolts in all holes of the connection and initially tighten sufficiently to bring all plies of the joint into firm contact, but without yielding or fracturing the control or indicator element of the fasteners. Continue to tighten systematically from the most rigid part of the connection to the free edges in a manner that will minimize relaxation of previously tightened fasteners.

Proper tensioning of the bolts may require more than a single cycle of systematic partial tightening before final twist-off or pull-off of the control or indicator element of individual fasteners.

(7) Inspection. Inspect the tightened bolts in the presence of the CO. Use an inspection torque wrench to verify tightening of threaded fasteners. For nonthreaded fasteners, ping each fastener with a hammer to test for soundness. Replace or retighten any loose or relaxed fastener. Cutting with a torch will not be permitted for removal of bolts.

Individually place three bolts of the same grade, size, and condition as those under inspection in a device calibrated to measure bolt tension. Perform this calibration operation at least once each inspection day. Permit the CO full opportunity to witness calibration tests.

Use a washer under the part turned in tightening each bolt if washers are used on the structure. If washers are not used on the structure, use the same specification material that abuts the part turned in the tension-measuring device as used on the structure. In the calibrated device, tighten each bolt by any convenient means to the specified tension. Apply the inspecting wrench to the tightened bolt to determine the torque required to turn the nut or head 5°, approximately 30 mm at a 300-mm radius, in the tightening direction. Use the average of the torque required for all three bolts as the job-inspection torque.

Select at random in each connection 10 percent (at least two) of the tightened bolts on the structure represented by the test bolts, and apply the job-inspection torque to each selected bolt with the inspecting wrench turned in the tightening direction. If this torque turns no bolt head or nut, the bolts in the connection will be considered to be properly tightened. If the torque turns one or more bolt heads or nuts, apply the job-inspection torque to all bolts in the connection. Tighten and reinspect any bolt whose head or nut turns at this stage. As an option, retighten all bolts in the connection and resubmit for inspection.

555.18 Welding. Ensure that welding, welder qualifications, prequalification of weld details, and inspection of welds conform to ANSI/AASHTO/AWS Bridge Welding Code D 1.5. Delete the provisions of section 9.25.1.7. Do not underrun the nominal fillet weld size.

Do not weld or tack brackets, clips, shipping devices, or other material not required to any member unless SHOWN ON THE DRAWINGS.

555.19 Erection. Ensure that falsework and forms conform to Section 562.

(a) Handling & Storing Material. Place material stored at the jobsite on skids above ground. Keep material clean and properly drained. Place and shore girders and beams upright. Support long members, such as columns and chords, on skids placed near enough together to prevent damage due to deflection.

(b) Bearings & Anchorages. Furnish and install bridge bearings in accordance with Section 564. If the steel superstructure is to be placed on a substructure that was built under a separate contract, verify that the masonry has been correctly constructed before ordering material.

(c) Erection Procedures. Follow the procedures shown below.

(1) Conformance to Drawings. Erect as SHOWN ON THE DRAWINGS. Modifications to or deviations from the approved erection procedure will require revised drawings and verification of stresses and geometry.

(2) Erection Stresses. Allow for erection stresses induced in the structure as a result of the use of a method of erection or equipment that differs from that previously approved, and that will remain in the finished structure as locked-in stresses. Provide additional material, as needed, to keep both temporary and final stresses within the allowable limits used in the design.

Provide temporary bracing or stiffening devices to accommodate handling stresses in individual members or segments of the structure during erection.

(3) Maintaining Alignment & Camber. During erection, support segments of the structure in a manner that will produce the proper alignment and camber in the completed structure. Install cross frames and diagonal bracing as necessary during erection to provide stability and assure correct geometry. As necessary, provide temporary bracing at any stage of erection.

(d) Field Assembly. Accurately assemble as SHOWN ON THE DRAWINGS and required by match-marks. Carefully handle the material. Do not hammer, damage, or distort the members. Clean bearing surfaces and permanent contact surfaces before assembly.

Assemble splices and field connections with at least two cylindrical erection pins per part (a minimum of four per splice or connection). Use cylindrical erection pins 1 mm larger than the bolts to be used. A plate girder splice requires, for example, at least four cylindrical erection pins for the top flange splice, four pins for the web splice, and four pins for the bottom flange splice. (These provide two pins for each part.) Place the pins in the corner holes of the splice plates.

Install more cylindrical erection pins, if necessary, to accurately align the parts. Fill the remaining holes in the connection with bolts, and tighten systematically in accordance with Subsection 555.17 from the most rigid part of the connection to the free edges. Remove cylindrical erection pins and replace with tightened bolts.

Release temporary erection supports at a splice or connection only after all bolts are installed and tightened. Special assembly and support situations are SHOWN ON THE DRAWINGS or approved submittals.

Fitting-up bolts may be the same high-strength bolts used in the installation. If other fitting-up bolts are required, use the same nominal diameter as the high-strength bolts.

(e) ***Pin Connections.*** Use pilot and driving nuts in driving pins. Drive the pins so that the members will fully bear on the pins. Screw pin nuts tight, install nut retaining devices as SHOWN ON THE DRAWINGS, and burr the threads at the face of the nut with a pointed tool.

(f) ***Misfits.*** Correction of minor misfits involving minor amounts of reaming, cutting, and chipping may be done, if approved. Any error in the shop fabrication or deformation resulting from handling and transporting will be cause for rejection.

Measurement

555.20 Method. Use the method of measurement that is DESIGNATED IN THE SCHEDULE OF ITEMS.

Measure structural steel by the kilogram or lump sum computed in accordance with the AASHTO "Standard Specifications for Highway Bridges." The quantity will include metal items incidental to the structure, such as castings, steel plates, anchor bolts and nuts, bearings, rockers, rollers, pins and nuts, expansion dams, roadway drains and scuppers, weld metal, bolts embedded in concrete, cradles and brackets, posts, conduits and ducts, and structural shapes.

Changes in quantities resulting from alternative details proposed by the Contractor and approved by the CO are not subject to price adjustment.

Payment

555.21 Basis. The accepted quantities will be paid for at the contract unit price for each PAY ITEM DESIGNATED IN THE SCHEDULE OF ITEMS.

Payment will be made under:

Pay Item		Pay Unit
555 (01)	Structural steel, ________________, furnished, *Description* fabricated, and erected	Kilogram
555 (02)	Structural steel, ________________, furnished, *Description* fabricated, and erected	Lump Sum

Section 556—Bridge Railing

Description

556.01 Work. Furnish and erect and/or remove and reset bridge railing and bridge approach railing.

Bridge railing is designated as concrete, steel, aluminum, or timber in accordance with the predominant material contained in the railing.

Materials

556.02 Requirements. Furnish material that conforms to specifications in the following sections and subsections:

Aluminum Bolt Heads and Nuts	717.14
Aluminum Alloy for Bridge Rail	717.13
Aluminum-Impregnated Caulking Compound	725.27
Aluminum Welding Wire	717.15
Box Beam Rail	710.07
Guardrail	606
Painting	563
Reinforcing Steel	554
Steel Structures	555
Structural Concrete	552
Timber Structures	557

Construction

556.03 General. Accurately place anchor bolts to provide correct and true alignment of the railing. Set anchor bolts so that they project not more than 10 mm beyond the nut when tightened. Chamfer or round by grinding or filing all sharp exposed metal edges.

Provide bridge rail shop drawings when SHOWN ON THE DRAWINGS or called for in the SPECIAL PROJECT SPECIFICATIONS.

Do not erect railing until centering or falsework for the supporting span is removed. Construct bridge railing so that it does not follow any unevenness in the curb, sidewalk, or wall that supports the railing. The railing shall present a smooth, uniform appearance in its final position. Set all posts vertical.

556.04 Concrete Railing. Construct in accordance with Section 552 and the following:

- Construct expansion joints that permit freedom of movement. After all other work is completed, use a sharp chisel to remove all loose or thin shells of concrete likely to spall under movement at expansion joints.

(a) Fixed Forms. Construct forms that are smooth and tight fitting, rigidly held in line and grade, and removed without damage to the concrete. Make form joints in vertical planes. Construct all moldings, panel work, and bevel strips as SHOWN ON THE DRAWINGS. Make corners in the finished work true and free from cracks, spalls, or other defects.

(b) Slipformed. Concrete rails may be slipformed if the DRAWINGS contain details for slipforming. Before slipforming any permanent rail, one or both of the following requirements shall be met, as directed by the CO:

(1) Cast a test section at least 6 m long that shall:

  (a) Be placed off the structure.

  (b) Have the same section and reinforcement as detailed for use on the structure.

  (c) Include one typical contraction or open joint.

  (d) Be removed and disposed of without compensation.

(2) Identify, for the purposes of evaluating work quality, at least two recent slipformed rail projects completed by the Contractor.

The CO will make the final decision about the use of slipforming on the project based on work quality. If slipforming is approved by the CO:

- Provide concrete with a slump of 25 mm ± 12 mm.

- Keep the top and faces of the finished rail free from sags, humps, and other irregularities.

- Maintain contraction joints, open joints, and expansion joints to the dimensions SHOWN ON THE DRAWINGS until the concrete sets.

- Use slipforming only for section of rail with constant dimensions. Use fixed forms where dimensions vary, as at luminaire or signal supports and at rail end transitions.

- Brush finish exposed rail surfaces with vertical strokes. Do not grind brush finished surfaces that are to receive a class 1 finish as specified in Subsection 552.18(a).

- Remove and replace any unsatisfactory work without compensation.

(c) Surface Finish. Apply a general surface finish using a class 2 finish to all exposed concrete surfaces as specified in Subsection 552.18(b).

556.05 Steel Railing. Construct in accordance with Section 555. Ensure that structural tubing conforms to AASHTO M 183 (ASTM A 500, grade B).

If required, galvanize in accordance with AASHTO M 111, and furnish nuts, bolts, and washers galvanized in accordance with AASHTO M 232. Repair minor abrasions with zinc-rich paint.

For exposed weathering steel, use railing fasteners, railing hardware, rail post anchor bolts, nuts, washers, and shims with the same atmospheric corrosion resistance and weathering characteristics as the railing and posts. Use hand methods to clean erected steel railing of all oil, dirt, grease, mortar, and other foreign substances. Use weld metal with similar atmospheric corrosion resistance and coloring characteristics as the base metal. Clean welds by power brushing or blast cleaning to remove welding flux, slag, and spatter.

Unless a coating is required, clean all weathering steel in accordance with Steel Structures Painting Council (SSPC) standard SSPC–SP 6 and remove all mill scale and other foreign substances so that the steel surface is uniformly exposed to the atmosphere.

556.06 Aluminum Railing. Construct in accordance with Section 555, except as amended by the following:

(a) Cutting. Material that is 13 mm thick or less may be cut by shearing, sawing, or milling. Saw or mill material that is more than 13 mm thick. Do not flame cut. Make cut edges true, smooth, and free from excessive burrs or ragged breaks. Fillet re-entrant cuts by drilling before cutting.

(b) Bending. Material may be heated to a maximum 200 °C for a period not to exceed 30 minutes to facilitate bending.

(c) Rivet & Bolt Holes. Drill rivet and bolt holes to finished size or subpunch smaller than the nominal diameter of the fastener and ream to size. Subpunch to a diameter that is smaller than that of the finished hole by at least one-quarter the thickness of the piece. Make the finished diameter of holes not more than 7 percent greater than the nominal diameter of the fastener, except:

(1) Fabricate slotted bolt holes as required.

(2) Fabricate anchor bolt holes up to 25 percent larger, not to exceed 15 mm larger than the nominal bolt diameter.

(d) Welding. Weld in accordance with AWS Structural Aluminum Welding Code D 1.2.

(e) Contact With Other Material. Do not place aluminum alloys in contact with copper, copper base alloys, lead, or nickel. Where aluminum alloys come in contact with other metals, coat the contacting surfaces thoroughly with an approved aluminum-impregnated caulking compound or place a neoprene gasket between the surfaces.

Where aluminum alloys come in contact with concrete or stone, coat the contacting surfaces with an aluminum-impregnated caulking compound. When bond between aluminum and concrete is required, coat the aluminum with zinc-chromate paint and allow to dry before installation.

Where aluminum alloys come in contact with wood, coat the contacting wood surface with three coats of paint in accordance with Section 563 and coat the contacting aluminum surface with an aluminum caulking compound.

556.07 Timber Railing. Construct in accordance with Section 557.

When SHOWN ON THE DRAWINGS or directed in the SPECIAL PROJECT SPECIFICATIONS, clean all exposed surfaces of timber railing treated with pentachlorophenol or creosote that are located where contact by people may occur. Seal these surfaces with two coats of urethane shellac, latex epoxy, enamel, or varnish.

556.08 Approach Railing. Construct in accordance with Subsection 556.05 and Section 606.

556.09 Remove & Reset Bridge Railing. Remove and store the existing bridge railings and appurtenances. Replace all railings, supports, and hardware damaged during removal, storage, or resetting.

556.10 Painting. Paint in accordance with Section 563.

Measurement

556.11 Method. Use the method of measurement that is DESIGNATED IN THE SCHEDULE OF ITEMS.

Measure bridge railing by the meter or by the lump sum. Measure removed and reset bridge railing by the meter. When bridge railing is measured by the meter, measure along the top of the railing center to center of end posts.

When bridge approach railing is measured by the meter, measure the total approach railing length along the face of the railing from the ends of the bridge railing, as SHOWN ON THE DRAWINGS, to the center of the end approach railing posts, unless otherwise SHOWN ON THE DRAWINGS.

Payment

556.12 Basis. The accepted quantities will be paid for at the contract unit price for each PAY ITEM DESIGNATED IN THE SCHEDULE OF ITEMS.

Payment will be made under:

Pay Item		Pay Unit
556 (01)	__________ bridge railing	Meter
556 (02)	__________ bridge railing	Lump Sum
556 (03)	Remove and reset bridge railing	Meter
556 (04)	Bridge approach railing, type_____ , class_____	Meter
556 (05)	Terminal section, ________________ *Description*	Each
556 (06)	Anchorage	Each

Section 557—Timber Structures

Description

557.01 Work. Furnish, fabricate, erect, and paint structural timber, including all required yard lumber and hardware.

Materials

557.02 Requirements. Furnish material that conforms to specifications in the following section and subsections:

Hardware & Structural Steel	716.02
Painting	563
Structural Glued Laminated Timber	716.04
Treated Structural Timber & Lumber	716.03
Treated Timber Piles	715.02
Untreated Structural Timber & Lumber	716.01

Furnish the following compliance certificates to the CO upon delivery of the materials to the jobsite:

(a) Verification of compliance with grading rules and species of timber and lumber. Provide certification by an agency accepted as competent by the American Lumber Standards Committee (ALSC).

(b) Lot certification of each charge for preservative, penetration in millimeters, and retention in kilograms per cubic meter (assay method) by a qualified independent inspection and testing agency. In addition, have the producer of the treated products provide written certification that Best Management Practices (BMP's) in accordance with "Best Management Practices for Treated Wood in Western Aquatic Environments," published by the Western Wood Preservation Institute (WWPI) and Canadian Institute of Treated Wood, were followed, including a description and appropriate documentation of the applicable BMP's used.

(c) Certification from a qualified inspection and testing agency indicating that all glued laminated members are in accordance with the requirements of American National Standard for Wood Products, "Structural Glued Laminated Timber" (ANSI/AITC A190.1), modified as SHOWN ON THE DRAWINGS.

(d) Such other certifications as SHOWN ON THE DRAWINGS or called for in the SPECIAL PROJECT SPECIFICATIONS.

Incise all glued laminated and solid sawn members thicker than 50 mm in accordance with AWPA standard C1, unless otherwise SHOWN ON THE DRAWINGS.

Provide shop drawings for all timber 21 days in advance of fabrication when SHOWN ON THE DRAWINGS or in the SPECIAL PROJECT SPECIFICATIONS. Show all dimensions and fabrication details for all cut, framed, or bored timbers.

Construction

557.03 General. Perform the work under Section 206. Furnish structural lumber and timber of the required stress grade.

Clear stacks of weeds, rubbish, or other objectionable material from the ground under and in the vicinity of all stored material. Place the bottom layer of material at least 200 mm above the ground level. Provide sufficient support to prevent sagging.

Open-stack untreated material to shed water. Stack material in layers on spacers (stickers) that extend across the full width of the stack to allow for free air circulation. Align all stickers vertically and space them at regular intervals.

Close-stack treated material to shed water.

Protect material from the weather. If covered, use sheet material such as water-resistant paper or opaque polyethylene film. Do not cover with impervious membranes, such as polyethylene film, during dry weather. Slit individual wrappings full length or puncture on the lower side to permit drainage of water.

Store and protect glued laminated timber in accordance with the recommendations for Loading and Handling, Job Site Storage, and Erection in "Recommended Practice for Protection of Structural Glued Laminated Timber During Transit, Storage, and Erection," published by the American Institute of Timber Construction, AITC 111.

Use slings or other devices to protect corners of heavy construction timbers and banded packages of lighter construction timber.

557.04 Treated Timber. Fabricate timbers before treatment. Handle treated timber according to the Consumer Information Sheet published by AWPA. Do not cut, frame, or bore treated timber after treatment unless approved by the CO. Handle treated timbers carefully and do not drop, damage outer fibers, or penetrate the surface with tools. Do not use cant dogs, hooks, or pike poles. In coastal waters, do not cut or bore timber below the highwater mark.

For timbers originally treated with pentachlorophenol, creosote, creosote solutions, or waterborne preservatives, field treat all cuts, abrasions, bolt holes, and recesses that occur after treatment with two liberal applications of a compatible preservative in accordance with the requirements specified in AWPA standard M4, Standard for the Care of Pressure-Treated Wood Products.

Unless otherwise specified, copper naphthenate solutions may be used for field treatments of material originally treated with copper naphthenate, pentachlorophenol, creosote, creosote solution, or waterborne preservatives. Prepare the preservative solution by blending copper naphthenate preservative that meets P8 requirements with a solvent conforming to AWPA standard P9. Ensure that the resulting preservative solution concentration contains a minimum of 2 percent copper metal.

Plug all unused holes with preservative-treated plugs. Perform all field-applied preservative treatment with necessary precautions so as to prevent any soil and/or water contamination.

557.05 Untreated Timber. Coat the following untreated timber surfaces in accordance with AWPA standard M4:

(a) All ends and tops, and all contact surfaces of posts, sills, and caps.

(b) All ends, joints, and contact surfaces of bracing and truss members.

(c) All surfaces of timber bumpers and the back faces of bulkheads.

(d) All other timber that will be in contact with earth.

557.06 Workmanship. Cut and form all lumber and construction timber so all joints will have even bearing over the entire contact surface. Do not use shims in making joints. Construct all joints to be closed. Drive nails and spikes to set the heads flush with the wood surface. Use the same end, face, and edge of the timber member for all layout dimensions. Bore all holes from mating faces.

557.07 Holes for Bolts, Dowels, Rods, & Lag Screws. Bore all holes before preservative treating the wood. Bore holes for round driftbolts and dowels 2 mm smaller in diameter than that of the bolt or dowel to be used. Ensure that the diameter of holes for square driftbolts or dowels is equal to the side dimension of the bolt or dowel.

Bore holes for machine bolts with a bit 1.5 mm larger than the diameter, except when galvanized bolts are specified. In this case, drill all holes 3 mm greater than the bolt size. Bore holes for lag screws with a bit not larger than the body of the screw at the base of the thread. Drill the depth of lag screw bolt holes 25 mm less than the length under the screw head and with a diameter approximately 75 percent of the shank diameter.

557.08 Hardware. Furnish the hardware as SHOWN ON THE DRAWINGS, as specified below.

(a) Bolts & Washers. Finally tighten all nuts to provide proper bearing, and cut off excess bolt lengths of more than 25 mm. After final tightening, check or burr all bolts effectively with a pointing tool to prevent loosening of the nuts.

Use malleable iron washers with a diameter approximately three times the bolt diameter under all bolt heads or nuts in contact with wood. Use cast-iron washers when the timber is in contact with the ground. Use square washers only when SHOWN ON THE DRAWINGS or with the approval of the CO.

(b) Galvanizing. Unless otherwise SHOWN ON THE DRAWINGS, ensure that all hardware for timber structures is galvanized, except for the glued laminated deck panel dowels. Ensure that all fasteners, including nails, spikes, bolts, washers, and timber connectors, other than malleable iron, are galvanized.

557.09 Countersinking. Countersink nuts and bolt heads where SHOWN ON THE DRAWINGS. Paint recesses formed for countersinking with an approved preservative, except in railing. After bolts or screws are in place, fill the holes with hot pitch or other approved filler.

557.10 Framing. Do not slab or trim treated piles for fitting sway or sash braces. Fill all gaps that occur between braces and piles with treated blocks so that the bracing is securely fastened to the piles.

557.11 Framing Bents. Bed mud sills firmly, evenly, and level to solid bearing, and tamp in place.

When concrete is cast and dowels are used for anchoring sills and posts, install dowels (18 mm minimum diameter) that project at least 150 mm above the tops of the pedestals. Carefully finish concrete pedestals supporting framed bents so that sills or posts bear evenly on the pedestals.

Provide firm, uniform bedding for mud sills. Make sills bear true and even on mud sills, piles, or pedestals. Drift bolt sills with bolts that extend into the mud sills or piles for at least 150 mm. Where possible, remove all earth in contact with sills for circulation of air around the sills.

557.12 Posts. Fasten posts to pedestals with dowels not less than 18 mm in diameter that extend at least 150 mm into the posts, or with other types of connectors as SHOWN ON THE DRAWINGS. Fasten posts to sills using one of the following methods, as SHOWN ON THE DRAWINGS:

(a) With dowels not less than 18 mm in diameter that extend at least 150 mm into posts and sills.

(b) With drift bolts not less than 18 mm diameter driven diagonally through the base of the post and extending at least 175 mm into the sill. Drive drift bolts into holes at a 45° angle to enter the post at least 150 mm above the post base.

(c) With other types of connectors as SHOWN ON THE DRAWINGS.

557.13 Pile Bents. Treat, furnish, and drive piles in accordance with Section 551.

557.14 Caps for All Bents. Make timber caps bear even and uniform over the tops of the supporting posts or piles, with their ends in alignment. Secure all caps with drift bolts and set approximately at the center of and extending into the posts or piles at least 230 mm.

557.15 Bracing. Bolt the ends of bracing through the pile, post, cap, or sill. Brace intermediate intersections with posts or piles with bolts or spikes, as required. In all cases, use galvanized spikes in addition to bolts.

Make all bracing bear firmly against the pile or cap to which it is bolted. Provide and place shims as necessary to prevent bending the bracing more than 25 mm out of line when bracing bolts are tightened.

Where the space between the bracing and cap or pile is less than 25 mm, shims need not be used.

Where the space between the bracing and the cap or pile is 40 mm ± 15 mm, place two ogee washers, with their narrow faces together, or other approved washers on each bolt that passes through the space.

Where the space between the bracing and the cap or pile is over 55 mm, use wooden shims of the proper thickness. Fabricate the wooden shims from White Oak or from other approved hardwood. Do not use built-up wooden shims. Make wooden shims from a single piece of lumber with the width not less than 100 mm and the length not less than the width of the bracing measured along the cap or pile. Do not adze, trim, or cut any treated member to avoid the use of shims.

557.16 Stringers. Place solid sawn stringers in position so that knots near edges are in the top portions of the stringers.

Outside stringers may have butt joints with the ends cut on a taper. Lap interior stringers to take bearing over the full width of the floor beam or cap at each end. Separate the lapped ends of untreated stringers by at least 15 mm for air circulation. Securely fasten the lapped ends with drift bolts, as required. Stagger the joints where stringers are two panels in length.

Install cross-bridging between stringers as SHOWN ON THE DRAWINGS. If timber cross-bridging members are used, cut for a full bearing at each end against the sides of the stringers. Place cross-bridging at the center of each span or as SHOWN ON THE DRAWINGS.

557.17 Plank Floors. Use plank that is surfaced on four sides (S4S).

Single-ply timber floors consist of a single thickness of planks supported on stringers. Lay the planks heart side down with 5 mm space between them for seasoned material, and with tight joints for unseasoned material. Spike each plank securely to each stringer. Carefully grade the planks as to thickness and lay so that no two adjacent planks vary in thickness by more than 2 mm.

Two-ply timber floors consist of two layers of flooring supported on stringers. Pressure treat the lower layer with creosote oil or with another preservative as SHOWN ON THE DRAWINGS. Lay the top layer either diagonal or parallel to the centerline of roadway as required. Securely fasten each floor piece to the lower layer. Stagger joints at least 1 m. Where the top layer is placed parallel to the centerline of the roadway, use special care to securely fasten the ends of the flooring. Bevel the ends of top layer members at each end of the structure.

557.18 Transversely Nail-Laminated Decks. Use 50-mm nominal thickness laminations; surface one edge hit or miss 3 mm scant (SIE–H or M 3 mm scant), and one side hit or miss 3 mm scant (SIS–H or M 3 mm scant).

Place the laminations on edge and at right angles to the centerline of the roadway. Spike each piece to the preceding piece at each end and at approximately 450-mm intervals, with the galvanized spikes driven alternately near the top and bottom edges. Use spikes of sufficient length to pass through two pieces and at least halfway through the third piece.

Where timber stringers are used, toenail every other piece to every other stringer. Use the size spikes specified. When steel stringers are used, securely attach the pieces using approved galvanized metal clips.

Use pieces of sufficient length to bear on at least four stringers. Do not splice pieces between stringers. Space end joints on any one stringer no closer than every third piece. Space end joints in adjoining pieces no closer than every second stringer.

557.19 Glued Laminated Panel Decks. Do not drag or skid panels. When lifted, support panels in the weak-moment plane at a sufficient number of points to avoid overstressing, and protect the edges from damage.

When dowels are SHOWN ON THE DRAWINGS between deck panels, use a template or drilling jig to ensure that dowel holes are accurately spaced and drilled parallel to one another and to the horizontal surfaces of the panel. Drill holes to a depth 6 mm greater than one-half the dowel length, and of a diameter that is 2 mm greater than the dowel, unless otherwise SHOWN ON THE DRAWINGS. Use a temporary dowel as a check for snug fit prior to production drilling. Use dowels of the size SHOWN ON THE DRAWINGS, with the tips slightly tapered or rounded. Use an approved lubricant to facilitate the connection process.

Start the tips of all dowels partially and equally into the holes of the two panels being joined. Draw the panels together keeping the edges parallel, until the panels abut tightly. Securely fasten each panel to each stringer as SHOWN ON THE DRAWINGS.

Assemble and match-mark panels prior to delivery to the construction site when SHOWN ON THE DRAWINGS or called for in the SPECIAL PROJECT SPECIFICATIONS. Follow erection procedures given in FPL–263, Forest Service, Forest Products Laboratory (FPL), Madison, Wisconsin.

557.20 Wheel Guards & Railings. Surface (S4S) wheel guards, rails, and posts. Place wheel guards in sections not less than 4 m in length. Squarely butt-joint all rails at posts.

557.21 Trusses. Fabricate trusses to show no irregularities of line when completed. Fabricate chords straight and true from end to end in horizontal projection. In vertical projection, fabricate chords to a smooth chorded curve through panel points conforming to the correct camber. Do not make uneven or rough cuts at the points of bearing.

557.22 Drains. Hot-dip galvanize drains, including anchorages, after fabrication.

557.23 Painting. Paint in accordance with Section 563.

Measurement

557.24 Method. Use the method of measurement that is DESIGNATED IN THE SCHEDULE OF ITEMS.

Measure untreated and treated structural timber and lumber by the cubic meter of lumber and timber in place in the completed structure. Compute the quantities from nominal dimensions and actual lengths, except for transversely nail-laminated decks. Measure transversely nail-laminated decks in place after dressing.

Measure timber piles under Subsection 551.16.

Measure timber bridge rail under Subsection 556.11.

Measure structural excavation under Subsection 206.12.

Payment

557.25 Basis. The accepted quantities will be paid for at the contract unit price for each PAY ITEM DESIGNATED IN THE SCHEDULE OF ITEMS.

Payment will be made under:

Pay Item		Pay Unit
557 (01)	Untreated structural timber and lumber	Cubic Meter
557 (02)	Treated structural timber and lumber	Cubic Meter
557 (03)	Untreated structural timber and lumber	Lump Sum
557 (04)	Treated structural timber and lumber	Lump Sum
557 (05)	Treated structural timber, glued laminated	Cubic Meter
557 (06)	Treated structural timber, glued laminated	Lump Sum

Section 558—Prefabricated, Modular Bridge Superstructure

Description

558.01 Work. Design, fabricate, deliver, and install a prefabricated, modular bridge superstructure, or transport and install Government-furnished prefabricated, modular superstructure and components as DESIGNATED IN THE SCHEDULE OF ITEMS. Construct the length, width, and capacity of the structure, including curbs and railings and the horizontal and vertical alignment, as SHOWN ON THE DRAWINGS.

Also furnish material for, and construct, bridge railing as SHOWN ON THE DRAWINGS or on approved manufacturer's drawings. Unless components are furnished by the Government, furnish prefabricated, modular bridge superstructure components complete and in place, including deck and railing, when required, to form a bridge superstructure capable of supporting traffic as soon as construction of approaches is complete. Include all incidental materials required to provide a completed structure ready for use.

When there are specific requirements for design, materials, appearance, and/or construction, they shall be SHOWN ON THE DRAWINGS or in the SPECIAL PROJECT SPECIFICATIONS.

Materials

558.02 Requirements. Furnish materials that meet the requirements specified in the following sections and subsections:

Bridge Railing	556
Hardware & Structural Steel	716.02
Precast Concrete Structures	553A
Prestressed Concrete	553
Reinforcing Steel	554
Steel Structures	555
Structural Concrete	552
Timber Structures	557

Concrete compressive strength, structural steel tensile strength, finish, designation, timber species, grade, treatment, and other material specifications shall be as SHOWN ON THE DRAWINGS or in the SPECIAL PROJECT SPECIFICATIONS. If material specifications are not in the contract documents, take them from the manufacturer's drawings, and have them approved by the CO prior to fabrication.

558.03 Design Requirements. Design in accordance with the AASHTO "Standard Specifications for Highway Bridges," latest edition and interims, for the HS20–44 loading, including impact, unless otherwise SHOWN ON THE DRAWINGS or in the SPECIAL PROJECT SPECIFICATIONS.

When design of the structure is required, provide on all the drawings and calculations that are submitted for review the signature and seal of a professional engineer who is currently licensed in the State where the bridge will initially be located.

Use materials that are durable enough to allow removal, transportation, and re-installation using typical forest logging or construction equipment. Use design techniques and fabrication methods to minimize field erection difficulties. Fabricate primary components from steel unless otherwise SHOWN ON THE DRAWINGS or in the SPECIAL PROJECT SPECIFICATIONS.

Rig main superstructure components with permanent lifting devices to facilitate efficient installation and removal of these items with equipment common to logging or construction operations. Place lifting devices so as not to interfere with traffic utilizing the structure.

558.04 Design Drawings. When furnishing a prefabricated bridge superstructure, submit design drawings, calculations, and/or shop drawings sufficiently in advance of the start of fabrication to allow time for review by the CO and correction of any changes. Such time shall be proportional to the work, but not less than 21 days. Include plan, elevation, and section views of the modular bridge superstructure, dimensions of all components, welding and connection details, and general and specific notes regarding design and construction.

When Government-furnished prefabricated bridge superstructure components are specified, material lists, erection information, and manufacturer's instructions will be furnished by the Government.

Construction

558.05 General. Perform excavation, backfill, and embankment work under Sections 203 and 206.

Dispose of all debris resulting from operations in accordance with Section 202.

Perform all construction of substructures, riprap, and signs under Sections 206, 206A, 251, 551, 552, 553A, 554, 555, 557, 564, 602, and 633, as applicable.

558.06 Performance. Provide 2 weeks' notice prior to delivery and/or installation.

If the prefabricated superstructure is not installed immediately upon delivery to the project site, provide appropriate equipment and labor to unload and stack, support, and store all material at the delivery point. Support and stack all components to prevent damage. Furnish and install blocking such that all components are supported at least 300 mm above the ground.

Furnish all tools, devices, special equipment, and material needed for installation in well-marked watertight containers suitable for long-term, outdoor storage.

558.07 Contractor-Furnished Prefabricated Bridge Superstructure. As applicable, furnish the CO with the following items for approval prior to delivery of the bridge component:

(a) Supplier or inspection agency certification of wood species and grade of all timber and a conformance certificate for all sawn and glued laminated members.

(b) Certification by an approved inspection and testing agency of wood treatment, listing method of treatment, type of preservative, retention, and penetration. Supplier certification is permitted if each piece is stamped or branded with a legible American Wood Preservers Bureau quality mark.

(c) Certification of structural steel, fasteners, and hardware.

(d) Certification of galvanizing process used.

(e) Steel fabricator certification that steel fabrication and quality control meet the requirements of the AISC Code of Standard Practice; and that all welding meets the requirements of ANSI/AASHTO/AWS D 1.5 Bridge Welding Code.

(f) A complete list of all bridge components, hardware, and fasteners.

(g) Complete erection instructions and drawings. Provide drawings that are black line, on a reproducible mylar media, on ANSI sheet size B or D.

As appropriate to the type of modular bridge, mark each major component of the bridge superstructure with the same serial number. Ensure that the marking is permanent and clearly visible on each component, both when stacked in storage and when erected on a bridge site.

When called for in the SPECIAL PROJECT SPECIFICATIONS, assemble each bridge superstructure prior to delivery to ensure proper fit-up of all components. Notify the CO of the assembly 2 weeks in advance so that inspection of the assembly can be arranged.

558.08 Government-Furnished Prefabricated Bridge Superstructure. When Government-furnished prefabricated bridge units are specified, transport all designated material from the storage site(s) designated in the SPECIAL PROJECT SPECIFICATIONS or SHOWN ON THE DRAWINGS to the bridge site, and install the superstructure complete and in place, including connection of all girders, diaphragms, railings, panels, transoms, and other elements.

Upon taking possession of the Government-furnished units at the storage site, assume all liability for damage resulting from handling, transporting, and/or erecting the units in place, until final acceptance of the project.

Measurement

558.09 Method. Use the method of measurement that is DESIGNATED IN THE SCHEDULE OF ITEMS.

Measure prefabricated bridge superstructures on a lump sum basis. Include all materials and work necessary to furnish, transport, and install the superstructure, including the deck and railing, as SHOWN ON THE DRAWINGS.

Payment

558.10 Basis. The accepted quantities will be paid for at the contract unit price for each PAY ITEM DESIGNATED IN THE SCHEDULE OF ITEMS.

Payment will be made under:

Pay Item		Pay Unit
558 (01)	Prefabricated bridge superstructure—design, fabricate, deliver, and install	Each
558 (02)	Government-furnished prefabricated bridge superstructure—transport and install	Each

Section 559—Log Bridges

Description

559.01 Work. Furnish, fabricate, and install the logs and timber for constructing log bridges, including abutments, piers, and superstructure. In addition, furnish and install all hardware and other required material.

Materials

559.02 Requirements. Furnish materials that conform to specifications in the following section and subsections:

Geotextiles	714.01
Reinforcing Steel	554
Structural Concrete	552
Timber Structures	557

559.03 Logs. Furnish logs used for stringers within the dimensional tolerance and of the species SHOWN ON THE DRAWINGS. They must be of high quality, straight, sound, and free of wind shake, decay, or excessive twist (spiral grain with a slope of grain relative to the longitudinal axis of the log exceeding 1 in 8). Ensure that knots in the middle half of the stringer length do not significantly affect structural capacity.

If SHOWN ON THE DRAWINGS, peel logs and provide preservative treatment as SHOWN ON THE DRAWINGS. Obtain written approval from the CO for all logs to be used in the structure.

559.04 Timber & Lumber. Furnish structural lumber and timber in accordance with the species, grades, and dimensions SHOWN ON THE DRAWINGS and in accordance with Section 557.

559.05 Aggregate. When required, furnish aggregate for decking or surfacing to meet the requirements SHOWN ON THE DRAWINGS.

Construction

559.06 General. Perform excavation, foundation, backfill, and embankment work specified in Sections 203 and 206, as applicable.

Handle all logs and timber carefully to prevent damage to the wood and/or preservative treatment.

Dispose of all debris resulting from operations in accordance with Section 202.

Construct abutments and pier as SHOWN ON THE DRAWINGS.

559.07 Performance. Construct bridge superstructure and substructures as SHOWN ON THE DRAWINGS, with attention paid to the details of erection, fit-up, and connection. Obtain written approval for all deviations from the CO.

Place timber caps to obtain even and uniform bearing over the tops of supporting posts or piles and with post and pile ends in true alignment. Secure all caps as SHOWN ON THE DRAWINGS.

Match stringers for size at the bearings and place them in position so that the crown is up. Alternate stringers butt to tip. Locate any knots that may affect the strength of the member in the top portion of the stringer.

Cut stringers to length with a square cut. Remove sufficient material from the top surface of the log stringer to provide an adequate bearing area for the decking as SHOWN ON THE DRAWINGS. Do not allow hewing to exceed 19 mm in depth at the small end of the log. Do not allow hewing of the top of the butt end to exceed 75 mm in depth for a distance not to exceed one-fourth span length.

Cut or hew the bottom surface of the small end of the stringer logs only to the depth necessary to achieve the required bearing area. Block or shim tip ends that are smaller than the largest tip. Cut or dap butt ends to the depth of the largest top end. Allow the maximum slope of any dap to be 1 to 10. Make top and bottom cuts parallel. Require shims or blocks used under small ends to cover the entire bearing area.

Notch all logs together, including face logs, tie logs, mud sills, and anchor logs as SHOWN ON THE DRAWINGS, and drift pin all connections.

Use an approved type of suitable granular, free-draining material and/or rock for backfill when crib abutments are to be constructed.

Use tiebacks or other abutment anchoring devices as SHOWN ON THE DRAWINGS or as approved in writing by the CO.

Measurement

559.08 Method. Use the method of measurement that is DESIGNATED IN THE SCHEDULE OF ITEMS.

When untreated and treated timber and lumber is measured, measure by the cubic meter of timber and lumber in place in the completed structure. Compute the quantities from nominal cross section dimensions and actual lengths.

When bridge railing is measured, measure under Subsection 556.11. When concrete is measured, measure under Subsection 552.21.

Measure log bridges on a lump sum basis, including all work necessary to furnish, prepare, and install the log portions of the bridge superstructure and substructure units.

Payment

559.09 Basis. The accepted quantities will be paid for at the contract unit price for each PAY ITEM DESIGNATED IN THE SCHEDULE OF ITEMS.

Payment will be made under:

Pay Item	Pay Unit
559 (01) Log bridge	Lump Sum

Section 561—Structural Concrete Bonding

Description

561.01 Work. Repair cracks in concrete structures by pressure injecting epoxy.

Materials

561.02 Requirements. Furnish material that conforms to specifications in the following subsections:

Epoxy Resin Adhesives	725.21
Low-Strength Grout	701.03(b)
Polymer Grout	701.05

Construction

561.03 Crack Preparation. Provide notice of crack sealing at least 14 days before beginning work. The work areas will be identified and the locations of the cracks to be repaired will be marked.

Remove all dirt, laitance, and other debris from the exterior and interior of cracks. Apply a temporary surface seal material to the face of cracks. Use surface seal material with sufficient strength and adhesion to confine the injected epoxy material until cured.

Provide openings (entry ports) in the surface seal along the crack. Make the distance between entry ports at least the thickness of the concrete member being repaired.

After the injection adhesive has cured, remove the surface seal. Finish the face of the crack and entry ports flush with the adjacent surface.

561.04 Injection Procedures. Begin injecting epoxy at the lowest entry port. Continue injection at the first port until epoxy begins to flow out of the next highest port. Plug the first port and inject epoxy in the second port until the epoxy flows from the next highest port. Continue this sequence until the entire crack is filled. Use a two-component epoxy system. Maintain the mix ratio for the epoxy as prescribed by the manufacturer within 5 percent by volume at any discharge pressure not to exceed 1.4 MPa. Do not use solvents to thin the epoxy.

Use positive inline displacement-type equipment to meter, mix, and inject the epoxy at pressures not to exceed 1.4 MPa.

(a) Test for Proper Ratio. Perform this test for each injection unit at the beginning and end of every day that the unit is used. Disconnect the mixing head of the injection equipment and pump the two adhesive components through a ratio check device with two independent valved nozzles capable of controlling flow rate and back pressure by opening or closing valves on the check device. Use a pressure gage capable of sensing the back pressure behind each valve to adjust the discharge pressure to 1.4 MPa for both epoxy components. Simultaneously discharge both epoxy components into separate calibrated containers. Compare the discharged amounts to determine the mix ratio.

After the test is completed at 1.4 MPa discharge pressure, repeat the procedures for zero MPa discharge pressure.

(b) Test for Pressure Check. Perform this test for each injection unit at the beginning and end of every day that the unit is used.

Disconnect the mixing head of the injection equipment and attach the two adhesive component delivery lines to a pressure check device with two independent valved nozzles capable of controlling flow rate and pressure by opening or closing the valves. Use a pressure gage capable of sensing the pressure buildup behind each valve. Close the valves on the pressure check device and operate the equipment until the gage pressure on each line reads 1.4 MPa. When the pumps are stopped, the gage pressure must not drop below 1.3 MPa within 3 minutes.

(c) Records. Maintain and make available complete and accurate records of the ratio check tests and the pressure check tests. Additional ratio and pressure check tests may be required.

561.05 Coring. Take one 50-mm diameter test core, in accordance with AASHTO T 24, for every 15 m of repaired crack at designated locations. The crack repair is acceptable if the core sample indicates that 90 percent or more of the crack has been successfully bonded.

When a test core shows that the epoxy bonding has penetrated less than 90 percent of the crack volume within the core sample, redo that 15-m crack segment, or the segment that the core represents, and resample. Repeat this procedure until acceptable crack repair is achieved.

Fill all sample core holes with polymer grout and finish the surface to match the adjacent concrete.

Measurement

561.06 Method. Use the method of measurement that is DESIGNATED IN THE SCHEDULE OF ITEMS.

Measure crack preparation by the meter or lump sum. Measure structural concrete bonding by the meter, liter, or lump sum. When measurement is by the meter, measure the actual meters of surface crack acceptably repaired.

When measurement is by the liter, measure the actual number of liters of bonding material injected in the marked cracks that are acceptably repaired.

Payment

561.07 Basis. The accepted quantities will be paid for at the contract unit price for each PAY ITEM DESIGNATED IN THE SCHEDULE OF ITEMS.

Payment will be made under:

Pay Item		Pay Unit
561 (01)	Structural concrete bonding	Meter
561 (02)	Structural concrete bonding	Liter
561 (03)	Structural concrete bonding	Lump Sum
561 (04)	Crack preparation	Meter
561 (05)	Crack preparation	Lump Sum

Section 562—Forms & Falsework

Description

562.01 Work. Design, construct, and remove forms and falsework to temporarily support concrete, girders, and other structural elements until the structure is completed to the point where it can support itself.

Design & Construction

562.02 Drawings. When complete details for forms and falsework are not shown, prepare and submit drawings as SHOWN ON THE DRAWINGS or as directed in the SPECIAL PROJECT SPECIFICATIONS. Perform the following, as applicable:

(a) Design and show the details for constructing safe and adequate forms and falsework that provide the necessary rigidity, support the loads imposed, and produce the required lines and grades in the finished structure. See Subsection 562.03 for design loads; Subsection 562.04 for design stresses, loadings, and deflections; and Subsection 562.05 for manufactured assemblies.

(b) Show the maximum applied structural load on the foundation material. Include a drainage plan or description of how foundations will be protected from saturation, erosion, and/or scour. See Subsection 562.06.

(c) Precisely describe all proposed material. Describe the material that is not describable by standard nomenclature (such as AASHTO or ASTM specifications) based on manufacturer's tests, and recommended working loads. Evaluate falsework material and ascertain whether the physical properties and conditions of the material are such that the material can support the loads assumed in the design.

(d) Furnish design calculations and material specifications showing that the proposed system will support the imposed concrete pressures and other loads. Provide an outline of the proposed concrete placement operation listing the equipment, labor, and procedures to be used for the duration of each operation. Include proposed placement rates and design pressures for each pour. Include a superstructure placing diagram showing the concrete placing sequence and construction joint locations.

(e) Provide design calculations for proposed bridge falsework. Appoint a licensed professional engineer proficient in structural design to design, sign, and seal the drawings. Ensure that the falsework design calculations show the stresses and deflections in load supporting members.

(f) Show anticipated total settlements of falsework and forms. Include falsework footing settlement and joint takeup. Design for anticipated settlements not to exceed 25 mm. Design and detail falsework supporting deck slabs and overhangs on girder bridges so there is no differential settlement between the girders and the deck forms during placement of deck concrete. Design and construct the falsework to elevations that include anticipated settlement during concrete placement and required camber to compensate for member deflections during construction.

(g) Show the support systems for form panels supporting concrete deck slabs and overhangs on girder bridges.

(h) Show details for strengthening and protecting falsework over or adjacent to roadways and railroads during each phase of erection and removal. See Subsection 562.07.

(i) Include intended steel erection procedures with calculations in sufficient detail to substantiate the girder geometry. See Subsection 562.08.

(j) Submit details of proposed anchorage and ties for void forms. See Subsection 562.10 for void form requirements.

Submit separate falsework drawings for each structure, except for identical structures with identical falsework design and details. Do not start construction of any unit of falsework until the drawings for that unit are reviewed and accepted.

562.03 Design Loads. Conform to the following:

(a) Vertical Design Loads. Dead loads include the weight of concrete, reinforcing steel, forms, and falsework. Consider the entire superstructure, or any concrete mass being supported by falsework, to be a fluid dead load with no ability to support itself. If the concrete is to be prestressed, design the falsework to support any increased or readjusted loads caused by the prestressing forces.

Assume that the density of concrete, reinforcing steel, and forms is not less than 2,600 kg/m^3 for normal concrete, and not less than 2,100 kg/m^3 for lightweight concrete.

Consider live loads to be the actual mass of equipment to be supported by falsework applied as concentrated loads at the point of contact plus a uniform load of not less than 1,000 Pa applied over the area supported, plus 1,100 N/m applied at the outside edge of deck falsework overhangs.

The total vertical design load for falsework is the sum of vertical dead and live loads. Use a total vertical design load of not less than 4,800 Pa.

(b) Horizontal Design Loads. Use an assumed horizontal design load on falsework towers, bents, frames, and other falsework structures to verify lateral stability. The assumed horizontal load is the sum of the actual horizontal loads due to equipment, construction sequence, or other causes and an allowance for wind. However, in no case is the assumed horizontal load to be less than 2 percent of the total supported dead load at the location under consideration.

The minimum wind allowance for each heavy-duty steel shoring with a vertical load carrying capacity exceeding 130 kN per leg is the sum of the products of the wind impact area, shape factor, and applicable wind pressure value for each height zone. The wind impact area is the total projected area of all the elements in the tower face normal to the applied wind. Assume that the shape factor for heavy-duty shoring is 2.2. Determine design wind pressure values from table 562-1.

Table 562-1.—Design wind pressure—heavy duty steel shoring.

Height Zone Above Ground (m)	Wind Pressure Value (Pa)	
	Adjacent to Traffic	At Other Locations
0	960	720
9–15	1,200	960
15–30	1,450	1,200
Over 30	1,675	1,450

The minimum wind allowance on all other types of falsework, including falsework supported on heavy-duty shoring, is the sum of the products of the wind impact area and the applicable wind pressure value for each height zone. The wind impact area is the gross projected area of the falsework and unrestrained portion of the permanent structure, excluding the areas between falsework posts or towers where diagonal bracing is not used. Determine design wind pressure values from table 562-2.

Table 562-2.—Design wind pressure—other types of falsework.

Height Zone Above Ground (m)	Wind Pressure Value (Pa)	
	For Members Over and Bents Adjacent to Traffic Openings	At Other Locations
0	320 Q	240 Q
9–15	400 Q	320 Q
15–30	480 Q	400 Q
Over 30	560 Q	480 Q

Note: Q = 0.3 + 0.2W, but not more than 3. W is the width of the falsework system in meters measured in the direction of the wind force being considered.

Design the falsework to have sufficient rigidity to resist the assumed horizontal load without vertical dead load. Neglect the effects of frictional resistance.

*(c) **Lateral Fluid Pressure.*** For concrete with retarding admixture, fly ash, or other pozzolan replacement for cement, design forms, form ties, and bracing for a lateral fluid pressure based on concrete with a density of 2,400 kg/m^3. For concrete containing no pozzolans or admixtures, which affect the time to initial set, determine the lateral fluid pressure based on concrete temperature and rate of placement in accordance with ACI standard 347R, "Guide for Formwork for Concrete."

562.04 Design Stresses, Loads, & Deflections. The allowable maximum design stresses and loads listed in this section are based on the use of undamaged, high-quality material. If lesser quality material is used, reduce the allowable stresses and loads. Do not exceed the following maximum stresses, loads, and deflections in the falsework design:

*(a) **Timber.*** For timber, use the following values:

Compression perpendicular to the grain = 3,100 kPa
Compression parallel to the grain[1] = 3,309 MPa/$(L/d)^2$

[1]Compression parallel to the grain is not to exceed 11 MPa.

where

L = unsupported length.
d = least dimension of a square or rectangular column or the width of a square of equivalent cross-sectional area for round columns

Flexural stress[1] = 12.4 MPa

[1]Reduce flexural stress to 10 MPa for members with a nominal depth of 200 mm or less.

Horizontal shear = 1,300 kPa
Axial tension = 8.3 MPa

Deflection due to the weight of concrete may not exceed 1/500 of the span, even if the deflection is compensated for by camber strips.

The modulus of elasticity (E) for timber = 11.7 GPa
Maximum axial loading on timber piles = 400 kN

Design timber connections in accordance with the stresses and loads allowed in the "National Design Specification for Wood Construction," published by the National Forest Products Association, except:

(1) Reductions in allowable loads required for high moisture condition of the lumber and service conditions do not apply.

(2) Use 75 percent of the tabulated design value as the design value of bolts in two member connections (single shear).

(b) Steel. For identified grades of steel, do not exceed the design stresses (other than stresses due to flexural compression) specified in the "Manual of Steel Construction," Allowable Stress Design, as published by the AISC.

When the grade of steel cannot be positively identified, do not exceed the design stresses, other than stresses due to flexural compression, specified in the AISC Manual for ASTM A 36 steel or the following:

Tension, axial and flexural = 150 MPa
Compression, axial[1] = 110,000 – $2.6(L/r)^2$ kPa

[1]L/r is not to exceed 120.

Shear on the web gross section of rolled shapes = 100 MPa
Web crippling for rolled shapes = 185 MPa

For all grades of steel, do not exceed the following design stresses and deflection:

Compression, flexural[1] = 82,750 MPa/(Ld/bt)

[1]Not to exceed 150 MPa for unidentified steel or steel conforming to ASTM A 36. Not to exceed 0.6 F_y for other identified steel.

where

L = unsupported length
d = least dimension of a square or rectangular column or the width of a square of equivalent cross sectional area for round columns or depth of beams
b = width of the compression flange
t = thickness of the compression flange
F_y = specified minimum yield stress for the grade of steel used

Deflection due to the mass of concrete may not exceed 1/500 of the span, even if the deflection is compensated for by camber strips

The modulus of elasticity (E) for steel = 210 GPa

*(c) **Other Requirements.*** Limit falsework spans supporting T-beam girder bridges to 4.3 m plus 8.5 times the overall depth of the T-beam girder.

562.05 Manufactured Assemblies. For jacks, brackets, columns, joists, and other manufactured devices, do not exceed the manufacturer's recommendations or 40 percent of the ultimate load-carrying capacity of the assembly based on the manufacturer's tests or additional tests ordered. Limit the maximum allowable dead load deflection of joists to 1/500 of their spans.

Furnish catalog or equivalent data showing the manufacturer's recommendations, or perform tests, as necessary, to demonstrate the adequacy of any manufactured device proposed for use. Do not substitute other manufacturers' components unless the manufacturer's data encompass such substitutions, or field tests reaffirm the integrity of the system.

If a component of the falsework system consists of a steel frame tower more than two or more tiers high, the differential leg loading within the steel tower unit shall not exceed 4 to 1. An exception may be approved if the manufacturer of the steel frame certifies, based on manufacturer's tests, that the proposed differential loadings are not detrimental to the safe load-carrying capacity of the steel frame.

562.06 Falsework Foundations. Field-verify all ground elevations at proposed foundation locations before design.

Where spread footing type foundations are used, determine the bearing capacity of the soil. The maximum allowable bearing capacity for foundation material, other than rock, is 190 kPa.

Do not locate the edge of footings closer than 300 mm from the intersection of the bench and the top of the slope. Unless the excavation for footings is adequately supported by shoring, do not locate the edge of the footings closer than 1.2 m or the depth of excavation, whichever is greater, from the edge of the excavation.

When a pile type foundation is used, use in accordance with Section 551. When falsework is supported by footings placed on paved, well-compacted slopes of berm fills, do not strut the falsework to columns unless the column is founded on rock or supported by piling.

Size spread footings to support the footing design load at the assumed bearing capacity of the soil without exceeding anticipated settlements. Provide steel reinforcement in concrete footings.

When individual steel towers have maximum leg loads exceeding 130 kN, provide for uniform settlement under all legs or each tower under all loading conditions.

Protect the foundation from adverse effects for the duration of its use. Advise the CO of actions that will be taken to protect the foundation.

562.07 Falsework Over or Adjacent to Roadways & Railroads. Design and construct the falsework to be protected from vehicle impact. This includes falsework posts that support members crossing over a roadway or railroad and other falsework posts if they are located in the row of falsework posts nearest to the roadway or railroad and if the horizontal distance from the traffic side of the falsework to the edge of pavement or to a point 3 m from the centerline of track is less than the total height of the falsework.

Provide additional features to ensure that this falsework will remain stable if subjected to impact by vehicles. Use vertical design loads for these falsework posts, columns, and towers (but not footings) that are not less than either of the following:

(a) 150 percent of the design load calculated in accordance with Subsection 562.03, but not including any increased or readjusted loads caused by prestressing forces.

(b) The increased or readjusted loads caused by prestressing forces.

Install temporary traffic barriers before erecting falsework towers or columns adjacent to an open public roadway. Locate barriers so that falsework footings or pile caps are at least 75 mm clear of concrete traffic barriers, and all other falsework members are at least 300 mm clear. Do not remove barriers until approved.

Use falsework columns that are steel with a minimum section modulus about each axis of 156,000 mm^3 or sound timbers with a minimum section modulus about each axis of 4,100,000 mm^3.

Mechanically connect the base of each column or tower frame supporting falsework over or immediately adjacent to an open public road to its supporting footing or provide other lateral restraint to withstand a force of not less than 9 kN applied to the base of the column in any direction. Mechanically connect such columns or frames to the falsework cap or stringer to resist a horizontal force of not less than 4.5 kN in any direction. Neglect the effects of frictional resistance.

For exterior girders upon which overhanging bridge deck falsework brackets are hung, brace or tie them to the adjacent interior girders as necessary to prevent rotation of the exterior girders or overstressing of the exterior girder web.

Mechanically connect all exterior falsework stringers and stringers adjacent to the end of discontinuous caps, the stringer or stringers over points of minimum vertical clearance, and every fifth remaining stringer to the falsework cap or framing. Provide mechanical connections capable of resisting a load in any direction, including uplift on the stringer, of not less than 2.2 kN. Install connections before traffic is allowed to pass beneath the span.

Use 16-mm-diameter or larger bolts to connect timber members used to brace falsework bents located adjacent to roadways or railroads.

Sheath falsework bents within 6 m of the centerline of a railroad track solid in the area between 1 and 5 m above the track on the side facing the track. Construct sheathing of plywood not less than 16 mm thick or lumber not less than 25 mm nominal thickness. Provide adequate bracing on such bents so that the bent resists the required assumed horizontal load or 22 kN, whichever is greater, without the aid of sheathing.

Provide at least the minimum required vertical and horizontal clearances through falsework for roadways, railroads, pedestrians, and boats.

562.08 Falsework for Steel Structures. Conform to the following:

(a) Use falsework design loads consisting of the mass of structural steel, the load of supported erection equipment, and all other loads supported by the falsework.

(b) Design falsework and forms for concrete supported on steel structures so that loads are applied to girder webs within 150 mm of a flange or stiffener. Distribute the loads in a manner that does not produce local distortion of the web. Do not use deck overhang forms that require holes to be drilled in the girder webs.

(c) Strut and tie exterior girders supporting overhanging deck falsework brackets to adjacent interior girders to prevent distortion and overstressing of the exterior girder web.

(d) Do not apply loads to existing, new, or partially completed structures that exceed the load-carrying capacity of any part of the structure in accordance with the Load Factor Design methods of the AASHTO "Standard Specifications for Highway Bridges" using Load Group IB.

(e) Build supporting falsework that will accommodate the proposed method of erection without overstressing the structural steel, and will produce the required final structural geometry, intended continuity, and structural action.

562.09 Falsework Construction. Construct falsework as SHOWN ON THE DRAWINGS.

When welding is required, submit a welder certification for each welder, in accordance with Subsection 555.18.

Build camber into the falsework to compensate for falsework deflection and anticipated structure deflection. Camber as SHOWN ON THE DRAWINGS or specified by the CO is for anticipated structure deflection only.

Attach tell-tales to soffit of concrete forms in enough systematically placed locations to be able to determine from the ground the total settlement of the structure while concrete is placed.

Do not apply dead loads, other than forms and reinforcing steel, to any falsework until authorized.

When the falsework installation is complete and when SHOWN ON THE DRAWINGS or specified in the SPECIAL PROJECT SPECIFICATIONS, have the falsework inspected by a licensed professional engineer proficient in structural engineering. Certify in writing that the falsework installation conforms to accepted falsework drawings, specifications, and acceptable engineering practices. Provide a copy of the certification to the CO prior to concrete placement.

Discontinue concrete placement and take corrective action if unanticipated events occur, including settlements that cause a deviation of more then 10 mm from those SHOWN ON THE DRAWINGS. If satisfactory corrective action is not taken before initial set, remove all unacceptable concrete.

562.10 Forms. For exposed concrete surfaces, use U.S. Product Standard 1 for Exterior B–B (Concrete Form) class I plywood or other approved material that will produce a smooth and uniform concrete surface. Use only form panels in good condition and free of defects on exposed surfaces. If form panel material other than plywood is used, ensure that it has flexural strength, modulus of elasticity, and other physical properties equal to or greater than the physical properties for the type of plywood specified.

Furnish and place form panels for exposed surfaces in uniform widths of not less than 1 m and in uniform lengths of not less than 2 m, except where the width of the member formed is less than 1 m.

Arrange panels in symmetrical patterns conforming to the general lines of the structure. Place panels for vertical surfaces with the long dimension horizontal and with horizontal joints level and continuous. For walls with sloping footings that do not abut other walls, panels may be placed with the long dimension parallel to the footing.

Precisely align form panels on each side of the panel joint by means of supports or fasteners common to both panels. Provide 19-mm triangular fillets at all sharp edges of the concrete, unless otherwise SHOWN ON THE DRAWINGS.

Devices may be cast into the concrete for later use in supporting forms or for lifting precast members. Do not use driven devices for fastening forms or form supports to concrete. Use form ties consisting of form bolts, clamps, or other devices necessary to prevent spreading of the forms during concrete placement.

Do not use form ties consisting of twisted wire loops. Use form ties and anchors that can be removed without damaging the concrete surface. Construct metal ties or anchorages within the forms to permit their removal to a depth of at least 25 mm from the face without damage to the concrete. Fill cavities with cement mortar in accordance with Subsection 701.04, and finish to a sound, smooth, uniform colored surface.

Construct all exposed concrete surfaces that will not be completely enclosed or hidden below the permanent ground surface so the formed surface of the concrete does not undulate more than 2.5 mm or 1/360 of the center-to-center distance between studs, joists, form stiffeners, form fasteners, or wales. Interior surfaces of underground drainage structures are considered to be completely enclosed surfaces. Form all exposed surfaces for each element of a concrete structure with the same forming material or with material that produces similar surface textures, color, and appearance.

Support forms for cast-in-place concrete bridge decks on the girders upon which the deck is to be cast. Do not shore deck forms to the ground or to the substructure.

Support roadway slab forms of box girder type structures on wales or similar supports fastened, as nearly as possible, to the top of the web walls.

Construct concrete forms mortar-tight, true to the dimensions, lines, and grades of the structure, and of sufficient strength to prevent appreciable deflection during placement of concrete. Place all material required to be embedded in the concrete before concrete placement. Clean inside surfaces of forms of all dirt, mortar, and foreign material. Remove all loose material before the completion of forming for the roadway deck slab of cast-in-place box girders or cells or voids of other members in which the forms are to either remain in place or be removed.

Form exposed curved surfaces to follow the shape of the curve. However, on any retaining walls that follow a horizontal curve, the wall stems may be a series of short chords if all of the following conditions apply:

- The chords within the panel are the same length.
- The chords do not vary from a true curve by more than 15 mm at any point.

- All panel points are on the true curve.

When architectural treatment is required, make the angle points for chords in wall stems fall at vertical rustication joints.

Coat with form oil all forms to be removed. Use commercial-quality form oil or an equivalent coating that permits release of the forms and does not discolor the concrete. Do not place concrete in forms until the forms have been inspected and approved.

(a) Stay-in-Place Deck Forms. Use permanent or stay-in-place forms only when SHOWN ON THE DRAWINGS.

Fabricate permanent steel bridge deck forms and supports from steel conforming to ASTM A 653M, coating designation 2600, any grade except 340, class 3.

Install forms in accordance with approved fabrication and erection drawings. Do not rest form sheets directly on the top of stringer or floor beam flanges. Securely fasten sheets to form supports. Place form supports in direct contact with the stringer flange or floor beam. Make all attachments with permissible welds, bolts, or clips. Do not weld form supports to flanges of steels not considered weldable or to portions of flanges subject to tensile stresses.

Clean, wire brush, and paint with two coats of zinc dust zinc-oxide primer (FSS TT–P–641 type II, no color added) any permanently exposed form metal where the galvanized coating has been damaged. Minor heat discoloration in areas of welds need not be touched up.

Locate transverse construction joints in slabs at the bottom of a flute. Field-drill 6-mm-diameter weep holes at not less than 300 mm on center along the line of the joint.

(b) Void Forms. Store void forms in a dry location to prevent distortion. Secure the forms using anchors and ties that leave a minimum of metal or other supporting material exposed at the bottom of finished slab.

Make the outside surface of the forms waterproof. Cover the ends with waterproof mortar-tight caps. Use a premolded 6-mm-thick rubber joint filler around the perimeter of the caps to permit expansion.

Provide a PVC vent near each end of each void form. Construct vents so the vent tube does not extend more than 13 mm below the bottom surface of the finished concrete after form removal. Protect void forms from the weather until concrete is placed.

(c) Metal Forms. The specifications for forms relative to design, mortar tightness, filleted corners, beveled projections, bracing, alignment, removal, reuse, and oiling also apply to metal forms.

562.11 Removal of Forms & Falsework. Remove all forms except:

(a) Interior soffit forms for roadway deck slabs of cast-in-place box girders.

(b) Forms for the interior voids of precast members.

(c) Forms for abutments or piers when no permanent access is available into the cells or voids.

To facilitate finishing, when approved by the CO, the removal of forms that do not support the dead load of concrete members and of forms for railings and barriers may begin 24 hours after the concrete for the member has been placed. Protect exposed concrete surfaces from damage. Cure all exposed concrete surfaces in accordance with Subsection 552.17, if forms are removed less than 7 days after concrete placement.

Do not remove forms and falsework until the concrete strength and time requirements in table 562-3 have been met.

Do not remove falsework under concrete that has been cured at a temperature continuously under 10 °C without first determining if the concrete has gained the specified strength, no matter how much time has passed.

Ensure that substructure concrete has reached the required 28-day compressive strength prior to erecting any superstructure or additional substructure elements, unless approved otherwise by the CO.

Do not release falsework in any span in continuous structures until the first and second adjoining spans on each side have reached the strength specified herein or in the SPECIAL PROJECT SPECIFICATIONS.

Uniformly and gradually remove falsework for arch bridges, beginning at the crown and working toward the springing. Remove falsework for adjacent arch spans simultaneously.

Completely release the falsework under all spans of continuous structures before concrete is placed in curbs, railings, and parapets.

Remove forms from columns before releasing supports from beneath beams and girders in order to determine the condition of column concrete.

Table 562-3.—Minimum form/support release criteria.

Structural Element	% of Specified 28-Day Strength (f_c')	Minimum Number of Days Since Last Pour	
		Standard Concrete	Type III Concrete
Columns and wall faces (not yet supporting loads)	50	3	2
Mass piers and mass abutments (not yet supporting loads) except pier caps	50	3	N/A
Box girders	80	14	7
Simple span girders, T-beam girders, slab bridges, cross beams, caps, pier caps not continuously supported, struts, and top slabs of concrete box culverts	80	14	7
Trestle slabs where supported on wood stringers	70	10	4
Slabs and overhangs where supported on steel stringers or prestressed concrete girders	70	10	4
Pier caps continuously supported	60	7	3
Arches, continuous span bridges, and rigid frames	90	21	10

Remove all forms from the cells of concrete box girders unless otherwise SHOWN ON THE DRAWINGS or permitted by the CO. Leave no forms that might jeopardize drainage or enclosed utilities.

Do not release falsework for cast-in-place prestressed portions of structures until after the prestressing steel has been tensioned.

Do not remove falsework supporting the deck of rigid frame structures, excluding box culverts, until compacted backfill material has been placed against vertical legs of the frame.

Install a reshoring system if the falsework supporting the sides of girder stems with slopes steeper than 1:1 are removed before placing deck slab concrete. Design a reshoring system with lateral supports that resist all rotational forces acting on the stem, including those caused by the placement of deck slab concrete. Install the lateral supports immediately after each form panel is removed and before release of supports for the adjacent form panel.

Completely remove falsework material. Remove falsework piling at least 0.5 m below the surface of the original ground or stream bed. Where falsework piling is driven within the limits of ditch or channel excavation, remove the piling to at least 0.5 m below the bottom and side slopes of the excavated areas.

Leave the forms for footings constructed within a cofferdam or crib in place when their removal would endanger the safety of the cofferdam or crib, and where the forms will not be exposed to view in the finished structure.

Remove all other forms, whether above or below groundline or water level.

Measurement & Payment

562.12 Method & Basis. Use the method of measurement that is DESIGNATED IN THE SCHEDULE OF ITEMS.

Do not measure forms and falsework for payment.

Section 563—Painting

Description

563.01 Work. Apply protective coatings to metal, timber, or concrete surfaces to control corrosion and deterioration.

Materials

563.02 Requirements. Furnish material that conforms to specifications in the following section and subsections:

Boiled Linseed Oil	725.14(a)
Paint	708
Petroleum Spirits (Mineral Spirits)	725.14(b)
Water	725.01

Construction

563.03 Protection of Public, Property, & Workers. Comply with the SSPC's "SSPC–PA Guide 3—A Guide to Safety in Paint Application" and with OSHA requirements. If the paint being removed is a hazardous material containing lead chromium, comply with all of the following:

- SSPC Guide 6I(CON)—"Guide for Containing Debris Generated During Paint Removal Operations."
- SSPC Guide 7I(DIS)—"Guide for the Disposal of Lead-Contaminated Surface Preparation Debris."
- 29 CFR 1926.62—"OSHA Construction Industry Standards for Lead."
- 40 CFR 50.6—"EPA National Primary and Secondary Ambient Air Quality Standards for Particulate Matter."
- 40 CFR, 50.12—"EPA National Primary and Secondary Ambient Air Quality Standards for Lead."
- 40 CFR, parts 260–268—"Resource Conservation and Recovery Act (RCRA)."

At least 28 days before beginning surface preparation, submit a written plan for approval that details the measures to be used for the protection of the environment, public, adjacent property, and the workers while performing the work. Include in the plan the following:

(a) Manufacturer's material safety data sheets and product sheets for all cleaning and painting products.

(b) A detailed containment plan for removed material, cleaning products, and paint debris. Include details of attachment to the structure.

(c) A detailed disposal plan for removal, cleaning products, and paint debris.

(d) Specific safety measures to protect workers from site hazards, including falls, fumes, fires, or explosions.

(e) If paint being removed is hazardous material, include specific safety measures to comply with 29 CFR 1962.26, 40 CFR 50.6, 40 CFR 50.12, and 40 CFR, parts 260–268. Document compliance upon request.

(f) A written plan for emergency spill procedures.

(g) A competent person responsible for ensuring that all necessary health, safety, and containment measures are enacted and maintained.

After acceptance, perform work according to the plan. If the measures fail to perform as intended, immediately stop work and take corrective action. Collect and properly dispose of all material, including wastewater that is used in preparing, cleaning, or painting.

563.04 Protection of the Work. Use tarps, screens, paper, cloth, or other suitable means to protect adjacent surfaces that are not to be painted. Prevent contamination of freshly painted surfaces by dust, oil, grease, or other harmful and deleterious material.

563.05 Surface Preparation, General. Notify the CO in writing at least 7 days before beginning operations. Immediately before painting, prepare the surface according to the following:

(a) Clean the surface to the specified cleanliness level.

(b) Remove dirt, dust, and other contaminants from the surface using methods recommended by the paint manufacturer.

(c) Thoroughly dry the surface.

(d) Determine that the surface temperature is between 10 °C and 40 °C.

(e) Determine that the surface temperature is 3 °C or more above the dew point according to ASTM 337.

(f) Determine that the humidity is 85 percent or less, unless specified otherwise on the manufacturer's product data sheet.

Suitable engineering control, such as enclosures and dehumidification, may be used to provide the conditions required above.

563.06 Paint Application, General. Use safe handling practices that conform to the manufacturer's safety data sheet and instructions. Mix and apply paint according to the product instructions. Mix paint with mechanical mixers for a sufficient length of time to thoroughly blend the pigment and vehicle together. Continue the mixing during application. Do not thin paint that is formulated ready for application.

Paint in a neat and workmanlike manner that does not produce excessive paint buildup, runs, sags, skips, holidays, or thin areas in the paint film. Measure the wet film thickness during application, and adjust the application rate such that, after curing, the desired dry film thickness is obtained. Apply paint by brush, spray, roller, or any combination thereof if permitted by the manufacturer's product data sheet.

Use brushes that have sufficient bristle body and length to spread the paint in a uniform film. Use round, oval-shaped, or flat brushes no wider than 120 mm. Evenly spread and thoroughly brush out the paint as it is applied.

Use airless or conventional spray equipment with suitable traps, filters, or separators to exclude oil and water from the compressed air. Use the spray gun tip sizes and pressures recommended by the manufacturer. Use compressed air that is free from oil or moisture and does not show black or wet spots when tested in accordance with ASTM D 4285.

Use rollers only on flat, even surfaces. Do not use rollers that leave a stippled texture in the paint film.

Use sheepskin daubers, bottle brushes, or other acceptable methods to paint surfaces that are inaccessible for painting by regular means.

Cure each coat of paint according to the manufacturer's recommendations. Correct all thin areas, skips, holidays, and other deficiencies before the next application of paint. Tint succeeding applications of paint to contrast with the paint being covered. The CO will approve the color for the finish coat before application.

Coat structures with the total thickness of undercoats before erection. Coat any surfaces that will be inaccessible after erection with the full number of required applications before erection. After erection and before applying the final coat, thoroughly clean all areas where coating has been damaged or has deteriorated, or where there are exposed unpainted surfaces, and spot coat with the specified undercoats to the specified thickness.

563.07 Structural Iron & Steel. Conform to the following:

(a) Paint Systems. Conform to the following:

(1) New Surfaces or Surfaces With All Existing Paint Removed. Furnish a paint system shown in table 563-1.

(2) Surfaces With Existing Sound Paint. Furnish a paint system that is compatible with the existing paint. Any of the systems listed in table 563-2 or any system that is approved for use on steel structures by the State department of transportation in the State in which the structure is located is acceptable if the proposed system is compatible with the existing system.

At least 14 days before ordering paint, verify compatibility of the proposed system with the existing system as follows:

(a) Select a test area of at least 3 m^2 in a condition representative of the condition of the structure. Perform the specified level of surface preparation and apply the proposed system to the existing topcoat and primer. Watch for lifting, bleeding, blistering, wrinkling, cracking, flaking, or other evidence of incompatibility.

(b) Verify that no indication of incompatibility exists at least 14 days after the application of each product. Perform adhesion tests according to ASTM D 3359, method A. Notify the CO immediately if adhesive testing fails at the interface of the existing finish coat and primer. An adhesion failure indicates incompatibility. Choose a more compatible paint system.

(b) Surface Preparation. Do not remove sound paint unless SHOWN ON THE DRAWINGS.

(1) New Surfaces or Surfaces With All Existing Paint Removed. Remove all dirt, mill scale, rust, paint, and other foreign material from exposed surfaces by blast cleaning to near white metal in accordance with SSPC–SP 10.

Use compressed air that is free from oil or moisture and does not show black or wet spots when tested in accordance with ASTM D 4285. Do not use unwashed sand or abrasives that contain salts, dirt, oil, or other foreign matter. Before blast cleaning

Table 563-1.—Structural iron and steel coating systems for new surfaces and surfaces with all existing paint removed.

	Paint System[a]				
	1	2	3	4	5
Coat	Aggressive Environments (Salt)	Aggressive Environments (Salt)	Aggressive Environments (Salt)	Less Aggressive Environments (No Salt)	Less Aggressive Environments (No Salt)
Primer	Inorganic zinc, type II, 75–100 µm dry	Zinc-rich epoxy, 75–100 µm dry	Moisture-cured urethane, 50–75 µm dry	Acrylic latex, 50–75 µm dry	Low-VOC alkyd, 50–75 µm dry
Intermediate	Epoxy, 75–100 µm dry	Epoxy, 75–100 µm dry	Moisture-cured urethane, 50–75 µm dry	Acrylic latex, 50–75 µm dry	Low-VOC alkyd, 50–75 µm dry
Top	Aliphatic urethane, 50–75 µm dry	Aliphatic urethane, 50–75 µm dry	Moisture-cured or aliphatic urethane, 50–75 µm dry	Acrylic latex, 50–75 µm dry	Low-VOC alkyd, 50–75 µm dry
Total thickness	200–275 µm dry	200–275 µm dry	150–225 µm dry	150–225 µm dry	150–225 µm dry

a. System 1, 2, or 3 is for the corrosion protection of iron and steel in aggressively corrosive atmospheric environments, such as marine, industrial, or high-humidity environments, and in structures exposed to deicing salts. System 4 or 5 is for use in environments that are free from high concentrations of salts or pollutants that cause aggressive corrosion.

Table 563-2—Structural iron and steel coating systems for surfaces with existing sound paint.

Coat	Paint System[a] 6 Aggressive Environments (Salt)	Paint System[a] 7 Less Aggressive Environments (No Salt)	Paint System[a] 8 Less Aggressive Environments (No Salt)
Primer	Moisture-cured urethane, 50–75 μm dry	Low-VOC alkyd, 50–75 μm dry	Low-viscosity epoxy sealer, 25–50 μm dry
Intermediate	Moisture-cured urethane, 50–75 μm dry	Low-VOC alkyd, 50–75 μm dry	Epoxy, 75–100 μm dry
Top	Moisture-cured or aliphatic urethane, 50–75 μm dry	Low-VOC silicone-alkyd, 50–75 μm dry	Aliphatic urethane, 50–75 μm dry
Total thickness	150–225 μm dry	150–225 μm dry	150–225 μm dry

a. System 6 is for the corrosion protection of iron and steel in aggressively corrosive atmospheric environments, such as marine, industrial, or high-humidity environments, and in structures exposed to deicing salts. System 7 or 8 is for use in environments that are free from high concentrations of salts or pollutants that cause aggressive corrosion.

near machinery, seal all bearings, journals, motors, and moving parts against entry of abrasive dust.

Blast clean with clean, dry sand, mineral grit, steel shot, or steel grit. Use a suitable gradation to produce a dense, uniform anchor pattern. Produce an anchor profile height of 25 to 50 μm, but not less than that recommended by the manufacturer's product data sheet for the paint system specified. Measure anchor profile height using the tape method in accordance with ASTM D 4417.

The same day cleaning is performed, remove dirt, dust, and other debris from the surface by brushing, blowing with clean, dry air, or vacuuming and apply the first coat of paint to the blast-cleaned surfaces. If the cleaned surfaces rust or become contaminated before painting, repeat blast cleaning.

(2) Surfaces With Existing Sound Paint. Wash all areas to be painted with pressurized water to remove dirt, surface chalking, loose rust, and contaminants such as chlorides. Maintain a washwater pressure of at least 3.5 MPa. Capture all washwater and removed waste according to appropriate regulations.

Clean according to SSPC–SP 2, Hand Tool Cleaning; SSPC–SP 3, Power Tool Cleaning; or SSPC–SP 6, Commercial Blast Cleaning, to remove dirt, loose mill scale, loose rust, or paint that is not firmly bonded to the underlying surface. Clean small areas that show pinhole corrosion, stone damage from traffic, or minor scratches. Clean at least 50 mm beyond the damaged areas. Feather edges of remaining old paint to achieve a reasonably smooth surface.

The same day hand- or power-tool cleaning is performed, remove dirt, dust, and other contaminants from the surface with solvent cleaning methods according to SSPC–SP 1, and spot paint all bare steel areas cleaned with the first coat of paint. If the cleaned surfaces rust or become contaminated before painting, repeat solvent cleaning. Repair all damage to sound paint by applying the entire system.

(c) Application of Paints. Apply each coat to the wet film thickness as recommended by the paint manufacturer to obtain the specified dry film thickness. Verify the application rate of each coat with a wet film paint thickness gauge immediately after applying paint to the surface. Confirm the application rate by measuring the dry film thickness after the solvent has evaporated from the surface.

For example, if 75 μm of dry thickness is desired and the volatile content of the paint is 50 percent, the wet film paint thickness gauge must read at least 150 μm immediately after application of the paint to achieve the desired dry coat thickness of 75 μm.

563.08 Painting Galvanized Surfaces. Clean and prepare the surface to be painted by washing with a mineral spirit solvent to remove all oil, grease, or other contaminants on the surface, in accordance with SSPC–SP 1.

Apply the coating system shown in table 563-3 for other metals.

563.09 Painting Timber Structures. Dry timber to a moisture content of 20 percent or less. On previously painted timber, remove all cracked or peeled paint, loose chalky paint, dirt, and other foreign material by wire brushing, scraping, or other approved methods. On timber treated with creosote or oilborne pentachlorophenol preservative, wash and brush away visible salt crystals on the wood surface and allow to dry. Remove all dust or other foreign material from the surface to be painted.

Apply the coating system shown in table 563-3. The primer may be applied before erection. After the primer dries and the timber is in place, fill all cracks, checks, nail holes, or other depressions flush with the surface using approved putty. Evenly spread and thoroughly work the paint into all corners and recesses. Allow the full thickness of the applied coat of paint to dry before applying the next coat.

Table 563-3—Coating systems for other structures.

Substrate	Paint Coatings			
	Primer	Intermediate	Finish	Total Thickness
Smooth wood	Exterior wood primer,[a] 60–70 μm dry	Exterior latex or alkyd, 35–50 μm dry	Exterior latex or alkyd, 35–50 μm dry	130–170 μm dry
Rough lumber	Exterior latex or alkyd,[a] 35–50 μm dry	Exterior latex or alkyd, 35–50 μm dry	Exterior latex or alkyd, 35–50 μm dry	105–150 μm dry
Concrete	Epoxy single coat, 80–100 μm dry. For gloss finish, finish with aliphatic polyurethane (50 μm dry).			80–150 μm dry
Masonry block	Masonry block filler, 50–60 μm dry	Exterior latex or alkyd, 35–50 μm dry	Exterior latex or alkyd, 35–50 μm dry	120–160 μm dry
Aluminum	Metal primer, 30–40 μm dry	Exterior latex or alkyd, 35–50 μm dry	Exterior latex or alkyd, 35–50 μm dry	100–140 μm dry
Other metals	Metal primer,[b] 35–45 μm dry	Exterior latex or alkyd, 35–50 μm dry	Exterior latex or alkyd, 35–50 μm dry	105–145 μm dry

a. For untreated wood, thin the primer with up to 0.1 L of turpentine and 0.1 L of linseed oil per liter of paint.

b. For galvanized surfaces, use an epoxy primer (35–45 μm dry thickness) or a vinyl wash primer (7–13 μm dry thickness).

563.10 Painting Concrete Structures. Clean and prepare the concrete surface to be painted by removing all laitance, dust, foreign material, curing compound, form oil, grease, or other deleterious material. If form oil, grease, or curing compound is present, wash the surface clean with a 5 percent solution of trisodium phosphate. After washing, thoroughly rinse the surface with clean water and allow to dry completely.

Give the cleaned surface a light abrasive sweep to remove mortar wash or other contaminants. Remove all residue and dust by hand, broom, compressed air, or other approved methods.

Apply the coating system shown in table 563-3. Evenly spread and thoroughly work the paint into all corners and recesses. Allow the full thickness of the applied coat of paint to dry before applying the succeeding coat.

Measurement

563.11 Method. Use the method of measurement that is DESIGNATED IN THE SCHEDULE OF ITEMS.

Measure painting by the square meter or lump sum.

When measurement is by the square meter, measure the visible surface area painted.

Payment

563.12 Basis. The accepted quantities will be paid for at the contract unit price for each PAY ITEM DESIGNATED IN THE SCHEDULE OF ITEMS.

Payment will be made under:

Pay Item	Pay Unit
563 (01) Painting, ____________ structure	Lump Sum
Description	
563 (02) Painting, _____________ structure	Square Meter
Description	

Section 564—Bearing Devices

Description

564.01 Work. Furnish and install bridge bearings. Bearing devices are designated as elastomeric, rocker, roller, and sliding plate.

Materials

564.02 Requirements. Furnish material that conforms to specifications in the following subsections:

Elastomeric Bearing Pads 717.10
TFE Surfaces for Bearings 717.11

Construction

564.03 General. Conform to the fallowing:

(a) ***Drawings.*** Prepare and submit drawings for the bearings in accordance with section 18, AASHTO "Standard Specifications for Highway Bridges," division II, volume II, when SHOWN ON THE DRAWINGS or in the SPECIAL PROJECT SPECIFICATIONS. Show all details of the bearings, including the material proposed for use. Obtain approval before beginning fabrication.

(b) ***Fabrication.*** Fabricate bearings in accordance with section 18 of the AASHTO "Standard Specifications for Highway Bridges," division II, volume II. Ensure that the surface finish of bearing components in contact with each other or with concrete, but not embedded in concrete, conforms to Subsection 555.08(e).

Preassemble bearing assemblies in the shop and check for proper completeness and geometry. Galvanize steel bearing components and anchor bolts in accordance with Subsection 717.07. Do not galvanize stainless steel bearing components or anchor bolts.

(c) ***Packaging, Handling, & Storage.*** Before shipping from the manufacturer, clearly identify each bearing component, and mark on its top the location and orientation in the structure. Securely bolt, strap, or otherwise fasten the bearings to prevent any relative movement.

Package bearings so they are protected from damage due to shipping, handling, weather, or other hazards. Do not dismantle bearing assemblies at the site except for inspection or installation.

Store all bearing devices and components at the worksite in a location that provides protection from environmental and physical damage.

(d) Construction & Installation. Clean the bearings of all deleterious substances. Install the bearings at the positions SHOWN ON THE DRAWINGS. Set bearings and bearing components to the dimensions SHOWN ON THE DRAWINGS or as prescribed by the manufacturer. Adjust in accordance with the manufacturer's instructions to compensate for installation temperature and future movements of the bridge.

Set bridge bearings level at the exact elevation and position. Provide full and even bearing on all external bearing contact surfaces. If bearing surfaces are at improper elevations or not level, or if bearings cannot otherwise be set properly, notify the CO and submit a written proposal to modify the installation for approval.

Bed metallic bearing assemblies that are not embedded in concrete on concrete with an approved filler or fabric material.

Set elastomeric bearing pads directly on properly prepared concrete surfaces without bedding material.

Machine all bearing surfaces that are seated directly on steel to provide a level and planar surface upon which to place the bearing.

564.04 Elastomeric Bearings. The bearings include nonreinforced pads (consisting of elastomer only) and reinforced bearings with steel or fabric laminates.

Reinforce elastomeric bearings that are more than 15 mm thick with laminates every 15 mm through the entire thickness.

If not specified, use 50 durometer elastomer that is capable of sustaining an average compressive stress of 7 MPa.

Fabricate elastomeric bearings in accordance with AASHTO M 251. Use material that meets the flash tolerance, finish, and appearance requirements of the "Rubber Handbook" published by the Rubber Manufacturer's Association Incorporated, RMA F3 and T.063 for molded bearings and RMA F2 for extruded bearings. Determine compliance with AASHTO M 251, Level I acceptance criteria.

Mark each reinforced bearing with indelible ink or flexible paint. Mark the order number, lot number, bearing identification number, and elastomer type and grade number. Unless otherwise specified, mark on a face that is visible after erection of the bridge. Furnish a list of all individual bearing numbers.

Place bearings on a level surface. Correct any misalignment in the support to form a level surface. Do not weld steel girders or base plates to the exterior plates of the bearing unless there is more than 40 mm of steel between the weld and elastomer. Do not expose the elastomer or elastomer bond to instantaneous temperatures greater than 200 °C.

564.05 Rocker, Roller, & Sliding Bearings. When TFE coatings are required, use coatings that conform to Subsection 564.07.

Fabricate rocker, roller, and sliding bearings in accordance with the details SHOWN ON THE DRAWINGS and with Section 555. Perform fabrication in accordance with standard practice in modern commercial shops. Remove burrs, rough and sharp edges, and other flaws. Stress-relieve rocker, roller, and other bearings that are built up by welding sections of plate together before boring, straightening, or finished machining.

Thoroughly coat all contact surfaces with oil and graphite just before placing roller bearings. Install rocker, roller, and sliding bearings so they are vertical at the specified mean temperature after release of falsework and after any shortening due to prestressing forces. Take into account any variation from mean temperature of the supported span at time of installation and any other anticipated changes in length of the supported span.

Make sure the superstructure has full and free movement at movable bearings. Carefully position cylindrical bearings so that their axes of rotation align and coincide with the axis of rotation of the superstructure.

564.06 Masonry, Sole, & Shim Plates for Bearings. Provide metal plates used in masonry, sole, and shim plates that conform to AASHTO M 270M, grade 250.

Fabricate and finish steel in accordance with Section 555. Form holes in bearing plates by drilling, punching, or accurately controlled oxygen cutting. Remove all burrs by grinding.

Accurately set bearing plates in level position as SHOWN ON THE DRAWINGS and provide a uniform bearing over the bearing contact area. When plates are embedded in concrete, make provision to keep them in correct position as the concrete is placed.

564.07 TFE Surfaces for Bearings. Furnish TFE material that is factory bonded, mechanically connected, or recessed into the backup material, as SHOWN ON THE DRAWINGS.

Bond or mechanically attach the fabric containing TFE fibers to a rigid substrate. Use a fabric capable of carrying unit loads of 70 MPa without cold flow. Use a fabric-substrate bond capable of withstanding, without delamination, a shear force equal to 10 percent of the perpendicular or normal application loading, plus any other bearing shear forces.

Determine compliance using approved test methods and procedures in accordance with section 18, subsection 18.8.3, AASHTO "Standard Specifications for Highway Bridges," division II, volume II. If the test facility does not permit testing completed bearings, manufacture extra bearings and prepare samples of at least 450 kN capacity at normal working stresses.

Determine static and dynamic coefficient of friction at first movement of the test bearing at a sliding speed of less than 25 mm per minute. Ensure that the coefficient of friction does not exceed the coefficient of friction specified in table 564-1 or by the manufacturer.

Table 564-1.—Coefficient of friction.

Material	Bearing Pressure (MPa)	Friction Coefficient
Unfilled TFE, fabric containing TFE fibers, or TFE-perforated metal composite	3.5	0.08
	14	0.06
	24	0.04
Filled TFE	3.5	0.12
	14	0.10
	24	0.08
Interlocked bronze and filled TFE structures	3.5	0.10
	14	0.07
	24	0.05

Furnish a listing of all individual bearing numbers.

564.08 Anchor Bolts. Furnish wedge or thread anchor bolts that conform to ASTM A 307 or as SHOWN ON THE DRAWINGS.

Drill holes for anchor bolts and set them in Portland cement nonshrink grout or preset them before placing the concrete.

Adjust bolt locations for superstructure temperature as required. Do not restrict free movement of the superstructure at movable bearings through anchor bolts or nuts.

564.09 Bedding of Masonry Plates. Place filler or fabric as bedding material under masonry plates, as SHOWN ON THE DRAWINGS. Use the type of filler or fabric specified and install to provide full bearing on contact areas. Thoroughly clean the contact surfaces of the concrete and steel immediately before placing the bedding material and installing bearings or masonry plates.

Measurement

564.10 Method. Use the method of measurement that is DESIGNATED IN THE SCHEDULE OF ITEMS.

Measure bearing devices by the each.

Payment

564.11 Basis. The accepted quantities will be paid for at the contract unit price for each PAY ITEM DESIGNATED IN THE SCHEDULE OF ITEMS.

Payment will be made under:

Pay Item	Pay Unit
564 (01) _______________ bearing device	Each
Description	

DIVISION 600
Incidental Construction

Section 601—Mobilization

Description

601.01 Work. Move personnel, equipment, material, and incidentals to the project, and perform all activities necessary to accomplish work at the project site. Obtain permits, insurance, and bonds.

Measurement

601.02 Method. Measure mobilization by the lump sum.

Payment

601.03 Basis. The accepted quantity, measured as provided above, will be paid at the contract price per unit of measurement for the PAY ITEM listed below that is DESIGNATED IN THE SCHEDULE OF ITEMS.

The mobilization lump sum will be paid as follows:

(a) If applicable, bond premiums will be reimbursed according to FAR clause 52.232–5, Payment Under Fixed-Price Construction Contracts, after receipt of evidence of payment.

(b) Fifty percent of the lump sum, not to exceed 5 percent of the original contract amount, will be paid following completion of 5 percent of the original contract amount, not including mobilization.

(c) Payment of the remaining portion of the lump sum, up to 10 percent of the original contract amount, will be paid following completion of 10 percent of the original contract amount, not including mobilization.

(d) Any portion of the lump sum in excess of 10 percent of the original contract amount will be paid after final acceptance.

Payment will be made under:

Pay Item	Pay Unit
601 (01) Mobilization	Lump Sum

Section 602—Minor Concrete Structures

Description

602.01 Work. Construct reinforced or unreinforced minor concrete structures.

Materials

602.02 Requirements. Furnish materials that meet the requirements specified in the following subsections:

Air-Entraining Admixtures	711.02
Cement	701.01
Chemical Admixtures	711.03
Coarse Aggregate for Portland Cement Concrete	703.02
Curing Material	711.01
Fine Aggregate for Portland Cement Concrete	703.01
Fly Ash	725.04
High-Strength Nonshrink Grout	701.02
Latex Modifier	711.04
Reinforcing Steel	709.01
Water	725.01

602.03 Concrete Composition. Use the concrete composition method DESIGNATED IN THE SCHEDULE OF ITEMS.

(a) Method A. Furnish to the CO a mix design showing the proposed weights of aggregate, water, and cement per cubic meter of concrete a minimum of 7 days prior to beginning placement. Proportion the cement, aggregate, and water to obtain concrete with good workability. Ensure that slump is 100 mm or less, as determined by AASHTO T 119. Ensure that air-entrainment is 6 ± 1 percent, as determined by AASHTO T 152 or T 196.

Ensure that the concrete develops a 28-day minimum compressive strength of 20 MPa, unless otherwise SHOWN ON THE DRAWINGS. Furnish concrete for specimens. Strength will be determined by test cylinders made and cured in accordance with AASHTO T 23 and tested in accordance with AASHTO T 22.

Failure of any test cylinder to meet the required strength, for any structural element tested, will be considered evidence of noncompliance with the strength requirement of this specification.

(b) Method B. Submit for approval the following information a minimum of 7 days prior to beginning placement:

(1) Type, grading, and sources of aggregate.

(2) Type and source of cement, blended cement, or fly ash.

(3) Saturated surface dry weights of the fine and coarse aggregate in kilograms per cubic meter of concrete.

(4) Weight of mixing water in kilograms per cubic meter of concrete.

(5) Weight of cement in kilograms per cubic meter of concrete.

(6) Admixture type, quantity, and certification by manufacturer.

(7) Air content.

(8) Slump.

(9) 28-day compressive strength.

Ensure that the concrete contains not less than 310 kg of cement per cubic meter. Ensure that slump is 100 mm or less, as determined by AASHTO T 119.

When a commercial supplier is used, furnish a certification with each truckload of concrete certifying that the material and mix proportions used are in conformance with the approved mixture.

(c) Method C. Make the concrete using a dry, preproportioned, blended, and bagged mix meeting the requirements of ASTM C 387 and mixed at the jobsite in accordance with the manufacturer's recommendations.

(d) Fly Ash- or Pozzolan-Modified Concrete. Fly ash may be substituted for cement at the rate of 550 g of fly ash per 450 g of Portland cement. After substitution, reduce the design aggregate volumes by an amount equal to the net increase in volume of the combined cement and fly ash. Replace no less than 10 percent and no more than 20 percent of the weight of Portland cement required with fly ash at the above rate. For purposes of controlling the maximum water/cement ratio of 0.49, make the water/cement ratio for fly-ash-modified concrete the ratio of the weight of water to the combined weights of Portland cement and 60 percent of the weight of the fly ash.

Extend the standard 28-day curing period for compressive-strength tests for fly-ash-modified concrete by 1 day (rounded to the nearest whole day) for each 1.5 percent of

Portland cement replaced with fly ash at the selected rate. (Example: If the maximum of 20 percent cement is replaced, the curing period for cylinders is 41 days.)

Construction

602.04 Forms. Design and construct forms so they can be removed without damaging the concrete. Make them free of bulge and warp, and constructed so that the finished concrete has the form and dimensions SHOWN ON THE DRAWINGS and is true to line and grade. Concrete may be placed without forms where SHOWN ON THE DRAWINGS.

Design forms for concrete that contains a retarding admixture, fly ash, or other pozzolan replacement for cement so that the lateral pressures exerted by the full anticipated height of fluidized concrete are contained, unless documented information in regard to initial set is provided by the manufacturer.

602.05 Placing Concrete. Place all reinforcing steel in position as SHOWN ON THE DRAWINGS, and ensure that it is securely held in place by approved supports during placing of concrete. Do not place concrete until the grading, forms, and steel reinforcements have been inspected and approved by the CO. Give the CO 24 hours written notice prior to placement of any concrete.

Ensure that reinforcing steel material and construction requirements are in accordance with Section 554.

Discharge all concrete prepared using methods A and B into the forms within the time limits shown in table 602-1. These time limits are based on jobsite ambient air temperature, cement type, and admixture used. Begin counting time from when the cement is introduced into the aggregate. Discharge concrete prepared using method C into the forms within 1-1/2 hours after introducing water to the mixture. Do not retemper concrete. Cement must be added to the mixer at the jobsite when required in the SPECIAL PROJECT SPECIFICATIONS. Do not mix or place concrete when the daily minimum atmospheric temperature is, or is expected to be, less than 5 °C unless adequate provisions are made to protect the concrete.

Place concrete to avoid segregation. Use high-frequency internal vibrators for consolidating concrete in the forms. Operate vibrators to produce concrete free of voids, but do not hold them in one place long enough to result in segregation or formation of laitance on the surface.

Method C concrete may be rodded instead of internally vibrated as necessary to remove voids.

Table 602-1.—Concrete discharge time limits.

Cement Type With and Without Admixtures	Time Limit (hour) < 30 °C[a]	Time Limit (hour) ≥ 30 °C[a]
Type I, IA, II, or IIA	2.0	1.5
Type I, IA, II, or IIA with water-reducing or -retarding admixture	3.0	2.0
Type III	1.5	1.0
Type III with water-reducing or -retarding admixture	2.0	1.5

a. Ambient air temperature.

Do not use aluminum pipe, conduit, or troughs for transporting concrete. When concrete is pumped, take samples from the discharge stream at the point of placement.

602.06 Finishing. Perform finishing of concrete surfaces as follows:

(a) Formed Surfaces. Unless otherwise SHOWN ON THE DRAWINGS, remove all fins and irregular projections exceeding 6 mm from the exposed surfaces. Fill holes produced by removing form ties with dry-pack mortar or other approved patching compounds.

(b) Unformed Surfaces. Strike off unformed surfaces with a straightedge, and finish them to a smooth uniform texture by floating and troweling. Prepare final finish of the surface as SHOWN ON THE DRAWINGS.

602.07 Curing Concrete. Beginning immediately after finishing, cure all concrete a minimum of 7 days or, if high-early-strength cement is used, a minimum of 3 days. For fly-ash-modified concrete placed in structures, the required moisture-controlled curing period shall be:

Percentage of Cement Replaced by Weight	Required Curing Period
10%	9 days
11–15%	10 days
16–20%	11 days

For cold weather concreting, maintain a controlled temperature for the required curing period. The above requirement for an extended curing period may be waived if a compressive strength of 65 percent of the specified 28-day design strength is achieved in 6 days.

Cure by maintaining a minimum concrete temperature of 5 °C and keeping the concrete continuously moist. Keep moist by supplying additional moisture or preventing moisture loss.

Acceptable methods of supplying additional moisture are ponding or sprinkling, and covering with burlap cloth that is kept saturated. Surfaces SHOWN ON THE DRAWINGS may be covered with saturated sand or 150 mm of saturated hay or straw to retain moisture.

Acceptable methods of preventing moisture loss are applying liquid membrane-forming compounds, or waterproof paper or polyethylene sheet materials. Apply liquid membrane-forming compounds by spraying at the coverage rates and patterns recommended by the manufacturer. Ensure that sheet material has overlapped sealed joints and forms a complete waterproof cover over the entire concrete surface.

602.08 Backfilling. Backfill in accordance with Subsection 206A.10. Do not backfill concrete until it has completed the required curing period.

Measurement

602.09 Method. Use the method of measurement that is DESIGNATED IN THE SCHEDULE OF ITEMS.

Payment

602.10 Basis. The accepted quantities will be paid for at the contract unit price for each PAY ITEM DESIGNATED IN THE SCHEDULE OF ITEMS.

Payment will be made under:

Pay Item	Pay Unit
602 (01) Concrete, method____	Cubic Meter
602 (02) Concrete, method____	Lump Sum

Section 603—Metal Pipe

Description

603.01 Work. Furnish and install, or install only, metal pipe and pipe appurtenances, including all bedding and backfilling required to complete the work. The term "metal" refers to aluminum and steel.

Materials

603.02 Requirements. Furnish materials that meet the requirements specified in the following subsections:

Aluminum-Alloy Corrugated Pipe	707.03
Aluminum-Alloy Spiral Rib Pipe	707.12
Asphalt-Coated Pipe	707.04
Concrete-Lined Corrugated Steel Pipe	707.13
Ductile Iron Culvert Pipe	707.01
Fiber-Bonded Bituminous-Coated Steel Pipe	707.09
Invert-Paved Corrugated Steel Pipe	707.14
Metallic-Coated Corrugated Steel Pipe	707.02
Metallic-Coated Spiral Rib Pipe	707.11
Polymer-Coated Steel Pipe	707.08
Repair of Damaged Coatings	707.15
Slotted Drain Pipe	707.10
Watertight Gaskets	712.03

Furnish bedding material that meets the requirements specified in Subsection 603.04.

Furnish backfill materials that meet the requirements specified in Subsection 603.08.

Clean and paint damaged spelter coating caused by welding, field cutting, or mishandling, as specified in Subsection 707.15.

To prevent electrolysis or physical failure, use materials in each pipe installation that are compatible with each other.

Either annular or helical pipe corrugations will be acceptable. Helical corrugated pipe containing annular rerolled ends may be used in conjunction with annular pipe of like or compatible materials.

Provide fabricator's certification that the sheet and pipe fabrication are in accordance with AASHTO M 36, M 196, and M 245, as applicable. Submit the certification before installing the pipe.

The lengths and locations of individual pipe SHOWN ON THE DRAWINGS are approximate. Do not order pipe until culvert locations are DESIGNATED ON THE GROUND and a written list of the correct lengths is approved by the CO.

Construction

603.03 Excavation. Excavate in accordance with the requirements specified in Section 206A.

Specific pipe installation time restrictions and installation plan requirements are SHOWN ON THE DRAWINGS.

603.04 Bedding. Bed the pipe to a depth of not less than 10 percent of its total height. After excavating in accordance with Subsection 206A.04(b), compact the foundation surface in accordance with Subsection 603.08 and shape it to fit the pipe.

As bedding material, provide selected mineral soil that meets the requirements for backfill specified in Subsection 603.08. When SHOWN ON THE DRAWINGS, ensure that completed bedding has a longitudinal camber.

603.05 Laying Pipe. Lay the lower segment of the pipe so that it is in contact with the bedding for the required depth throughout its length. Place outside circumferential laps facing upstream.

Lay paved or partially lined pipe so the longitudinal centerline of the paved segment coincides with the flowline. Place elliptical pipe with the major axis within 5° of a vertical plane through the longitudinal axis of the pipe.

Ensure that the final installed alignment allows no reverse grades, and does not permit any point to vary from a straight line drawn from inlet to outlet by more than 2 percent horizontally and vertically of the culvert length, or 300 mm, whichever is less.

Do not place any pipe in service until a suitable outlet is provided.

Install helically corrugated lock-seam pipe with the seam at the inlet end placed below the horizontal centerline. This requirement applies to the outlet end when the outlet is less than 1.5 m below subgrade.

Position longitudinal laps on riveted or spot-welded pipe at any location between 45° above or below horizontal.

603.06 Joining Pipes. Firmly join pipe using form-fitting coupling bands. Attach end sections to the pipe using connecting bands or other means, as recommended by the manufacturer. Install gaskets at each joint to form a watertight connection when SHOWN ON THE DRAWINGS. Do not use dimpled bands when the slope of the pipe is greater than 15 percent.

Ensure that coupling bands meet the strength requirements of field joints for Nonerodible Soil Condition—Special Joint Type, according to division II, section 26, of the "Standard Specifications for Highway Bridges" by AASHTO.

When aluminum alloys come in contact with other metals, coat the contacting surfaces with an asphalt mastic or other impregnated caulking compound approved by the CO.

603.07 Shop Elongation. When SHOWN ON THE DRAWINGS, increase the vertical diameter of round pipe 5 percent by shop elongation.

603.08 Backfilling. Do not place or backfill pipe that meets any of the following conditions until the excavation and foundation have been approved by the CO:

- Embankment height greater than 3 m at subgrade centerline.
- Installation in a live stream.
- Round pipe with a diameter of 1,200 mm or greater.
- Pipe arches with a span of 1,270 mm or greater.

After the bedding is prepared and the pipe is placed, place selected material in layers not exceeding 150 mm loose thickness, and compact the material under the haunches and alongside the pipe. Use material that is readily compactible and free of frozen lumps, chunks of highly plastic clay (with a plasticity index greater than 10), or other objectionable material. Do not use rocks larger than 75 mm in greatest dimension within 300 mm of the pipe. On each side of the pipe, place an area of compacted material at least as wide as the diameter of the pipe. Compact the backfill without damaging or displacing the pipe.

Continue backfilling and compacting until the backfill is a minimum of 300 mm above the top of the culvert.

After bedding and backfilling the pipe, protect it with an adequate cover of embankment before heavy equipment is permitted to cross during roadway construction.

Replace any pipe that is distorted by more than 5 percent of nominal dimensions, or that is ruptured or broken.

Compact backfill using method A, B, or C, as DESIGNATED IN THE SCHEDULE OF ITEMS.

(a) Method A. Ensure that backfill density exceeds the density of the surrounding embankment.

(b) Method B. Ensure that backfill density exceeds 95 percent of the maximum density as determined by AASHTO T 99, method C or D.

Determine density of the compacted material during the process of the work in accordance with AASHTO T 191, T 205, or T 238; and AASHTO T 217, T 239, or T 255. Corrections for coarse particles may be made in accordance with AASHTO T 224.

(c) Method C. Ensure a moisture content suitable for obtaining compaction. Compact each layer using compaction equipment designed for this purpose until visual displacement ceases.

Measurement

603.09 Method. Use the method of measurement that is DESIGNATED IN THE SCHEDULE OF ITEMS.

When DESIGNATED IN THE SCHEDULE OF ITEMS, measure backfill material adjacent to the pipe 300 mm horizontally and vertically from the outside dimensions of the pipe, with a deduction for the volume of the pipe along the full length of the backfill.

Payment

603.10 Basis. The accepted quantities will be paid for at the contract unit price for each PAY ITEM DESIGNATED IN THE SCHEDULE OF ITEMS.

Payment will made under:

Pay Item		Pay Unit
603 (01)	________-mm corrugated metal pipe, ______-mm thickness for steel or _______-mm thickness for aluminum, method ___________	Meter

603 (02) ________-mm span, ___________-mm rise corrugated metal pipe arch, _________-mm thickness for steel or ________ -mm thickness for aluminum, method __________ .... Meter

603 (03) ________-mm metal end section .... Each

603 (04) ________-mm span, _______-mm rise metal end section .... Each

603 (05) _______-mm corrugated steel pipe, ________-mm thickness, method _____________ .... Meter

603 (06) _______-mm span, __________-mm rise corrugated steel pipe arch, _______-mm thickness, method _______________ .... Meter

603 (07) _______-mm steel end section .... Each

603 (08) _______-mm span, _______-mm rise steel end section .... Each

603 (09) _______-mm ________ -type _________ -coated corrugated steel pipe, ______-mm thickness, method ____________ .... Meter

603 (10) _______-mm _______ -type _________ -coated paved invert corrugated steel pipe, _____-mm thickness, method __________ .... Meter

603 (11) _______-mm span, ______-mm rise ________-type _________ -coated corrugated steel pipe arch, _______-mm thickness, method _____________ .... Meter

603 (12) ______-mm _______ -type _______ -coated steel section end .... Each

603 (13) ______-mm span, ______-mm rise _______-coated steel end section .... Each

603 (14) ______-mm corrugated aluminum pipe, ______-mm thickness, method _____________ .... Meter

603 (15) _______-mm paved invert corrugated aluminum pipe, ______-mm thickness, method ____________ .... Meter

603 (16) _______-mm span, _______-mm rise corrugated aluminum pipe arch, _______-mm thickness, method _____________ .... Meter

603 (17) _______-mm aluminum end section .... Each

603 (18) _______-mm span, ______-mm rise aluminum end section .... Each

603 (19) Pipe elbow, ______-mm diameter, _______-mm thickness .... Each

603 (20) Branch connection, ______-mm diameter, _____-mm thickness .... Each

603 (21) Furnishing and placing backfill material for pipe .... Cubic Meter

Section 603A—Concrete Pipe

Description

603A.01 Work. Furnish and install, or install only, concrete pipe and pipe appurtenances, including all bedding and backfilling required to complete the work.

Materials

603A.02 Requirements. Furnish materials that meet the requirements specified in the following subsections:

Nonreinforced Concrete Pipe 706.01
Precast Reinforced Concrete Box Sections 706.07
Reinforced Arch-Shaped Concrete Pipe 706.04
Reinforced Concrete Pipe 706.02
Reinforced D-Load Concrete Pipe 706.06
Reinforced Elliptical-Shaped Concrete Pipe 706.05
Watertight Gaskets 712.03

Furnish end sections constructed of the same material as the main section of the pipe.

Furnish bedding material that conforms to the requirements of Subsection 603A.04. Furnish backfill material that conforms to the requirements of Subsection 603A.06, or as SHOWN ON THE DRAWINGS.

The lengths and locations of individual pipe SHOWN ON THE DRAWINGS are approximate. Do not order pipe until culvert locations are designated on the ground and a written list of the correct lengths is approved by the CO.

Construction

603A.03 Excavation. Conduct excavation in accordance with the requirements are specified in Section 206A. Excavate the trench a minimum of 100 mm below grade.

Specific pipe installation time restrictions and installation plan requirements are SHOWN ON THE DRAWINGS.

603A.04 Bedding. Unless otherwise SHOWN ON THE DRAWINGS, backfill the trench with bedding material to grade. Extend bedding material to a minimum height of one-sixth the pipe diameter above the bottom of the pipe, and compact it in accordance with Subsection 603A.06.

Bed pipe with select excavated material from the roadway in the vicinity of the pipe, or with material from the source SHOWN ON THE DRAWINGS. Use material that contains no rocks greater than 25 mm in size. Ensure that the bedding surface provides a foundation of uniform density and support throughout the entire length of the pipe; provides for camber as SHOWN ON THE DRAWINGS; and has recesses shaped to receive the bell, of the bell and spigot pipe.

603A.05 Placing & Joining. Do not place or backfill any pipe until the excavation and foundation have been approved by the CO and a suitable outlet has been constructed. Ensure that the bell or groove ends face upstream. Join the pipe section so that the inner surfaces are reasonably flush and even, and the ends are entered as required. Make joints with a cold applied bituminous mastic, with rubber, or with plastic ring gaskets, as SHOWN ON THE DRAWINGS. When using mastic material, fill the joints with the material prior to joining the pipe.

603A.06 Backfilling. Furnish readily compactible backfill material that is free from frozen lumps and chunks of highly plastic clay or other objectionable material. Use no rock larger than 75 mm in greatest dimension within 300 mm of the pipe.

Place backfill material at or near optimum moisture content, and compact it in layers not exceeding 150 mm loose thickness on both sides, and to an elevation of 300 mm above the top of the pipe. Thoroughly compact the backfill under the haunches of the pipe. Bring the backfill up evenly on both sides of the pipe for the full length. Make the width of backfill on each side of the pipe equal to the diameter of the pipe.

Compact the backfill to at least 95 percent of the maximum density, as determined by AASHTO T 99, method C or D, unless otherwise SHOWN ON THE DRAWINGS.

Determine density of the compacted material during the process of the work in accordance with AASHTO T 191, T 205, or T 238; and AASHTO T 217, T 239, or T 255. Corrections for coarse particles may be made in accordance with AASHTO T 224.

Ensure that the final installed alignment of all pipe allows no reverse grades, and does not permit horizontal and vertical alignments to vary from a straight line drawn from center of inlet to center of outlet by more than 2 percent of pipe center length or 300 mm, whichever is less.

Measurement

603A.07 Method
Use the method of measurement that is DESIGNATED IN THE SCHEDULE OF ITEMS.

When DESIGNATED IN THE SCHEDULE OF ITEMS, measure backfill material adjacent to the pipe 300 mm horizontally and vertically from the outside dimensions of the pipe, with a deduction for the volume of the pipe along the full length of the backfill.

Payment

603A.08 Basis. The accepted quantities will be paid for at the contract unit price for each PAY ITEM DESIGNATED IN THE SCHEDULE OF ITEMS.

Payment will be made under:

Pay Item		Pay Unit
603A (01)	______-mm reinforced concrete pipe, class ____________	Meter
603A (02)	______-mm span, _______-mm rise reinforced concrete pipe, class __________________	Meter
603A (03)	______-mm reinforced concrete end section	Each
603A (04)	______-mm span, ______-mm rise reinforced concrete end section	Each
603A (05)	Furnishing and placing backfill material for pipe	Cubic Meter

Section 603B—Plastic Pipe

Description

603B.01 Work. Furnish and install, or install only, plastic pipe and pipe appurtenances, including all bedding and backfilling required to complete the work.

Materials

603B.02 Requirements. Furnish materials that meet the requirements specified in the following subsections:

Plastic Pipe	706.08
Watertight Gaskets	712.03

Furnish bedding material that meets the requirements specified in Subsection 603B.04, or as SHOWN ON THE DRAWINGS.

Furnish backfill materials that meet the requirements specified in Subsection 603B.06, or as SHOWN ON THE DRAWINGS.

The lengths and locations of individual pipe SHOWN ON THE DRAWINGS are approximate. Do not order pipe until culvert locations are designated on the ground and a written list of the correct lengths is approved by the CO.

Construction

603B.03 Excavation. Conduct excavation in accordance with the requirements specified in Section 206A. Excavate a minimum of 200 mm below the designed invert elevation.

Specific pipe installation time restrictions and installation plan requirements are SHOWN ON THE DRAWINGS.

603B.04 Bedding. Place bedding material in the excavated section, and compact the material to ensure a uniform foundation bed for the pipe.

As bedding material, use selected mineral soil that meets the requirements for backfill specified in Subsection 603B.06.

When SHOWN ON THE DRAWINGS, ensure that the completed bedding has a longitudinal camber.

603B.05 Placing & Joining. Join to form a watertight connection, when SHOWN ON THE DRAWINGS.

Protect portions of the pipe that will be exposed, when SHOWN ON THE DRAWINGS.

Ensure that the final installed alignment allows no reverse grades, and does not permit any point to vary from a straight line drawn from center of inlet to center of outlet by more than 2 percent horizontally and vertically of the culvert length, or 300 mm, whichever is less, unless otherwise SHOWN ON THE DRAWINGS.

603B.06 Backfilling. Furnish readily compactible backfill material that is free of frozen lumps and chunks of highly plastic clay (with a plasticity index greater than 10) or other objectionable material. Do not use rocks larger than 25 mm in greatest dimension within 300 mm of the pipe.

Place backfill material that is at or near optimum moisture content, and compact it in layers not exceeding 150 mm loose thickness on both sides, and to an elevation of 300 mm minimum above the top of the pipe. Thoroughly compact the backfill under the haunches of the pipe. Bring the backfill up evenly on both sides of the pipe for the full length. Extend the width of the compacted backfill a minimum of 300 mm on each side of the pipe.

Continue backfilling and compacting until the backfill is a minimum of 300 mm above the top of the culvert.

After bedding and backfilling the pipe, protect it with an adequate cover of embankment before heavy equipment is permitted to cross during roadway construction.

Replace any pipe that is distorted by more than 5 percent of nominal dimensions, or that is ruptured or broken.

Compact backfill using method A, B, or C, as DESIGNATED IN THE SCHEDULE OF ITEMS.

(a) Method A. Ensure that backfill density exceeds the density of the surrounding embankment.

(b) Method B. Ensure that backfill density exceeds 95 percent of the maximum density, as determined by AASHTO T 99, method C or D.

Determine density of the compacted material during the process of the work in accordance with AASHTO T 191, T 205, or T 238; and AASHTO T 217, T 239, or T 255. Corrections for coarse particles may be made in accordance with AASHTO T 224.

(c) Method C. Ensure a moisture content suitable for obtaining compaction. Compact each layer using compaction equipment designed for this purpose until visual displacement ceases.

Measurement

603B.07 Method. Use the method of measurement that is DESIGNATED IN THE SCHEDULE OF ITEMS.

When DESIGNATED IN THE SCHEDULE OF ITEMS, measure backfill material adjacent to the pipe 300 mm horizontally and vertically from the outside dimensions of the pipe, with a deduction for the volume of the pipe along the full length of the backfill.

Payment

603B.08 Basis. The accepted quantities will be paid for at the contract unit price for each PAY ITEM DESIGNATED IN THE SCHEDULE OF ITEMS.

Payment will be made under:

Pay Item		Pay Unit
603B (01)	_______ -mm plastic pipe, method ____	Meter
603B (02)	Furnishing and placing backfill material for pipe	Cubic Meter

Section 605—Underdrains

Description

605.01 Work. Furnish and install underdrains, sheet drains, and pavement edge drains.

The term "metal" refers collectively to aluminum and steel.

Materials

605.02 Requirements. Furnish materials that meet the requirements specified in the following subsections:

Aluminum-Alloy Corrugated Pipe	707.03
Asphalt-Coated Pipe	707.04
Geocomposite Drains	714.02
Geotextile, Type I (A, B, C, D, E, or F)	714.01
Granular Backfill	703.03
Metallic-Coated Corrugated Steel Pipe	707.02
Perforated Concrete Pipe	706.03
Plastic Pipe	706.08
Repair of Damaged Coatings	707.15
Structural Backfill	704.04

Construction

605.03 General. Use the same material and coating on all contiguous drain sections, extensions, elbows, branch connections, and other special sections.

Type of backfill material and underdrain pipe material, orientation of geosynthetic material, and approximate location are SHOWN ON THE DRAWINGS. Determine the final location and length in the field.

Conduct excavation in accordance with the requirements specified in Section 206A.

If geotextile or geocomposite is used, smooth the trench surfaces by removing all projections that may damage the geotextile or geocomposite. Replace geotextile or geocomposite damaged during installation. Make repairs to geocomposites in accordance with the manufacturer's recommendations.

Do not permit soil materials or other foreign matter to enter the drain systems. Plug the upgrade end of installations.

Backfill in 150-mm layers by first dampening the granular backfill and then compacting each layer with two or more passes of a mechanical tamper.

Furnish nonperforated pipe for outlet pipe. Install outlet pipe as specified in Sections 603, 603A, and 603B. Immediately place and secure a screen made of 1.4-mm-diameter galvanized wire with approximately 13 × 13-mm mesh openings over the outlet ends of all exposed pipes and weep holes.

605.04 Underdrain & Trench. Place a layer of granular backfill at least 100 mm in thickness in the bottom of the trench.

Furnish a collector pipe at least 125 mm in diameter with all underdrains.

Join pipe sections securely with coupling fittings or bands. Join PVC and acrylonitrile-butadiene-styrene (ABS) pipe using either a flexible elastomeric seal or solvent cement. Join polyethylene pipe with snap-on, screw-on, or wrap-around coupling bands, as recommended by the manufacturer.

When underdrain is placed in drainages, prevent infiltration of surface water by placing material conforming to AASHTO M 145 classifications A-4, A-5, A-6, or A-7 in the top 300 mm of the trench.

(a) Standard Underdrain. When geotextile is required, place the long dimension of the geotextile parallel to the centerline of the trench. Position the geotextile, without stretching, such that it lies smoothly in contact with the trench surface. Overlap the joints a minimum of 600 mm with the upstream geotextile placed over the downstream geotextile.

Place collector pipe with the perforations down. Firmly embed the underdrain pipe in granular backfill material.

Place granular backfill to a height of 300 mm above the top of the collector pipe and compact. Do not displace the collector pipe. Place and compact the remainder of the granular backfill material as specified in Subsection 603.08, method A or B.

Fold the geotextile over the top of the granular backfill with a minimum overlap of 300 mm.

(b) Geocomposite Underdrain or Sheet Drain. Extend the geotextile from the bottom of the drainage core around the collector pipe.

Construct splices, joints, and outlet fittings as recommended by the manufacturer and in a manner that prevents infiltration of soil into the geocomposite core and does not impede flow through the geocomposite core or damage the core.

Place the assembled geocomposite in the trench with the geocomposite placed against the inflow side of the trench. If the trench wall is irregular such that flow along or through the geocomposite may be impeded, smooth the trench or place a layer of granular backfill between the geocomposite and the trench wall.

Temporarily support the drain against the trench side while backfilling.

When the trench is less than 0.5 m in width, backfill the trench using fine granular backfill. Except as otherwise indicated, backfilling in layers and compacting are not required. After the backfill is in place, densify by wheel rolling, vibrating, tamping with a mechanical tamper, or flooding with water.

When the trench is 0.5 m or more in width, place granular coarse or fine backfill to a height of 300 mm above the top of the collector pipe and compact the material. Finish backfilling the trench as indicated in Subsection 605.03.

605.05 Geocomposite Sheet Drain. Do not place sheet drain against a mortar course less than 4 days old.

When a geocomposite is used in conjunction with a waterproof membrane, install drainage panels compatible with the membrane using methods recommended by the membrane manufacturer. Assemble and place the geocomposite drain against the surface to be backfilled according to the manufacturer's recommendations.

Splice geocomposite drains so the flow across the edges is continuous. Overlap the geotextile a minimum of 75 mm in the direction of waterflow. For vertical splices, overlap the geotextile in the direction that backfill proceeds.

Connect the drainage core to the collector pipe or weep holes so the flow is continuous through the system. Extend the geotextile from the bottom of the drainage core around the collector pipe.

Backfill with structural backfill material, and compact it as specified in Subsection 605.03.

605.06 Geocomposite Pavement Edge Drain. Assemble the geocomposite pavement edge drain and outlet material according to the manufacturer's recommendations and place them in the trench. If the trench is irregular such that flow along or through the geocomposite may be impeded, smooth the trench or place a layer of fine granular backfill between the geocomposite and the trench wall.

Temporarily support the drain against the trench while backfilling.

When the trench is less than 0.5 m in width, backfill the trench using fine granular backfill. Backfilling in layers and compacting are not required. After the backfill is in place, densify by wheel rolling, vibrating, tamping with a mechanical tamper, or flooding with water.

When the trench is 0.5 m or more in width, place coarse or fine backfill to a height of 300 mm above the top of the collector pipe and compact. Finish backfilling the trench as indicated in Subsection 605.03.

When underdrain is placed in drainages, prevent infiltration of surface water by placing material conforming to AASHTO M 145 classifications A-4, A-5, A-6, or A-7 in the top 300 mm of the trench.

Measurement

605.07 Method. Use the method of measurement that is DESIGNATED IN THE SCHEDULE OF ITEMS.

Measure geotextile material on surface area covered according to the dimensions, as SHOWN ON THE DRAWINGS.

Payment

605.08 Basis. The accepted quantities will be paid for at the contract unit price for each PAY ITEM DESIGNATED IN THE SCHEDULE OF ITEMS.

Payment will be made under:

Pay Item		Pay Unit
605 (01)	Standard underdrain system	Meter
605 (02)	Geocomposite underdrain system	Meter
605 (03)	Geocomposite sheet drain system	Square Meter
605 (04)	Geocomposite pavement edge drain system	Meter
605 (05)	_____-mm collector pipe	Meter
605 (06)	_____-mm outlet pipe	Meter

605 (07) Coarse granular backfill .... Cubic Meter

605 (08) Fine granular backfill .... Cubic Meter

605 (09) Geotextile material .... Square Meter

605 (10) Structural backfill .... Cubic Meter

Section 606—Guardrail

Description

606.01 Work. Construct guardrail systems, and/or modify, remove, reset, and/or raise existing guardrail systems.

(a) Guardrail systems are designated as follows:

G1	Cable guardrail
G2	W-beam (weak post)
G3	Box beam
G4	Blocked-out W-beam standard barrier
G9	Blocked-out thrie-beam standard barrier
MB4	Blocked-out W-beam median barrier
SBTA	Steel-backed timber guardrail/timber posts and blockout
SBTB	Steel-backed timber guardrail/timber posts and no blockout
CRT	W-beam guardrail and no blockout
STLG	Steel-backed log rail

(b) Steel guardrail types are designated as follows:

I Zinc-coated, 550 g/m^2
II Zinc-coated, 1,100 g/m^2
III Painted rails
IV Corrosion-resistant steel

(c) Steel guardrail classes are designated as follows:

A Metal thickness—2.67 mm
B Metal thickness—3.43 mm

(d) Terminal section types are designated as follows:

BCT	Breakaway cable terminal
CRT	Cable releasing terminal
MELT	Modified eccentric loader terminal
G4–BAT	Back slope anchor terminal

Materials

606.02 Requirements. Furnish materials that meet the requirements specified in the following section and subsections:

Box Beam Rail	710.07
Corrosion-Resistant Steel Rail	710.06(b)
Galvanized Steel Rail	710.06(a)
Guardrail Hardware	710.10
Guardrail Hardware (Reflector Tabs)	710.10
Guardrail Posts	710.09
Minor Concrete Structures	602
Precast Concrete Units (Precast Anchors)	725.11
Retroreflective Sheeting, Type I or Type II	718.01
Steel-Backed Timber Rail	710.08
Treated Structural Timber & Lumber	716.03
Welding	555.18
Wire Rope or Wire Cable	709.02

Construction

606.03 Posts. When pavement is within 1 m of the guardrail, set posts before placing the pavement.

Do not shorten guardrail posts unless the cut end is set in concrete. Do not shorten posts in terminal sections.

Drive posts into pilot holes that are punched or drilled. The dimensions of the pilot hole shall not exceed the dimensions of the post by more than 15 mm. Set posts plumb, backfill, and compact in accordance with Subsection 206A.10.

When longer posts are specified, do not use them in the terminal sections.

Alternate hole arrangements, when specified, do not apply to posts in the anchorage assembly.

Protect posts from traffic at all times by attaching rail elements and all associated hardware, or by other approved methods.

606.04 Rail Elements. Install the rail elements after the pavement adjacent to the guardrail is complete. Do not modify specified hole diameters or slot dimensions. Install guardrail systems of the type and class SHOWN ON THE DRAWINGS.

(a) Steel Rail. Shop bend all curved guardrail with a radius of 45 m or less.

Erect rail elements in a smooth continuous line with the laps in the direction of traffic flow. Use bolts that extend at least 6 mm, but not more than 25 mm beyond the nuts. Tighten all bolts.

Paint all scrapes on galvanized surfaces that are through to the base metal with two coats of zinc-oxide paint.

Where installation of the rail elements interferes with paving operation, rail elements may be temporarily attached directly to the posts without blockouts. Install blockouts within 15 days following the paving operation. Securely bolt a type 1 end section assembly to the last post at the end of each day on guardrail sections that have an exposed end toward oncoming traffic. Diaphragms are optional in the end section assembly.

(b) ***Timber Rail.*** Align timber guardrail along the top of the rail.

Field cut timber rails to produce a close fit at joints. Treat field cuts with two coats of the preservative originally used for treatment.

(c) ***Log Rail.*** Construct log rail as SHOWN ON THE DRAWINGS.

606.05 Terminal Sections. Construct terminal sections at the locations shown. Terminal sections consist of posts, railing, hardware, and anchorage assembly necessary to construct the type of terminal section specified.

Where concrete anchors are installed, construct either cast-in-place or precast units. Do not connect the guardrail to cast-in-place anchors until the concrete has cured 7 days. Install end anchor cables tightly, without slack.

Use either the steel tube anchor or the concrete anchor in the construction of the anchorage assembly for the type BCT terminal section.

When required, construct earth berms as specified in Section 203.13.

606.06 Connection to Structure. Construct connection to structure and, where required, reinforced concrete transition as SHOWN ON THE DRAWINGS.

606.07 Removing & Resetting Guardrail. Remove and store the existing guardrail, posts, and appurtenances. Remove and dispose of posts that are set in concrete. Replace all guardrail, posts, and hardware damaged during removal, storage, or resetting. Backfill all holes resulting from the removal of guardrail posts and anchors with granular material as specified in Subsection 206A.10.

606.08 Raising Guardrail. Remove the existing guardrail and appurtenances. Replace and reset posts as needed. Replace all guardrail, posts, and hardware damaged during the removal and raising.

Measurement

606.09 Method. Use the method of measurement that is DESIGNATED IN THE SCHEDULE OF ITEMS.

Measure guardrail, except steel-backed timber guardrail and steel-backed log rail, by the meter along the face of the rail, excluding terminal sections. Measure steel-backed timber guardrail and steel-backed log rail by the meter along the face of the rail, including terminal sections. Measure transition sections from G9 rail to G4 rail as G9 rail.

Measure terminal sections, except steel-backed timber guardrail terminal sections and steel-backed log rail terminal sections, by the each.

Measure removing and resetting guardrail and raising guardrail by the meter along the face of the rail, including reset terminal sections.

Measure replacement posts (except replacement posts for posts damaged by construction operations) used in removing, resetting, or raising guardrail by the each.

Measure reinforced concrete flared back parapet wall and safety shape transition by the each.

Payment

606.10 Basis. The accepted quantities will be paid for at the contract unit price for each PAY ITEM DESIGNATED IN THE SCHEDULE OF ITEMS.

Payment will be made under:

Pay Item		Pay Unit
606 (01)	Guardrail system ____, type ___, class ___	Meter
606 (02)	Terminal section, _____________ (*Description*)	Each
606 (03)	Removing and resetting guardrail	Meter
606 (04)	Raising guardrail	Meter
606 (05)	Replacement posts	Each
606 (06)	Connection to structure	Each
606 (07)	Reinforced concrete transition	Each

Section 607—Fences, Gates, & Cattleguards

Description

607.01 Work. Furnish and install, or install only, fences, gates, and cattleguards.

Materials

607.02 Requirements. Furnish materials that meet the requirements specified in the following sections and subsections:

Barbed Wire	710.01
Chain Link Fence	710.03
Fence Posts	710.04
Material for Timber Structures	716
Precast Concrete Units	725.11
Reinforcing Steel	709.01
Structural Metal	717
Timber Rails	710.13
Woven Wire	710.02

Furnish materials for gates and cattleguards that meet the requirements as SHOWN ON THE DRAWINGS. Concrete for cattleguard units may be cast-in-place or precast. Furnish concrete that meets the requirements specified in Subsection 602.03, method A or B, as SHOWN ON THE DRAWINGS.

Construction

607.03 Fences & Gates. Remove trees, brush, and other obstacles along the fence line that interfere with the fence. Do not perform continuous grubbing or grading along the fence line. Where possible, erect the fence on natural ground. Establish clearing width and dispose of materials as SHOWN ON THE DRAWINGS.

When drilling into solid rock is required to set a post, the post may be shortened, provided a minimum of 300 mm of post is grouted into the rock.

Where breaks in a run of fencing are required, or at intersections with existing fences, adjust post spacing to meet the requirements for the type of closure.

When posts, braces, or anchors are to be embedded in concrete, install temporary guys or braces as required to hold the posts in proper position until the concrete has set. Install no materials on posts, and place no strain on guys and bracing set in concrete, until 7 days have elapsed from the time the concrete was placed.

Set all posts vertically and to the grade and alignment SHOWN ON THE DRAWINGS. Do not cut tops of posts unless approved by the CO.

Stretch wire or fencing taut, and firmly attach it to the posts and braces, as SHOWN ON THE DRAWINGS.

At each location where a high-voltage overhead electric transmission line crosses a fence containing metal, ground the fencing by installing a galvanized or copper-coated steel grounding rod 2.5 m long, with a minimum diameter of 13 mm, directly below the point of crossing. Drive the rod vertically until the top is 150 mm below the ground surface. Use a number 6 solid copper conductor or equivalent to connect each metal fence element to the grounding rod. Braze the connections or fasten them with noncorrosive clamps approved by the CO.

When a powerline is within 150 m and runs parallel or nearly parallel to the fence, ground the fence at each end, at gate posts, and at intervals not to exceed 450 m.

When acceptable vertical penetration of the grounding rod cannot be obtained, submit an equivalent horizontal grounding system for approval by the CO.

Ensure that the bottom of the fence fabric generally follows the contour of the ground. Grade where necessary to provide a neat appearance. Where abrupt changes in the ground profile make it impractical to maintain the specified ground clearance, longer posts may be used and multiple strands of barbed wire stretched between them. Make the vertical spacing between strands of barbed wire 150 mm, unless otherwise SHOWN ON THE DRAWINGS. At grade depressions, where stresses tend to pull posts from the ground, install sag bracing, as SHOWN ON THE DRAWINGS.

Splice wire as SHOWN ON THE DRAWINGS.

Repair all posts in accordance with approved procedures after cutting or drilling.

607.04 Cattleguards. Complete work required under Section 203 or 306 at the location of the cattleguard before beginning excavation for the cattleguard. Install the cattleguard at the grade elevation that is SHOWN ON THE DRAWINGS or as staked on the ground. Provide drainage at time of installation so the cattleguard will drain. Construct the bypass and gate as SHOWN ON THE DRAWINGS.

Conduct excavation and backfill in accordance with Section 206A.

After cattleguard is bedded, place selected material in layers not exceeding 150 mm loose thickness, uniformly compacted on all sides, along the cattleguard. Use readily compactible backfill material that is free of frozen lumps, chunks of highly plastic clay, or other objectionable material. Compact the backfill without damaging or displacing the cattleguard. Continue backfilling and compacting to the top of the cattleguard foundation.

Raise cattleguards by modifying the cattleguard base as SHOWN ON THE DRAWINGS. Replace or recondition cattleground wings, posts, or decks as SHOWN ON THE DRAWINGS.

After bedding and backfilling, protect the cattleguard with adequate ramps on each side before heavy equipment is permitted to cross during roadway construction.

Measurement

607.05 Method. Use the method of measurement that is DESIGNATED IN THE SCHEDULE OF ITEMS.

Measure the length of the fence along the top of the fence between the outsides of the end posts for each continuous run of fence.

When brace panels and bypass gates are SHOWN ON THE DRAWINGS, payment for cattleguards will include these items.

Payment

607.06 Basis. The accepted quantities will be paid for at the contract unit price for each PAY ITEM DESIGNATED IN THE SCHEDULE OF ITEMS.

Payment will be made under:

Pay Item		Pay Unit
607 (01)	Cattleguard, ________ foundation, loading ____________, width ____________	Each
607 (02)	Fence ________, type ____________, height ________	Meter
607 (03)	Gate ________, type ____________, size ________	Each
607 (04)	Cattleguard modification	Each
607 (05)	Recondition	Each
607 (06)	Replace	Each

Section 609—Curb or Curb & Gutter

Description

609.01 Work. Construct or reset curb, gutter, or combination curb and gutter.

Materials

609.02 Requirements. Furnish materials that meet the requirements specified in the following sections and subsections:

Aggregate for Lean Concrete Backfill	703.13
Curing Material	711.01
Emulsified Asphalt (for Tack Coat)	702.03
Joint Mortar	712.02
Precast Concrete Curbing	725.06
Reinforcing Steel	709.01
Sealants, Fillers, Seals, & Sleeves	712.01
Stone Curbing	705.06

Furnish concrete that meets the requirements of Subsection 602.03, method A or B, as SHOWN ON THE DRAWINGS.

Furnish bituminous mixtures that meet the requirements as SHOWN ON THE DRAWINGS or in the SPECIAL PROJECT SPECIFICATIONS.

Concrete, bituminous mixes, and manufactured curbing materials will be subject to inspection and tests at the plants for compliance with quality requirements.

Construction

609.03 Cast-in-Place Portland Cement Concrete Curbing or Curb & Gutter. For cast-in-place Portland cement concrete curbing or curb and gutter, meet the requirements in the following subsections:

(a) Excavation. Excavate to the depth SHOWN ON THE DRAWINGS. Compact the foundation to a firm, even surface. Remove all soft, yielding material and replace it with acceptable material.

(b) Forms. Use forms of wood, metal, or other suitable material and extend them to the full depth of the concrete. Ensure that all forms are straight, free of warp, and of sufficient strength to resist the pressure of the concrete without displacement. Brace and stake the forms to keep them in both horizontal and vertical alignment until their removal. Clean all forms and coat them with an approved form-release agent before concrete is placed. Use divider plates made of metal. After the forms have been set to line and grade, bring the foundation to the grade required, and wet it well approximately 12 hours before placing the concrete. Machine slip forming may be used.

(c) Mixing & Placing. Proportion, mix, and place the concrete in accordance with the requirements specified in Subsection 602.03, method A or B, and as SHOWN ON THE DRAWINGS. Deposit the concrete without segregation in a single course. Use vibration or other acceptable methods to consolidate concrete placed in the forms. Leave forms in place for 24 hours or until the concrete has set sufficiently so that forms can be removed without damage to the curbing. Strike off the concrete to the cross section SHOWN ON THE DRAWINGS, then finish the concrete smooth and even by means of a wooden float.

For the purpose of matching adjacent concrete finishes or for other reasons, the CO may permit other methods of finishing. No plastering shall be permitted.

(d) Contraction Joints. Construct curbing in sections of a uniform length of 3 m, unless otherwise approved by the CO. Separate sections by open joints approximately 3 mm wide and at least 25 mm deep, except at expansion joints. Where the curb is constructed adjacent to concrete pavement, match the contraction or open joints in the curb to the contraction joints in the pavement.

(e) Expansion Joints. Form expansion joints at the intervals SHOWN ON THE DRAWINGS using a preformed expansion joint filler with a thickness of 13 mm. When the curb is constructed adjacent to or on concrete pavement, locate expansion joints at expansion joints in the pavement.

(f) Curing. Immediately upon completion of the finishing, moisten the curb and keep it moist for 3 days, or use membrane-forming material to cure the curbing. Ensure that all materials meet the requirements specified in Subsection 711.01.

(g) Backfilling. After the concrete has set sufficiently, backfill the curb to the required elevation with suitable material, and compact the material in accordance with Subsection 203.16(b), method 4, in layers of not more than 150 mm loose thickness.

(h) Curb Machine. The curb or curb and gutter may be constructed using a curb-forming machine that meets the requirements of Subsection 609.06(c).

(i) Curb Template. Exposed curb face may be constructed and finished using trowel-type templates shaped to produce the desired contours when operated along approved forms set to the established lines and grades.

While the concrete is green, float the top, front, or other exposed surfaces of the curb or combined curb and gutter with a moist wooden float. Remove form marks and any other irregularities.

609.04 Precast Concrete Curbing. Set the curb so that the top surfaces of adjoining sections are true and even. Fill all spaces under the curbing with material that meets the requirements of the material for bed course, and compact this material.

609.05 Reflecting Concrete Curbing. Use construction methods for this item that meet the requirements specified in Subsection 609.03, with the following exceptions:

(a) Use a mortar mix consisting of one part white Portland cement to 1.75 parts light-colored washed mortar sand to create the reflecting surface of the curbing. Make this mortar mix approximately 25 mm thick.

(b) Alternatively, construct the entire curbing of concrete with white Portland cement.

Use washed mortar sand that meets all the requirements for mortar sand and is light in color. Place the reflecting surface mortar immediately after the base concrete. Never let more than 20 minutes elapse between placing the base concrete and the reflecting surface.

Perform scoring or surface deformation and finishing of the reflecting surface in accordance with the details SHOWN ON THE DRAWINGS.

609.06 Bituminous Concrete Curbing. For bituminous concrete curbing, meet the requirements in the following specifications:

(a) Excavation. Excavate as specified in Subsection 609.03(a).

(b) Preparation of Bed. When curbing is to be constructed on a cured or aged Portland cement concrete base, on bituminous pavement, or on a bituminous-treated base, thoroughly sweep the bed and clean it using compressed air. Thoroughly dry the surface, and immediately before placing the bituminous mixture, apply a tack coat of bituminous materials of the type and grade SHOWN ON THE DRAWINGS. Apply the tack coat material at a rate between 0.23 and 0.68 L/m^2 of surface area. Prevent the tack coat from spreading to areas outside of the area to be occupied by the curb.

(c) Placing. Construct bituminous curbing using a self-propelled automatic curber or curb machine, or a paver with curbing attachments. Use an automatic curber or machine that meets the following requirements:

(1) The weight of the machine must provide compaction without the machine riding above the bed on which curbing is constructed.

(2) The machine must form curbing of uniform texture, shape, and density.

(3) The construction of curbing by means other than the automatic curber or machine is acceptable when short sections or sections with short radii are required. Ensure that the resulting curbing conforms in all respects to the curbing produced by using the machine.

Place the mixture only when the bed is dry and weather conditions are suitable for properly handling and finishing the mixture.

Place the bituminous mixture at a workable temperature of not less than 105 °C. Place the curbs to an accurate alignment and with a high density such that material is free of honeycombs. When joining to a section of curb that has become cold, give the contact surface of the cold curb a thin uniform tack coat of bituminous material prior to placing the fresh bituminous mixture against the cold joint. Protect the curb from traffic using barricades or other suitable methods until the curb has hardened.

(d) Painting & Sealing. Seal or paint only on a curbing that is clean and dry and has reached an ambient temperature.

(e) Backfilling. Backfill as specified in Subsection 609.03(g).

609.07 Resetting Curb. In resetting curb, meet the requirements in the following specifications.

(a) Salvage of Curbing. Carefully remove, store, and clean curbing that is specified for resetting. Replace any curbing to be reset that is lost from storage or damaged through improper handling.

(b) Excavation. Excavate and provide bedding as specified in Subsection 609.03(a).

(c) Resetting Curb. Set the curb on a firm bed with the top surface of adjoining sections true and even. Set all sections of curbing so that the maximum opening between adjacent sections is not more than 19 mm wide for the entire exposed top and face. Dress the ends of the curbing as necessary to meet this requirement.

After the curb has been set, completely fill the joints with mortar as SHOWN ON THE DRAWINGS.

(d) Backfilling. Backfill the curb with suitable material to the required elevation. Thoroughly tamp backfill material in layers of not more than 150 mm loose thickness.

(e) Cutting & Fitting. Cut or fit as necessary to install the curbing.

Measurement

609.08 Method. Use the method of measurement that is DESIGNATED IN THE SCHEDULE OF ITEMS.

Measure curbing along the front face of the section at the finished grade elevation. Measure the length of combination curb and gutter along the face of the curb. Make no deduction in length for drainage structures installed in the curbing section, or for driveway openings where the gutter is carried across the drive.

Payment

609.09 Basis. The accepted quantities will be paid for at the contract unit price for each PAY ITEM DESIGNATED IN THE SCHEDULE OF ITEMS.

Payment will be made under:

Pay Item		Pay Unit
609 (01)	Portland cement concrete curb, ______ -mm depth, type ______	Meter
609 (02)	Portland cement concrete gutter, type ____________	Meter
609 (03)	Portland cement concrete curb and gutter, ________ -mm depth, type ______________	Meter
609 (04)	Bituminous concrete curb, ________ -mm depth	Meter
609 (05)	Reset curb	Meter
609 (06)	Bed course material	Ton
609 (07)	Bed course material	Cubic Meter

Section 610—Stone Masonry Structures

Description

610.01 Work. Construct stone masonry structures and stone masonry portions of composite structures.

610.02 Classes of Masonry. The class of masonry required for each part of a structure will be SHOWN ON THE DRAWINGS.

(a) *Cement rubble masonry* shall consist of roughly dressed stones of various sizes and shapes laid in random courses in cement mortar.

(b) *Class A and class B masonry* shall consist of stones shaped, dressed, and laid broken-coursed in cement mortar.

(c) *Dimensioned masonry* shall consist of broken-coursed ashlar masonry composed of stones with two or more dimensions SHOWN ON THE DRAWINGS.

Materials

610.03 Stone. Furnish stone that is sound and durable. Use stone for dimensioned masonry that is free of reeds, rifts, seams, laminations, and minerals that, by weathering, would cause discoloration or deterioration.

(a) Sizes & Shapes. Furnish stones in the sizes and face areas necessary to produce the general characteristics and appearance SHOWN ON THE DRAWINGS.

In general, furnish stones with a thickness of not less than 125 mm, a width not less than 300 mm and one and one-half times the thickness of the stones, and a length not less than one and one-half times their width. Where headers are required, use stones with lengths not less than the width of the bed of the widest adjacent stretcher, plus 300 mm.

Ensure that at least 50 percent of the total volume of the masonry is made up of stones with a volume of at least 0.03 m^3 each.

(b) Dressing. Dress the stone to remove any thin or weak portions. Dress face stones to provide bed and joint lines with a maximum variation from true line as follows:

Type of Masonry	Maximum Variation
Cement rubble masonry	37 mm
Class A masonry	19 mm
Class B masonry	6 mm
Dimensioned masonry	Reasonably true

(c) Bed Surfaces. Ensure that bed surfaces of face stones are normal to the faces of the stones for about 75 mm. From this point, they may depart from normal, not to exceed 25 mm in 300 mm for dimensioned masonry, and 50 mm in 300 mm for all other classes.

(d) Joint Surfaces. In all classes of masonry except dimensioned masonry, construct joint surfaces of face stones to form an angle with the bed surfaces of not less than 45°.

In dimensioned masonry, ensure that joint surfaces are normal to the bed surfaces, and normal to the exposed faces of the stone for at least 50 mm. From this point, they may depart from normal by not more than 25 mm in 300 mm.

Do not round the corners at the meeting of the bed and joint lines in excess of the following radii:

Type of Masonry	Dimensions
Cement rubble masonry	37 mm
Class A masonry	25 mm
Class B masonry	No rounding
Dimensioned masonry	No rounding

(e) Arch Ring Stone Joint Surfaces. Ensure that arch ring stone joint surfaces are radial and at right angles to the front faces of the stones. Dress them for a distance of at least 75 mm from the front faces and soffits. From these points, they may depart from a plane normal to the face, not to exceed 19 mm in 300 mm. Ensure that the back surface is in contact with the concrete of the arch barrel parallel to the front face, and dress it for a distance of 150 mm from the intrados. Cut the top perpendicular to the front face, and dress it for a distance of at least 75 mm from the front.

When concrete is to be placed after the masonry has been constructed, place adjacent ring stones to vary at least 150 mm in depth.

(f) Stratification. Ensure that stratification in arch ring stones is parallel to the radial joints, and in other stones parallel to the beds.

(g) Finish for Exposed Faces. Pitch face stones to the line along all beds and joints. The kind of finish for exposed faces will be SHOWN ON THE DRAWINGS. The following symbols will be used, representing the type of surface or dressing specified below:

(1) Fine Pointed (F.P.). Make the point depressions approximately 10 mm apart with surface variation not to exceed 3 mm from the pitch line.

(2) Medium Pointed (M.P.). Make the point depressions approximately 15 mm apart with surface variations not to exceed 6 mm from the pitch line.

(3) Coarse Pointed (C.P.). Make the point depressions approximately 25 to 31 mm apart, with surface variations not to exceed 10 mm from the pitch line.

(4) Split or Seam Faced (S.). Make the surface present a smooth appearance without tool marks, without depressions below the pitch line, and with no projection exceeding 19 mm beyond the pitch line.

(5) Rock Faced (R.F.). Make the face an irregular, projecting surface without indications of tool marks, without concave surfaces below the pitch line, and with projections beyond the pitch line, when measured in millimeters, not exceeding the figure preceding the symbol as SHOWN ON THE DRAWINGS (for example, "37 R.F." means projections beyond the pitch line not exceeding 37 mm). Where a variable rock face is specified, distribute stones of the same height of projection.

Removal of drill and quarry marks from the faces of stones in cement rubble masonry shall not be required.

610.04 Quarry Operations. Organize quarry operations and delivery of stone to the point of use to ensure that deliveries are well ahead of masonry operations. Keep a sufficiently large stock of stone on the site at all times to permit adequate selection of stone by the masons.

610.05 Mortar. Use mortar that meets the requirements specified in Subsection 712.05.

Construction

610.06 Excavation & Backfill. Excavate and backfill as specified in Section 206A, modified as follows:

For filled spandrel arches, carefully place the backfill to load the ring uniformly and symmetrically. Use backfill material approved by the CO. Place it in horizontal layers, carefully tamp it, and bring it up simultaneously from both haunches. Do not place wedge-shaped sections of backfill material against spandrels, wings, or abutments.

610.07 Falsework. Construct arch centering in accordance with construction drawings submitted. Provide wedges for raising or lowering the forms to the exact elevation for taking up any settlement that occurs during loading. Lower the centering gradually and symmetrically to avoid overstresses in the arch.

Rest centering upon jacks in order to take up and correct any slight settlement that may occur after the placing of masonry has begun. In general, strike the centering and make the arch self-supporting before the railing or coping is placed. For filled spandrel arches, leave these portions of the spandrel walls for construction subsequent to the striking of centers as necessary to avoid jamming of the expansion joints.

610.08 Sample Section. When SHOWN ON THE DRAWINGS, build an L-shaped sample section of wall not less than 1.5 m high and 2.5 m long, showing examples of face wall, top wall, method of turning corners, and method of forming joints. The sample section will be subject to the CO's approval. Do not lay any masonry other than the foundation masonry before such samples are approved.

610.09 Arch Ring Template. Lay out a full-size template of the arch ring near the quarry site, showing face dimensions of each ring stone and thickness of joints. Do not begin shaping any ring stone before the template is approved by the CO, and do not place any ring stone in the structure that does not correspond to approved configuration.

610.10 Selection & Placing. When the masonry is to be placed on a prepared foundation bed, make the bed firm and normal to, or in steps normal to, the face of the wall. Do not place any stone before the bed is approved by the CO. When stone is to be placed on foundation masonry, thoroughly clean the bearing surface of this masonry thoroughly and wet it immediately before the mortar bed is spread.

Set face stones in random bond to produce the effect SHOWN ON THE DRAWINGS and to correspond with the sample section approved by the CO.

Prevent small stones or stones of the same size from bunching. When weathered or colored stones or stones of varying texture are used, uniformly distribute the various kinds of stones throughout the exposed faces of the work. Use large stones for the bottom courses and large selected stones in the corners. In general, ensure that the stones decrease in size from the bottom to the top of work.

Thoroughly clean all stones and wet them immediately before they are set. Clean and moisten the bed before the mortar is spread. Lay the stones with their longest faces horizontal in full beds of mortar, and flush the joints with mortar.

610.11 Beds & Joints. Ensure that the exposed faces of individual stones are parallel to the faces of the walls in which the stones are set.

Do not jar or displace stones already set. Provide equipment for setting stones larger than those that can be handled by two people. Do not roll or turn stones on the walls. If a stone is loosened after the mortar has taken initial set, remove it, clean off the mortar, and relay the stone with fresh mortar.

Carefully set arch ring stone to exact positions, and hold the stone in place with hardwood wedges until the joints are packed with mortar.

Ensure that the thickness of beds and joints of face stones are as follows:

Type of Masonry	Beds	Joints
Cement rubble masonry	12–62 mm	12–62 mm
Class A masonry	12–50 mm	12–50 mm
Class B masonry	12–50 mm	12–37 mm
Dimensioned masonry	(See note)	19–25 mm

Note: The thickness of beds in dimensioned masonry may vary from 19 to 25 mm from the bottom to the top of the work. However, make the beds of uniform thickness throughout in each course.

Do not allow beds to extend in an unbroken line across more than 5 stones, and joints across more than 2 stones.

Make joints in dimensioned masonry vertical. In all other masonry, joints may be at angles with the vertical from 0° to 45°.

Bond each face stone with all contiguous face stones at least 150 mm longitudinally and 50 mm vertically.

Do not make the corners of four stones adjacent to each other.

Make cross beds for vertical walls level. For battered walls, cross beds may vary from level to normal to the batter line of the face of the wall. Completely fill all arch ring joints with mortar.

610.12 Headers. Uniformly distribute headers throughout the walls of structures to form at least one-fifth of the faces.

610.13 Backing. Build the backing chiefly of large stones and in a workmanlike manner. Ensure that the individual stones composing the backing and hearting are well bonded with the stones in the face wall and with each other. Completely fill all openings and interstices in the backing with mortar, or with spalls completely surrounded by mortar.

610.14 Coping. Prepare copings as SHOWN ON THE DRAWINGS. If copings are not SHOWN ON THE DRAWINGS, finish the top of the wall with stones wide enough to cover the top of the wall, from 0.5 to 1.5 m in length, and of random heights, with a minimum height of 150 mm. Lay stones so the top course is an integral part of the wall. Align the tops of the top courses of stone in both vertical and horizontal planes.

610.15 Parapet Walls. Use selected stones, squared and pitched to line and with heads dressed, in the ends of parapet walls and in all exposed angles and corners. Interlock headers well, and extend as many as possible entirely through the wall. Ensure that both the headers and stretchers in the two faces of the wall are well interlocked in heart and comprise practically the whole volume of the wall. Completely fill all interstices in the wall with cement grout, or with spalls completely surrounded with mortar or grout.

610.16 Facing for Concrete. Construct the stone masonry before placing concrete. Concrete may be placed before constructing the stone masonry if approved by the CO.

(a) Stone Masonry Constructed Prior to Placing Concrete. Ensure that hooked steel anchors, consisting of number 4 bars each bent into an elongated "S" shape, are spaced 600 mm apart both horizontally and vertically, unless closer spacing is SHOWN ON THE DRAWINGS. To improve the bond between the stone masonry and the concrete backing, make the back of the former as uneven as the stones will permit. Rigidly embed each anchor in a horizontal joint of the masonry, with one end 50 mm from the faces of the stones. Project the other end approximately 250 mm into the concrete hacking.

When the stone facing has been laid and the mortar has attained sufficient strength, carefully clean all surfaces against which concrete is to be placed and remove all dirt, loose material, and accumulations of mortar droppings. If necessary, use picks, scrapers, and wire brooms for this purpose. If compressed air is available, use it to blow out the dust and dirt. Just before the concrete is placed, wash the surfaces thoroughly. Forcibly dash water against the stones and into the joints, preferably using a hose. In depositing concrete, hold the top surface immediately adjacent to the stones, slightly low, and carry a neat cement grout of the consistency of cream on top of the concrete and against the masonry at all times, coating the entire exposed areas of all the stones with grout. Fill all interstices of the masonry, and thoroughly spade the concrete, working it until it is brought into intimate contact with every part of the back of the masonry.

(b) Concrete Placed Before Constructing Masonry. Except where otherwise SHOWN ON THE DRAWINGS, allow a thickness of 225 mm for facing. Set galvanized metal slots with anchors for the stone work, or other approved type of metal anchor, vertically in the concrete face at a horizontal spacing of no more than 600 mm. Temporarily fill the slots with felt or other material to prevent them from being filled with concrete. During the setting of the stone facing, fit the metal anchors tightly in the slots at an average vertical spacing of 600 mm. The CO will mark on the concrete backing the approximate location of the anchors. Place the anchor in the stone joint nearest to the mark. Ensure that at least 25 percent of the metal anchors have a short right-angle bend to engage a recess to be cut into the stone. Extend the anchors to within 75 mm of the exposed face of the stone work.

Where the shape of the concrete face is unsuitable for the use of metal slots, place ties consisting of U.S. Standard Gauge number 9 galvanized iron wire, as approved by the CO, with not less than one wire tie for each 0.14 m^2 of exposed stone surface. In laying the stone, continuously keep the concrete face wet for 2 hours before placing the stone, and thoroughly fill all spaces between the stone and concrete with mortar. Immediately after laying, clean all exposed stone surfaces, and keep them clean of loose mortar and cement stains.

610.17 Pointing. Point or finish all joints as SHOWN ON THE DRAWINGS.

When raked joints are called for, rake out all mortar in exposed faced joints and beds to the depth SHOWN ON THE DRAWINGS. Clean stone faces in the joints free of mortar.

When weather joints are called for, weather strike the bed. Slightly rake the joints to conform to the bed weather joint. Never make the mortar flush with the faces of the stones.

To provide drainage, slightly crown the mortar in joints on top surfaces at the center of the masonry.

610.18 Weep Holes. Provide all walls and abutments with weep holes, as SHOWN ON THE DRAWINGS.

610.19 Cleaning Exposed Faces. Immediately after being laid and while the mortar is fresh, thoroughly clean all face stone of mortar stains, and keep it clean until the work is completed. Before the final acceptance, clean the surface of the masonry using wire brushes and acid, if necessary.

610.20 Weather Limitations. Do not lay stone in freezing weather unless the CO approves in writing, and then only using precautionary methods prescribed for doing the work and protecting it at all times. This permission and the use of the prescribed methods shall not release the Contractor from the obligation to build a satisfactory structure. Remove and replace all work damaged by cold weather. In hot or dry weather, use satisfactory means to protect the masonry from the sun, and keep it wet for at least 3 days after completion.

Measurement

610.21 Method. Use the method of measurement that is DESIGNATED IN THE SCHEDULE OF ITEMS.

Do not include sample sections of wall, unless they are permitted to be incorporated in the work.

When computing quantities, use the dimensions determined by the lines SHOWN ON THE DRAWINGS. Make no deductions for weep holes, drain pipes, or other openings of less than 0.2 m^2 in area, or for chamfers or other ornamental cuts that amount to 5 percent or less of the volume of the stone in which they occur.

Payment

610.22 Basis. The accepted quantities will be paid for at the contract unit price for each PAY ITEM DESIGNATED IN THE SCHEDULE OF ITEMS.

Payment will be made under:

Pay Item		Pay Unit
610 (01)	Cement rubble masonry	Cubic Meter
610 (02)	Class A masonry	Cubic Meter
610 (03)	Class B masonry	Cubic Meter
610 (04)	Dimensioned masonry	Cubic Meter

Section 610A—Simulated Stone Masonry Surface

Description

610A.01 Work. Design, furnish, and install textured form liners. Apply a surface finish (color/stain application) that will duplicate the unique coloring and mottled appearance of stone masonry. Prepare a simulated stone masonry test wall and demonstrate the surface finish before beginning production work.

In accordance with the intent of the contract, simulate the texture and color of native stone masonry. Construct the simulated stone masonry stone pattern as SHOWN ON THE DRAWINGS.

Materials

610A.02 Requirements. Furnish material as specified in the following subsections:

Form Liner	725.26
Low-Strength Grout (Plaster Mix)	701.03(b)
Penetrating Stain	708.05
Preformed Expansion Joint Fillers	712.01(b)

Construction

610A.03 Form Liner Fabrication. Take an impression of the stone shape, texture, and mortar joints from a designated location. Design form liners from the impressions according to the stone pattern, as SHOWN ON THE DRAWINGS.

610A.04 Form Liner Installation. Attach the form liners to the form. Attach adjacent form liners to each other, with less than a 3-mm seam. Do not repeat the form liner pattern between expansion joints or within 6-m intervals, whichever is greater.

Form expansion joints at the intervals as SHOWN ON THE DRAWINGS. Blend the butt joints into the pattern and the final concrete surface.

Coordinate the forms with wall ties. Place form tie holes in the high point of rustication or in the mortar joint.

Clean off buildup before reusing form liners. Visually inspect each liner for blemishes and tears. Repair the liner before installation.

610A.05 Top Surface. Emboss the plastic concrete in the exposed top surface by stamping, tooling, troweling, or hand shaping, or a combination thereof, to simulate the stone masonry texture and mortared joints. Match the side pattern of the formed mortared joints. Immediately after the free surface water evaporates and the finish embossing is complete, cure the concrete for 7 days, as specified in Subsection 552.17(b). Do not use liquid membrane curing compounds.

610A.06 Form Liner Removal. Within 24 hours after placing concrete, remove or break free the form liners without causing concrete surface deterioration or weakness in the substratum. Remove all form tie material to a depth of at least 25 mm below the concrete face without spalling and damaging the concrete.

Cure the concrete for 7 days, as specified in Subsection 552.17(b). Do not use liquid membrane curing compounds.

610A.07 Preparation of Concrete Surface. Finish all exposed formed concrete surfaces as specified in Subsection 552.18(a). Finish so that vertical seams, horizontal seams, and butt joint marks are not visible. Keep grinding and chipping to a minimum to avoid exposing aggregate.

Provide a completed surface free of blemishes, discolorations, surface voids, and conspicuous form marks. Make the finished texture and patterns continuous without visual interruption.

610A.08 Color/Stain Application. Age concrete, including patches, a minimum of 30 days. Use approved methods to clean the surface of all latency, dirt, dust, grease, and any foreign material.

Remove efflorescence with a pressure water wash. Use a fan nozzle held perpendicular to the surface at a distance from 0.6 to 1 m. Use a minimum 20 MPa water pressure at a rate of 12 to 16 L/min. Do not sandblast any surface that will receive color or stain.

Correct any surface irregularities created by the surface cleaning.

Maintain the concrete temperature between 4 °C and 30 °C when applying color or stain, and for 48 hours after applying color or stain.

Color or stain all exposed concrete surfaces. Use a color or stain application suitable to obtain the appearance of the native stone masonry. Use a minimum of three colors or stains.

When required at boundaries between two color tones or between surfaces that receive color at different times, take care to provide protection to avoid overspray and color overlap.

Apply grout of a natural cement color to each form joint. Use sufficient grout so the overspray of the color or stain is not visible. Give the form pattern grout joint the appearance of mortared joints in completed masonry.

Recoat any areas that lack a uniform appearance or are inconsistent in appearance with the approved test wall.

Treat expansion joints with caulk or grout to blend with the appearance of the adjacent stone or motor joint.

610A.09 Test Wall. Before beginning production work on the simulated stone masonry, construct a test wall with a minimum size of 1.0 m height by 0.5 m width by 3.0 m length, in accordance with Section 552 and these specifications.

Cast the test wall on site, using the same forming methods, procedures, form liner, texture configuration, expansion joint, concrete mixture, and color or stain application proposed for the production work. Demonstrate the quality and consistency of joint treatment, end treatment, top embossing methods, back treatment, and the color or stain application on the test wall. Construct a new test wall if results are not acceptable.

Begin production structural concrete work only after the test wall is approved. Begin production color or stain application only after the color or stain application on the test wall is approved. Dispose of the test wall after use as SHOWN ON THE DRAWINGS.

Measurement

610A.10 Method. Use the method of measurement that is DESIGNATED IN THE SCHEDULE OF ITEMS.

Measure simulated stone masonry surface treatment by the square meter.

Measure the simulated stone masonry test wall, including concrete and surface finish, by the each.

Do not measure form liners.

Payment

610A.11 Basis. The accepted quantities will be paid for at the contract unit price for each PAY ITEM DESIGNATED IN THE SCHEDULE OF ITEMS.

Payment will be made under:

Pay Item		Pay Unit
610A (01)	Simulated stone masonry surface treatment	Square Meter
610A (02)	Simulated stone masonry test wall	Each

Section 611—Development of Pits & Quarries

Description

611.01 Work. Clear, grub, strip topsoil, remove overburden, construct access roads, conduct restoration activities, and perform other incidental work required for pit or quarry development.

Construction

611.02 General. Perform all work in accordance with Sections 201, 203, and 625; landscape preservation requirements; and the pit and quarry development and/or restoration plan, as SHOWN ON THE DRAWINGS.

611.03 Source. Develop designated sources in accordance with requirements SHOWN ON THE DRAWINGS or in the SPECIAL PROJECT SPECIFICATIONS.

611.04 Clearing, Grubbing, & Slash Cleanup. Meet clearing, grubbing, and slash cleanup requirements as specified in Section 201 and as SHOWN ON THE DRAWINGS.

611.05 Access Roads. Construct or recondition access roads to the pit or quarry as specified in Section 203 or 306, and as SHOWN ON THE DRAWINGS.

611.06 Topsoil. Strip, stockpile, and place topsoil obtained from the site as specified in Section 203 and as SHOWN ON THE DRAWINGS.

611.07 Overburden. Remove overburden to expose rock material for aggregate production, and stockpile or place the overburden in the embankment within the limits of the pit or quarry, as specified in Section 203 and as SHOWN ON THE DRAWINGS.

611.08 Ground Control & Haulways. Perform the work in accordance with MSHA 30 CFR, part 56, as related to ground control and haulways. Immediately correct any deterioration of overburden slopes, safety benches, or protective berms, or any encroachment on clearing limits.

611.09 Oversize Material. Use all suitable material for aggregate, regardless of size, that is developed in stripping, overburden removal, and excavation of rock material, unless other disposition is SHOWN ON THE DRAWINGS.

611.10 Restoration. After excavation has been completed in part or all of the area, slope and grade the sides, and smooth the general pit area as SHOWN ON THE DRAWINGS.

Rip and drain access roads that are marked on the drawings for obliteration; block them to traffic; and seed them in accordance with Section 625.

Measurement

611.11 Method. Use the method of measurement that is DESIGNATED IN THE SCHEDULE OF ITEMS.

Payment

611.12 Basis. The accepted quantities will be paid for at the contract unit price for each PAY ITEM DESIGNATED IN THE SCHEDULE OF ITEMS.

Payment will be made under:

Pay Item	Pay Unit
611 (01) Pit development	Each
611 (02) Quarry development	Each

Section 617—Structural-Plate Structures

Description

617.01 Work. Furnish and install, or install only, structural-plate pipes, pipe arches, arches, boxes, and underpasses. Include joints and connections to pipes, catch basins, headwalls, stiffening ribs, thrust beams, and other appurtenances required to complete the structure.

Materials

617.02 Requirements. Ensure that materials meet the requirements specified in the following subsections:

Aluminum-Alloy Structural-Plate Structures	707.06
Asphalt-Coated Structural-Plate Structures	707.07
Backfill Material	704.03
High-Strength Nonshrink Grout	701.02
Reinforcing Steel	709.01
Repair of Damaged Coatings	707.15
Steel Structural-Plate Structures	707.05

Furnish concrete that meets the requirements specified in Subsection 602.03, method A or B, or as SHOWN ON THE DRAWINGS.

Construction

617.03 Excavation & Bedding. Excavate to the limits SHOWN ON THE DRAWINGS and in accordance with the requirements of Sections 206 and 206A.

Specific structure installation time restrictions and installation plan requirements are SHOWN ON THE DRAWINGS.

Provide bedding material that meets the same requirements as backfill material.

Bed the culvert in a compacted layer of bedding material with a thickness equal to at least 10 percent of the culvert height. Place the bedding material in layers that do not exceed 150 mm in depth when compacted. Shape and compact the bed to fit the culvert. Where applicable, recess the shaped foundation to receive the joints.

Do not place or backfill structure until the CO has approved the excavation and foundation.

617.04 Design & Fabrication. Submit four sets of shop drawings of the plate structure to the CO at least 21 days before planned construction. Accompany shop drawings with all calculations used to determine the size, shape, location, and spacing of stiffening ribs, thrust beams, or other special structural features.

Fabricate plates in accordance with AASHTO M 167 or M 219.

Form plates to provide lap joints. Stagger joints so that not more than three plates come together at any one point. Punch the bolt holes so that all plates with like dimensions, curvature, and the same number of bolts per meter of seam are interchangeable. Curve each plate to the proper radius so that the cross-sectional dimensions of the finished structure are as SHOWN ON THE DRAWINGS.

Cut plates for forming skewed or sloped ends to give the angle of skew or slope as SHOWN ON THE DRAWINGS. Place legible identification numerals on each plate to designate its proper position in the finished structure.

617.05 Erection. Provide a copy of manufacturer's assembly instructions before assembly. The instructions shall show the position of each plate and the assembly order.

Assemble the structural plates according to the manufacturer's instructions. Exercise care in the use of drift pins and pry bars to prevent damage to the structural plate and its coating. Torque all bolts before beginning the backfill. Repair damaged coatings according to Subsection 707.15. Ensure that the plates have a proper fit-up.

When aluminum alloys come in contact with other metals, coat the contacting surfaces with asphalt mastic according to Subsection 707.07 or a preapproved asphalt-impregnated caulking compound.

Torque steel bolts on steel plates to a minimum of 135 N•m and a maximum of 400 N•m.

Torque steel and aluminum bolts on 2.5-mm-thick aluminum plates to a minimum of 120 N•m and a maximum of 155 N•m.

Torque steel bolts and aluminum bolts on 3-mm-thick and heavier aluminum plates to a minimum of 155 N•m and a maximum of 180 N•m.

For long-span structures:

(a) Tighten the longitudinal seams when the plates are assembled unless the plates are held in shape by cables, struts, or backfill. Properly align plates circumferentially to avoid permanent distortion from the design shape. Before backfilling, do not exceed 2 percent variation from the design shape.

(b) Do not distort the shape of the structure by operating equipment over or near it.

(c) Provide survey control on the structure to check structure movement when SHOWN ON THE DRAWINGS.

(d) Check and control the deflection movements of the structure during the entire backfilling operation. Do not exceed the manufacturer's recommended limits.

(e) Provide a manufacturer's representative onsite to assist in the erection and backfilling of the structure when SHOWN ON THE DRAWINGS.

617.06 Backfilling. Provide backfill meeting the material and placement requirements of Section 206 or 206A.

After preparing bedding and placing the structure, place backfill in layers not exceeding 150 mm loose thickness, and compact it under the haunches and alongside the structure. The material shall be readily compactible and free of frozen lumps, chunks of highly plastic clay, or other objectionable material. Do not use rocks larger than 75 mm in greatest dimension within 300 mm of the structure. Ensure that there is an area of compacted material at least as wide as one diameter of the structure or 3.5 m, whichever is less, on each side of the structure. Compact the backfill without damaging or displacing the structure.

When filling around and over arches before headwalls are in place, place the first embankment material midway between the ends of the arch, forming as narrow a ramp as possible, until reaching the top of the arch. Build the ramp evenly from both sides, and compact the embankment material as it is placed. After building the two ramps to the top of the arch, deposit the remainder of the embankment material from the top of the arch in both directions from the center to the ends, and as evenly as possible on both sides of the arch.

If headwalls are built before any embankment material is placed around and over the arch, first place the embankment material adjacent to one headwall until reaching the top of the arch. Then dump the embankment material from the top of the arch toward the other headwall, depositing the material evenly on both sides of the arch.

Follow the procedures specified above in multiple installations. Bring up the embankment material evenly on each side of each arch to avoid unequal pressure.

Ensure that backfill density exceeds 95 percent of the maximum density as determined by AASHTO T 99, method C or D.

Continue backfilling and compacting until the backfill is 300 mm above the top of the structure.

After bedding and backfilling, protect the structure with an adequate cover of compacted embankment, as indicated by manufacturer's recommendation. Do this before permitting heavy equipment to cross during roadway construction.

Replace structure that is distorted by more than 5 percent of nominal dimensions, or that is ruptured or broken.

Ensure that backfilling materials, methods, and procedures meet the manufacturer's requirements.

Measurement

617.07 Method. Use the method of measurement that is DESIGNATED IN THE SCHEDULE OF ITEMS.

The quantity of concrete or reinforcing steel used in thrust beams or additional structural metal or plates used in stiffening ribs or other special structural features will not be included in the quantities for payment.

Payment

617.08 Basis. The accepted quantities will be paid for at the contract unit price for each PAY ITEM DESIGNATED IN THE SCHEDULE OF ITEMS.

Payment will be made under:

Pay Item		Pay Unit
617 (01)	________ structural-plate pipe, size ________, *Material* ________ -coated, ________ thickness	Meter
617 (02)	________ structural-plate pipe arch, _____ span, *Material* _____ rise, ________ -coated, ________ thickness	Meter
617 (03)	________ structural-plate arch, ______ span, *Material* _______ rise, ________ -coated, ________ thickness	Meter
617 (04)	________ structural-plate underpass, size ________, *Material* ________ -coated, ________ thickness.........................	Meter

617 (05) ________ structural-plate box culvert, _____ span,
Material
_____ rise, ________ -coated, _________ thickness ..... Meter

617 (06) ________ long-span structure, plate _______ -coated,
Material *Type*
________ span, _______ rise, ________ thickness .... Meter

617 (07) Installation only .... Each

Section 618—Cleaning & Reconditioning Existing Drainage Structures

Description

618.01 Work. Clean and recondition existing pipe and appurtenant structures.

Materials

618.02 Requirements. Ensure that the materials used for repair or replacement meet the applicable requirements of sections SHOWN ON THE DRAWINGS or specified in the SPECIAL PROJECT SPECIFICATIONS.

Construction

618.03 Pipe Removed & Cleaned. Carefully remove the pipe and appurtenant structures and clean foreign material out of the barrel and at the jointed ends.

618.04 Pipe Cleaned in Place. Remove all foreign material inside the barrel by methods that do not damage the pipe. Take adequate measures to protect the drainage and prevent stream siltation or increased turbidity when hydraulically cleaning pipe in place.

If approved by the CO, all or part of the pipe designated to be cleaned in place may be removed, cleaned, and relaid in accordance with Sections 603, 603A, and 603B. In these cases, furnish all material required to replace damaged pipe and joints, perform all excavation and backfill, and relay the pipe.

618.05 Relaying or Stockpiling Salvaged Pipe. The locations of pipe and appurtenant structures to be removed, cleaned, and relaid will be SHOWN ON THE DRAWINGS. Relay the pipe in accordance with Sections 603, 603A, and 603B. Furnish all jointing material, and replace pipe that has been damaged during removing or handling, in sufficient lengths to complete the designated length to be relaid, without added compensation. Place salvaged pipe designated to be stockpiled where SHOWN ON THE DRAWINGS. Carefully remove and handle all pipe to avoid breaking or damaging it. Do not place pipe that has been structurally damaged in stockpiles. Dispose of damaged pipe at an approved location in accordance with Subsection 202.04.

618.06 Reconditioning Drainage Structure. Remove all debris, repair leaks, and replace broken or missing metalwork on all structures, such as manholes and inlets, that are SHOWN ON THE DRAWINGS as needing reconditioning. Leave these structures in operating condition.

Measurement

618.07 Method. Use the method of measurement that is DESIGNATED IN THE SCHEDULE OF ITEMS.

The quantity of pipe and appurtenant structures removed, cleaned, and relaid is the length in final position.

The quantity of pipe and appurtenant structures removed, cleaned, and stockpiled is the total length of all pipe acceptably removed, cleaned, and placed in the stockpile.

The quantity of pipe and appurtenant structures cleaned in place is the length along the flow line.

No additional payment will be made for material to replace damaged pipe and joints, excavation, relaying pipes, or backfill if pipe is removed for cleaning, when damage is the result of operation.

Payment

618.08 Basis. The accepted quantities will be paid for at the contract unit price for each PAY ITEM DESIGNATED IN THE SCHEDULE OF ITEMS.

Payment will be made under:

Pay Item		Pay Unit
618 (01)	Removing, cleaning, and stockpiling salvaged ________	Meter
618 (02)	Removing, cleaning, and relaying salvaged ________	Meter
618 (03)	Cleaning ________ in place	Meter
618 (04)	Cleaning ________ in place	Each
618 (05)	Reconditioning drainage structures	Each

Section 621—Corrugated Metal Spillways

Description

621.01 Work. Furnish and install, or install only, corrugated metal spillway inlet assemblies, outlet pipes, half-round outlet pipe, rectangular flumes, and other appurtenances for downdrains.

Materials

621.02 Requirements. Ensure that spillway inlet assemblies, outlets, and connectors are of the type and thickness SHOWN ON THE DRAWINGS and are constructed of corrugated sheet metal that meets the requirements specified in Subsection 603.02. Fillet weld or rivet bulkheads and connections for outlet pipes to the inlet chamber to form watertight joints. Rivet or weld anchors, lips, and skirts so they are secure. Ensure that connections for outlet pipes meet the requirements specified in Subsection 603.06.

Ensure that outlets are the type, size, and arrangement SHOWN ON THE DRAWINGS, and that they meet the requirements for corrugated metal pipe specified in Subsection 603.02. Furnish half-round pipe with end sections punched so that joints can be riveted in the field. Furnish elbows of the full-circle type.

Ensure that anchor assemblies for the downdrains and other components are as SHOWN ON THE DRAWINGS.

Ensure that coating for spillway inlet assemblies and outlet pipes meets the requirements for coated corrugated pipe specified in Subsection 603.02.

Install a gasket or equivalent material on circular pipe at the joints on each side of elbows and at each joint on the downdrain to make the connections watertight. Install gaskets on the entire circumference. Use gasket material of sponge rubber or synthetic rubber compound specifically designed for such installations and recommended by the coupling band fabricator. Approved joint compounds, such as Thiocaulk or Plastiflex, may be used instead of gaskets.

Construction

621.03 Performance. Place spillway inlets where SHOWN ON THE DRAWINGS. Compact the earth backfill in accordance with Subsection 603.08, method B.

Install outlet pipes in accordance with Section 603. Place outside laps so they face upstream.

Repair damaged coating on the inlet assemblies or pipe and all field rivet heads as required in Subsection 707.15.

Ensure that there are no reverse grades, and that no point varies from a straight line drawn from inlet to outlet by more than 2 percent horizontally and vertically of the spillway length, or 300 mm on the final installed alignment, whichever is less, unless otherwise SHOWN ON THE DRAWINGS.

Measurement

621.04 Method. Use the method of measurement that is DESIGNATED IN THE SCHEDULE OF ITEMS.

The quantity of outlet pipes shall be the length from end to end of each outlet pipe, excluding elbows and spillway assemblies.

Payment

621.05 Basis. The accepted quantities will be paid for at the contract unit price for each PAY ITEM DESIGNATED IN THE SCHEDULE OF ITEMS.

Payment will be made under:

Pay Item		Pay Unit
621 (01)	Spillway inlet assemblies	Each
621 (02)	Spillway inlet assemblies with __________ coating	Each
621 (03)	_______-mm half-round outlet pipe	Meter
621 (04)	_______-mm half-round outlet pipe with ________ coating	Meter
621 (05)	_______-mm flexible downpipe	Meter
621 (06)	Anchors for downdrains for ________-mm pipe	Each
621 (07)	________-mm full-circle outlet pipe	Meter
621 (08)	________-mm full-circle outlet pipe with ________ coating	Meter
621 (09)	Pipe elbow __________	Each

621 (10) Pipe elbow, __________ -coated __________ .... Each

621 (11) ________ -mm pipe end section .... Each

621 (12) Starter section, type ____________ .... Each

621 (13) Flume, type ______________ .... Meter

621 (14) Anchor stakes _____________ .... Each

621 (15) Energy dissipator, type _______________ .... Each

Section 622—Paved Waterways

Description

622.01 Work. Pave ditches, gutters, spillways, and other similar waterways with concrete, grouted rubble, ungrouted rubble, mortared rubble, concrete and rubble, or a mixture of aggregate and bituminous material. Also construct a bed course.

Materials

622.02 Requirements. Furnish concrete that meets the requirements specified in Subsection 602.03, method A or B, as SHOWN ON THE DRAWINGS.

Ensure that materials meet the requirements specified in the following sections or subsections:

Bed Course	704.09
Bituminous Material	702
Cold Asphalt Concrete Pavement Aggregate	703.08
Low-Strength Grout	701.03
Reinforcing Steel	709.01

Ensure that materials and proportions for bituminous mixtures are as SHOWN ON THE DRAWINGS.

Provide rubble for pavement that is approved, sound, durable rock of the sizes SHOWN ON THE DRAWINGS. Inspect all rock before and after laying it, and remove all rejected material immediately.

Construction

622.03 Bed Course. Form the bed course to the required depth below and parallel with the finished surface of the waterway. Replace all soft, yielding, or otherwise unsuitable material with suitable material. Compact and finish the bed course to a smooth, firm surface.

Do not construct the paved waterway until the CO gives written approval of the bed.

622.04 Grouted Rubble. Place and compact bed course material to the required thickness when SHOWN ON THE DRAWINGS.

Place the pavement stones on the bed course with flat faces up and their longest dimensions at right angles to the centerline of the waterway.

Break joints so they are not wider than 25 mm. Ram the rocks until the surface is firm and reasonably true to the finished surface in grade, alignment, and cross section. Relay or replace any rock that causes an irregular or uneven surface. After the rocks are rammed into place and the surface is satisfactory, fill the spaces or voids between and around the stones with filler aggregate to within 100 mm of the surface. Then pour and broom cement grout into the spaces between the stones. Continue this operation until the grout is about 25 mm below the tops of the stones. Ensure that the grout flows readily into the spaces between the rocks, but is not so wet that solid matter separates from the water.

622.05 Ungrouted Rubble. Place the pavement rocks on the bed course with flat faces up and their longest dimensions at right angles to the centerline of the waterway. Break joints so they are not wider than 12 mm. Ram the rocks until the surface is firm and reasonably true to the finished surface in grade, alignment, and cross section. Relay or replace any rock that causes an irregular or uneven surface, or any rock that is not in reasonably close contact with adjacent rocks.

622.06 Mortared Rubble. Lay the pavement rocks with flat faces up and their longest dimensions parallel to the gutter line.

Break joints so they are not wider than 25 mm. After each rock is rammed into place and the surface is satisfactory, apply enough mortar on the exposed side so that when the adjacent rock is rammed into position, the mortar fills the interstices between the rocks to within 25 mm of the surface and does not protrude above their tops. Ensure the finished rock surface is free from mortar stain.

622.07 Reinforced Concrete & Rubble. Construct a reinforced concrete foundation upon a repaired foundation as SHOWN ON THE DRAWINGS. Construct it progressively by laying surface rocks and bedding them securely in the concrete before it hardens. Ensure that the faces of the rocks in contact with the concrete are clean and free of any defects that will impair their bond with the concrete.

Thoroughly wet rocks before laying them, allowing ample time for them to become nearly saturated. Fill joints between rocks with mortar. Keep the bedded reinforcement steel within the middle third of the depth of the concrete during construction.

622.08 Bituminous Mixture. Prepare the bituminous mixture, stake forms, and place the mixture as specified below.

(a) Preparing Mixture. Prepare the bituminous mixture by using either a rotary mixer or a pugmill, or by spreading the aggregate on a flat, firm surface off the area to be surfaced, and mix it using road-mix methods. Either batch- or continuous-type pugmills may be used.

Except when emulsified asphalt is used, ensure that the aggregate does not have a moisture content of more than 2 percent at the time it is mixed with the bituminous materials. However, if an approved additive is used, the aggregate may have a moisture content of up to 5 percent.

Apply bituminous material to the aggregate or introduce it into the mixture at the temperature at which the aggregate will be coated uniformly and completely.

When mixing is done in a mixer, mix for no less than 40 seconds from the time all materials are in the mixer until they are discharged. When road-mix methods are used, continue the mixing until all aggregate particles are uniformly coated with bituminous material.

(b) Forms. Securely stake all forms approved by the CO into position at the correct line and elevation.

(c) Placing Mixture. Place the mixture on the prepared bed only when the bed is sufficiently dry and weather conditions are suitable. Place and compact the mixture in one or more courses to the thickness SHOWN ON THE DRAWINGS. Smooth each course by raking or screeding and compact it thoroughly by rolling with a hand-operated roller weighing no less than 135 kg, or with an approved small power roller. Compact areas that cannot be reached with rollers by using hand tampers. Ensure that the surface is smooth and even, and that it has a dense texture after it is compacted.

622.09 Concrete Paving. Ensure that concrete paving is plain or reinforced, as SHOWN ON THE DRAWINGS, and that it meets the requirements specified in Section 602.

622.10 Finishing Work. Remove forms from paved waterways, and make necessary repairs to edges. Shape and compact the adjacent slopes and shoulders to the cross section SHOWN ON THE DRAWINGS.

Measurement

622.11 Method. Use the method of measurement that is DESIGNATED IN THE SCHEDULE OF ITEMS.

Base area computations upon surface measurements.

Payment

622.12 Basis. The accepted quantities will be paid for at the contract unit price for each PAY ITEM DESIGNATED IN THE SCHEDULE OF ITEMS.

Payment will be made under:

Pay Item		Pay Unit
622 (01)	Grouted rubble paved waterway	Square Meter
622 (02)	Ungrouted rubble paved waterway	Square Meter
622 (03)	Mortared rubble paved waterway	Square Meter
622 (04)	Concrete and rubble paved waterway	Square Meter
622 (05)	Bituminous paved waterway	Square Meter
622 (06)	Concrete paved waterway	Square Meter
622 (07)	Bed course material	Ton

Section 623—Monuments & Markers

Description

623.01 Basis. Furnish and install right-of-way monuments, milepost markers, underdrain markers, and culvert markers.

Materials

623.02 Requirements. Furnish materials for the various types of monuments and markers as SHOWN ON THE DRAWINGS.

Construction

623.03 Performance. Fabricate and install the various types of monuments and markers as SHOWN ON THE DRAWINGS, and also paint the posts, if required. Set each monument and post accurately at the required location and elevation.

Measurement

623.04 Method. Use the method of measurement that is DESIGNATED IN THE SCHEDULE OF ITEMS.

Payment

623.05 Basis. The accepted quantities will be paid for at the contract unit price for each PAY ITEM DESIGNATED IN THE SCHEDULE OF ITEMS.

Payment will be made under:

Pay Item		Pay Unit
623 (01)	Right-of-way monument	Each
623 (02)	Milepost marker	Each
623 (03)	Underdrain marker	Each
623 (04)	Culvert marker	Each

Section 624—Topsoiling

Description

624.01 Work. Furnish, excavate, or remove topsoil from stockpiles; then haul, deposit, and spread it.

Materials

624.02 Source. Obtain topsoil from sources SHOWN ON THE DRAWINGS or specified in the SPECIAL PROJECT SPECIFICATIONS.

624.03 Quality. Ensure that topsoil meets the requirements specified in Subsection 713.01.

Construction

624.04 Spreading. Spread the topsoil to the depth and at the locations SHOWN ON THE DRAWINGS.

Do not spread topsoil when the ground or topsoil is frozen, excessively wet, or in a condition detrimental to the work.

Remove and dispose of large clods, rocks larger than 50 mm in any dimension, roots, stumps, and other litter as SHOWN ON THE DRAWINGS.

624.05 Hauling. Keep the roadbed surfacing clean during hauling operations. Remove topsoil or other soil from the surfacing before traffic compacts it.

624.06 Source Area Other Than Roadway. After stripping operations have been completed, rough grade the source area and remove refuse. Leave the area in a neat condition. Leave a minimum 75 mm of topsoil on the source, and seed the area as SHOWN ON THE DRAWINGS.

Measurement

624.07 Method. Use the method of measurement that is DESIGNATED IN THE SCHEDULE OF ITEMS.

Measure the topsoil provided by the Contractor and paid for by cubic meter in the vehicles at the point of delivery. Measure the volume of topsoil from designated stockpiles in the original stockpile.

When topsoil is paid for by the square meter, compute its quantity along slope dimensions.

Payment

624.08 Basis. The accepted quantities will be paid for at the contract unit price for each PAY ITEM DESIGNATED IN THE SCHEDULE OF ITEMS.

Payment will be made under:

Pay Item		Pay Unit
624 (01)	Furnishing and placing topsoil	Square Meter
624 (02)	Furnishing and placing topsoil	Cubic Meter
624 (03)	Placing topsoil	Square Meter
624 (04)	Placing topsoil	Cubic Meter

Section 625—Seeding & Mulching

Description

625.01 Work. Prepare seedbeds, and furnish and place required seed, fertilizer, limestone, mulch, net, and blanket material.

Materials

625.02 Requirements. Ensure that materials meet the requirements specified in the following subsections:

Agricultural Limestone	713.02
Erosion Control Mats, Roving, & Geocell	713.07
Fertilizer	713.03
Mulch	713.05
Seed	713.04
Stabilizing Emulsion Tackifiers	713.12
Water	725.01(b)

Construction

625.03 Seeding Seasons. Observe the normal seasonal dates for seeding, as shown in the SPECIAL PROJECT SPECIFICATIONS. Do not apply seeding materials during windy weather or when the ground is excessively wet or frozen.

625.04 Soil Preparation. Finish the areas to be seeded, as required by other applicable sections, to the lines and grades SHOWN ON THE DRAWINGS. Restore areas that are damaged by erosion or other causes. Ensure that the surface soil is in a roughened condition favorable for germination and growth. When required, apply limestone uniformly at the rate given in the SPECIAL PROJECT SPECIFICATIONS, either before or after soil preparation.

625.05 Application Methods for Seed, Fertilizer, & Limestone. To control erosion, apply seed to disturbed soil and slopes within 30 days of disturbance. If the slopes have not been finished, apply the seed by the dry method as an interim erosion control measure. Apply fertilizer with the seed when specified in the SPECIAL PROJECT SPECIFICATIONS.

The following methods may be used to place material:

(a) Hydraulic Method. Mix the seed or seed and fertilizer with water in the amounts and mixtures shown in the SPECIAL PROJECT SPECIFICATIONS to produce a slurry, then apply it under pressure at the rates specified in the SPECIAL PROJECT SPECIFICATIONS. When wood cellulose or grass straw mulch materials are to be incorporated as an integral part of the slurry mix, add them after all other materials have been thoroughly mixed in the tank.

Use an inoculum for hydraulic seeding that is four times what is recommended for dry seeding.

(b) Dry Method. Use mechanical, landscape, or cultipacker seeders, seed drills, fertilizer spreaders, or other approved mechanical seeding equipment to apply the seed or seed and fertilizer in the amounts and mixtures shown in the SPECIAL PROJECT SPECIFICATIONS.

Spread dry fertilizer and ground limestone separately at the rates given in the SPECIAL PROJECT SPECIFICATIONS. Incorporate them in a single operation to the required depth in the areas SHOWN ON THE DRAWINGS.

Hand-operated seeding devices may be used to apply dry seed, fertilizer, and ground limestone.

625.06 Application of Mulch. The following methods may be used to apply mulch:

(a) Hydraulic Method. Hydraulic equipment that uses water as the carrying agent may be used to apply wood cellulose or grass straw fiber mulch, tackifier, and fertilizer in a single operation. Continuously agitate the materials to keep them in uniform suspension throughout the distribution cycle. Ensure that the discharge line evenly distributes the solution to the seedbed. Do not mulch where there is free surface water. Start application to areas SHOWN ON THE DRAWINGS at the top of the slope and work downward. If necessary, use extension hoses to reach the extremities of slopes. Apply at the rate specified in the SPECIAL PROJECT SPECIFICATIONS.

(b) Dry Method. After completion of seeding and fertilizing, unless otherwise indicated in the SPECIAL PROJECT SPECIFICATIONS, apply mulch uniformly at the rate specified in the SPECIAL PROJECT SPECIFICATIONS.

When a tackifier is used for mulch, apply the material at the rate specified in the SPECIAL PROJECT SPECIFICATIONS. Immediately distribute it evenly over the mulch. Prevent asphalt adhesive materials from marking or defacing structures, appurtenances, pavements, utilities, or plant growth.

625.07 Installation of Netting & Erosion Control Blankets. Install nettings and erosion control blankets as SHOWN ON THE DRAWINGS and in accordance with the manufacturer's recommendations.

625.08 Care During Construction. Protect and care for seeded areas until the work is finally accepted. Repair all damage to seeded areas caused by construction, without additional compensation.

Measurement

625.09 Method. Use the method of measurement that is DESIGNATED IN THE SCHEDULE OF ITEMS.

Base area computations on surface measurements.

Seed used for interim erosion control will be paid for as Seed Mix and will be measured by the kilogram. When fertilizer is used for interim erosion control, it will be measured by the ton.

Payment

625.10 Basis. The accepted quantities will be paid for at the contract unit price for each PAY ITEM DESIGNATED IN THE SCHEDULE OF ITEMS.

Payment will be made under:

Pay Item	Pay Unit
625 (01) Seeding, ________ method (without mulch)	Hectare
625 (02) Seeding, ________ method (with mulch)	Hectare
625 (03) Mulch (supplemental application)	Ton
625 (04) Fertilizer (supplemental application)	Ton
625 (05) Seed mix (supplemental application)	Kilogram
625 (06) Netting, type ________________	Square Meter
625 (07) Erosion control blanket, type _____________	Square Meter

Section 626—Trees, Shrubs, Vines, & Ground Cover

Description

626.01 Work. Furnish and plant trees, shrubs, vines, and ground cover plants.

Materials

626.02 Requirements. Ensure that materials meet the requirements specified in the following subsections:

Topsoil	713.01
Fertilizer	713.03
Mulch	713.05
Plant Materials	713.06
Miscellaneous Planting Material	713.08
Water	725.01(b)

Construction

626.03 Performance. Follow the specifications below.

(a) Planting Seasons. Plant during the seasons indicated in the SPECIAL PROJECT SPECIFICATIONS. Do not plant in frozen ground, when snow covers the ground, or when the soil is in an unsatisfactory condition for planting.

(b) Delivery & Inspection. Notify the CO not less than 15 days before plants are delivered. Inform the CO about the source of supply and the shipping dates for all plant material. Ensure that all plant materials comply with State and Federal laws applicable to inspection for plant diseases and insect infestations. Deliver all required certificates of inspection to the CO.

(c) Protection & Temporary Storage. Keep all plant material moist. Protect plants when they are in transit, in temporary storage, or on the project site awaiting planting.

(d) Layout. Designate the locations of plant material and bed outlines on the project site so they conform to the lines, grades, and elevations SHOWN ON THE DRAWINGS. The CO may adjust plant material locations to meet field conditions.

(e) Excavation for Plant Pits & Beds. Do not excavate plant pits and beds until the layout is approved. Remove and dispose of all sod, weeds, roots, and other objectionable material that is unsuitable for backfill.

For root spreads from 600 to 1,200 mm, make pit diameters 600 mm greater than the root spread. For root spreads over 1,200 mm, make pit diameters one and one-half times the root spread.

Ensure that all pits are deep enough to permit a minimum of 150 mm of loam–humus backfill under all roots or balls. Excavate to sufficient depth to plant at the root collar.

Loosen the soil at the bottom of the plant pit to a depth of at least 150 mm before backfilling or planting begins.

(f) Prepared Backfill Soil. Prepare the backfill soil to consist of a mixture of four parts topsoil, loam, or selected soil, and one part peat moss or peat humus.

(g) Setting Plants. Set all plants approximately plumb and at the same level, or not more than 25 mm lower than the depth at which they were grown in the nursery or collecting field.

(1) Bare Root Stock. Place prepared backfill soil in the plant pit to the required depth. Then place bare-rooted plants in the center of the plant pit and spread out the roots in a natural position. Cleanly cut back all broken or damaged roots to sound root growth.

Carefully work backfill soil around and over the roots, then settle it by firming or tamping. Thoroughly water or puddle around bare-rooted plants. Form earth saucers or water basins with a diameter equal to the plant, at least 100 mm deep for trees, and 75 mm deep for shrubs, around individual plants.

(2) Balled & Burlapped Stock. Carefully place balled and burlapped plants in the prepared pits on the required depth of tamped backfill soil so they are in a firm, upright position. Put backfill soil around the plant ball to half the depth of the ball, then tamp and thoroughly water. Cut away the burlap and remove it from the upper half of the ball, or loosen and fold it back. Place the remainder of the backfill. Provide earth saucers or water basins, and thoroughly water the plant.

(h) Fertilizing. Use the types and rates of fertilizer application for the different plants that are shown in the SPECIAL PROJECT SPECIFICATIONS. Uniformly apply fertilizer within 5 days after planting, and cultivate it into the top 50 mm of the plant pit area or shrub bed. Work the proper amount of fertilizer for each type of plant into the prepared backfill material. Apply fertilizer before mulching plant pits or shrub beds.

(i) Watering. Water all plants during and immediately after planting, and at intervals specified in the SPECIAL PROJECT SPECIFICATIONS. Ensure that water does not contain elements toxic to plant life. Thoroughly saturate the soil around each plant at each watering.

(j) Guying & Staking. Immediately after planting, guy and stake all trees as SHOWN ON THE DRAWINGS. Do not contact the roots when staking.

(k) Wrapping. Wrap deciduous trees only. Completely wrap tree trunks that are 50 mm in diameter or larger with burlap or other approved material. Begin wrapping at the base of the tree, extend it to the first branches, and tie it adequately. Do not wrap tree trunks until they have been inspected and approved. Finish wrapping tree trunks within 24 hours after approval.

(l) Antidesiccant Spray. An approved antidesiccant spray may be used in place of wrapping.

(m) Pruning. Prune plants before or immediately after planting in a manner that will preserve their natural character. Employ experienced personnel who use proper equipment and accepted horticultural practices when pruning. Paint cuts that are over 20 mm in diameter with an approved tree wound dressing.

(n) Mulching. Furnish and place mulch over all pit or saucer areas of individual trees and shrubs, and completely over shrub beds to the depth SHOWN ON THE DRAWINGS. Ensure that mulch is as SHOWN ON THE DRAWINGS or in the SPECIAL PROJECT SPECIFICATIONS. Put 3.5 kg of nitrogen per cubic meter of mulch material around plants to be mulched with wood chips, in addition to the normal dressing of commercial fertilizer. Mulch the plantings within 24 hours after fertilizing is completed.

626.04 Restoration & Cleanup. Restore grass areas that have been damaged or scarred during planting to their original condition. Clean up debris, spoil piles, containers, and so forth.

626.05 Plant Establishment Period & Replacement. During the plant establishment period, which is one full growing season, use all possible means to keep the plants in a healthy growing condition. There will be a semifinal inspection 15 days before the end of the full growing system. At the end of the establishment period, the CO will determine if the plantings are acceptable. Water, cultivate, prune, repair and adjust guys and stakes, and perform other maintenance during the establishment period. Promptly remove dead or unsatisfactory plants. During the next planting season, replace all dead and unsatisfactory plants in kind with robust, healthy plants. Use alternative or substitute varieties only if approved. There will be a final inspection of these plants within 15 days after planting is completed.

Measurement

626.06 Method. Use the method of measurement that is DESIGNATED IN THE SCHEDULE OF ITEMS.

Include in the quantities only living plants that are in healthy condition at the time of final inspection, as specified in Subsection 626.05.

Payment

626.07 Basis. The accepted quantities will be paid for at the contract unit price for each PAY ITEM DESIGNATED IN THE SCHEDULE OF ITEMS.

Payment will be made under:

Pay Item		Pay Unit
626 (01)	____________________, ________________ (*Name of plant*, *Size*)	Each
626 (02)	Plant materials	Lump Sum

Section 628—Sprigging

Description

628.01 Work. This work consists of furnishing and planting living grass plants. Sprigging is designated as broadcast, row, or spot, according to Subsection 628.06.

Materials

628.02 Requirements. Provide material that meets requirements specified in the following subsections:

Agricultural Limestone	713.02
Fertilizer	713.03
Mulch	713.05
Sprigs	713.09
Water	725.01(b)

Construction

628.03 General. Do not sprig during windy weather or when the ground is dry, excessively wet, frozen, or otherwise untillable.

628.04 Harvesting Sprigs. Provide at least 5 days notice before harvesting sprigs. Before harvesting, mow grass and weeds to a height of 65 ± 15 mm and remove all clippings.

Loosen sprigs by cross-disking, shallow plowing, or other acceptable methods. Gather the sprigs in small piles or windrows, water, and keep moist until planting. Dispose of sprigs that freeze or dry out.

628.05 Preparing the Soil. Clear and grade the area to be sprigged. Cultivate, disk, harrow, or otherwise loosen the grade to a depth of not less than 100 mm. Remove stones larger than 50 mm in any diameter, sticks, stumps, and other debris that might interfere with proper placement or subsequent growth.

Place topsoil according to Section 624.

Apply fertilizer and agricultural limestone uniformly over the sprigging area at the application rates SHOWN ON THE DRAWINGS. Mechanical spreaders or blower equipment may be used. Disk or till the fertilizer and limestone into the soil to a depth of 100 mm.

Moisten the prepared soil.

628.06 Planting Sprigs. Plant the sprigs within 24 hours after harvesting.

(a) Broadcast Sprigging. Broadcast sprigs by hand or using suitable equipment in a uniform layer over the prepared surface, with spacing between sprigs not to exceed 150 mm. Force the sprigs into the soil to a depth of 75 ± 25 mm with a straight spade, disk harrow, or other equipment.

(b) Row Sprigging. Open furrows along the approximate contour of slopes. Place sprigs in a continuous row in the open furrow, with successive sprigs touching. Cover the sprigs immediately.

(c) Spot Sprigging. Spot sprig according to Subsection 628.06(b), except that instead of planting in a continuous row, group four or more sprigs 500 mm apart in the rows.

After planting, clear the surface of stones larger than 25 mm, large clods, roots, and other litter brought to the surface during sprigging. Lightly compact the sprigged area within 24 hours. Do not compact when the soil is so wet that it is picked up by the equipment. Do not compact clayey soils.

When mulch is required, cover the sprigged area with mulch within 24 hours, in accordance with Subsection 625.06.

628.07 Maintaining Sprigged Areas. Keep the sprigged areas moist. Water carefully to avoid erosion. Erect warning signs and barriers to protect newly sprigged areas. Regrade and resprig all areas damaged.

Measurement

628.08 Method. Use the method of measurement that is DESIGNATED IN THE SCHEDULE OF ITEMS.

Measure sprigging by the hectare or square meter on the ground surface. Measure topsoil under Subsection 624.07.

Payment

628.09 Basis. The accepted quantities will be paid for at the contract unit price for each PAY ITEM DESIGNATED IN THE SCHEDULE OF ITEMS.

Payment will be made under:

Pay Item		Pay Unit
628 (01)	__________ sprigging	Hectare
628 (02)	__________ sprigging	Square Meter

Section 629—Sodding

Description

629.01 Work. Prepare the sod bed, and furnish, cut, haul, and lay live sod of perennial turf-forming grasses.

Materials

629.02 Requirements. Ensure that materials meet the requirements specified in the following subsections:

Agricultural Limestone	713.02
Fertilizer	713.03
Pegs for Sod	713.11
Sod	713.10
Water	725.01(b)

Construction

629.03 Season. Sod during the season SHOWN ON THE DRAWINGS or in the SPECIAL PROJECT SPECIFICATIONS.

629.04 Sources of Sod. The CO will approve sod obtained from other than commercial sources, in the original position, before cutting and delivery to the project. Notify the CO at least 5 days before cutting begins.

629.05 Soil Preparation & Cleanup. Bring areas to be sodded to the lines and grades SHOWN ON THE DRAWINGS, then plow, disk, harrow, or otherwise loosen them before sod is delivered. Remove stones larger than 50 mm in diameter, sticks, stumps, and other debris that might interfere with the proper laying or subsequent growth of sod.

629.06 Topsoiling. Place topsoil where SHOWN ON THE DRAWINGS. Remove and dispose of large clods, stones larger than 50 mm in any dimension, roots, stumps, and other litter at locations SHOWN ON THE DRAWINGS.

629.07 Applying Fertilizer & Ground Limestone. After soil preparation, cleanup, and topsoiling, uniformly spread fertilizer and ground limestone, when specified, at the rate SHOWN ON THE DRAWINGS or in the SPECIAL PROJECT SPECIFICATIONS. Use mechanical spreaders, blower equipment, or other approved methods to spread fertilizer and ground limestone. Incorporate these materials into the soil by disking or other tillage.

629.08 Laying Sod. Lay sod on the prepared bed within 24 hours after cutting, except when stored in stacks or piles, grass to grass and roots to roots, for not more than 5 days. Protect sod from drying by sun or wind and from freezing. Move and lay sod when weather conditions and soil moisture are favorable.

Lay sod according to one or more of the following methods, as DESIGNATED IN THE SCHEDULE OF ITEMS:

(a) Lay solid sodding when soils are moist. Thoroughly moisten dry sod bed areas before laying sod. Lay sections of solid sod edge to edge with staggered joints. Plug openings with sod or fill them with acceptable loamy topsoil. After laying and filling joints, roll or tamp the sod to eliminate air pockets and make an even surface. On slopes of 1:0.5 or steeper and in channels, peg sod on approximate 600-mm centers after tamping. Drive pegs flush with the sod bed surface.

(b) Lay strip sod in parallel rows of the width SHOWN ON THE DRAWINGS. Lay sod in a shallow trench and firmly roll or tamp it until the surface is level with or below the adjacent soils. If SHOWN ON THE DRAWINGS or in the SPECIAL PROJECT SPECIFICATIONS, seed the ground between strips of sod with grass seeds of the kind and at the rates specified. Then rake or drag the seeded areas to cover the seed.

(c) Perform spot sodding by laying sod blocks as SHOWN ON THE DRAWINGS. Firmly roll or tamp the pieces into the soil until the surfaces of sod blocks are slightly below the surrounding ground surface.

629.09 Care During Construction, Watering, & Temporary Maintenance of Sodded Areas. Water the sod when it is laid and keep it moist until final acceptance of the contract. Distribute water evenly at a measured rate per unit of area. Water to avoid erosion and prevent damage to sodded areas.

Erect necessary warning signs and barriers; mow sodded areas; repair or replace sodded areas that fail to show a uniform growth of grass or are damaged by construction operations; and otherwise maintain the sod until final acceptance of the contract.

Replace dried-out or damaged sod at no charge to the Government.

Measurement

629.10 Method. Use the method of measurement that is DESIGNATED IN THE SCHEDULE OF ITEMS.

Base area computations upon surface measurement.

Payment

629.11 Basis. The accepted quantities will be paid for at the contract unit price for each PAY ITEM DESIGNATED IN THE SCHEDULE OF ITEMS.

Payment will be made under:

Pay Item		Pay Unit
629 (01)	Solid sodding	Square Meter
629 (02)	Strip sodding	Square Meter
629 (03)	Spot sodding	Square Meter

Section 633—Signs

Description

633.01 Work. Install only, or furnish and install, delineators, markers, signs, sign supports, panels, and posts; or remove and dispose of existing signs, posts, and hardware.

633.02 Traffic Control Sign Details. Ensure that traffic control signs meet the requirements of the MUTCD and details as SHOWN ON THE DRAWINGS.

Materials

633.03 Requirements. Furnish materials that meet the requirements specified in the following subsections:

Aluminum Panels	718.05
Conventional Traffic Paint	718.16
Delineator & Object Marker Retroreflectors	718.15
Edge Film	718.10
Epoxy Markings	718.18
Epoxy Resin Adhesives	718.26
Extruded Aluminum Panels	718.07
Glass Beads	718.22
Hardware	718.13
Letters, Numerals, Arrows, Symbols, & Borders	718.14
Object Marker & Delineator Posts	718.12
Paint	718.08
Plastic Panels	718.06
Plywood Panels	718.03
Polyester Markings	718.19
Preformed Plastic Markings	718.21
Raised Pavement Markers	718.23
Retroreflective Sheeting	718.01
Signposts	718.11
Silk Screen Inks	718.09
Steel Panels	718.04
Temporary Pavement Markings	718.24
Temporary Traffic Control Devices	718.25
Test Procedures	718.02
Thermoplastic Markings	718.20
Waterborne Traffic Paint	718.17

Furnish certification to the CO that all materials comply with the specified requirements.

Ensure that all concrete meets the requirements specified in Section 602, or as SHOWN ON THE DRAWINGS.

Furnish reinforcing steel as SHOWN ON THE DRAWINGS that meets the requirements specified in Subsection 709.01.

Construction

633.04 Fabrication of Sign Panels. Fabricate all parts in a uniform manner. Complete all panel fabrication, including cutting, punching, and drilling of holes, prior to final surface preparation and application of reflective sheeting, except where required for the fabrication of diecut or sawed letters on processed and mounted signs. Ensure that workmanship is of high quality, and that there are no visible defects in the finished product.

(a) Sign Panel Preparation. Cut sign panel from the specified substrate material. Ensure that it is flat and free of warp and any defects that interrupt smooth continuity of the panel surface. Prepare all panels precisely as described in writing by the substrate and sheeting manufacturers.

For sign panels smaller than 1.2 × 2.4 m, cut them from a single sheet of substrate material without joints.

For high-density overlay (HDO) substrate sign panels larger than 1.2 × 2.4 m, fabricate in sections using 19-mm-thick material. Prepare individual panel sections to be joined using doweled butt joints. Use dowels that are 9-mm threaded bolt stock and 112 mm in length. Place them 50 mm from each side, and every 300 to 375 mm along the joint. Do not actually join the individual panels until sign installation.

Make panels of the dimensions that are SHOWN ON THE DRAWINGS, with a tolerance of ± 6 mm.

(b) Beveling. Slightly round or bevel all edges of sign panels to eliminate edge sharpness.

(c) Drilling. Drill holes at the locations and to the sizes that are SHOWN ON THE DRAWINGS. Debur all holes. Do not field drill holes in any part of the structural assembly without the approval of the CO.

(d) Surface & Edge Finishing. Fill all core-cap holes on HDO plywood signs with an exterior wood dough, sanded with medium grit (50–60) sandpaper to produce a smooth surface. Apply one coat of paint to the edges prior to application of background sheeting.

(e) Sheeting, Legend, Border, & Symbol Application. Apply all sheeting, legend, border, and symbols precisely as prescribed in writing by the sheeting manufacturer. Either pressure-sensitive or heat-activated material may be used, as SHOWN ON THE DRAWINGS. Cover the entire face of the sign panel with one unspliced sheet if the panel is less than 1.2 m in either dimension. On each section, use horizontal splicing only. Color match materials, and ensure that splices do not coincide with any legend. Ensure that the top piece overlaps the bottom piece by at least 12 mm.

Ensure that all letters, layout, and spacing are as specified in the Federal Highway Administration's (FHWA's) "Standard Alphabets for Highway Signs," current edition, and as SHOWN ON THE DRAWINGS. Apply the following tolerances:

(1) Horizontally align letters, numerals, and symbols to a tolerance of ± 3 mm.

(2) Vertically align letters, numerals, and symbols to a tolerance of ± 2 mm.

(3) Ensure that spacing between lines does not exceed a tolerance of ± 3 mm.

(f) Silk-Screening. Perform all silk-screening operations precisely as prescribed in writing by the manufacturers of the ink and the sheeting. Use direct screen process to apply messages and borders that are darker in color than the sign background. Use the reverse screen process to produce messages and borders that are lighter in color than the sign background. Screen to produce uniform colors and tone, with sharply defined edges of legend and border, and without blemishes on the sign background.

Color-match silk-screened inks to eliminate any visual difference between silk-screened material and applied material of the same color on the same sign.

(g) Trimming & Edge Finishing. After all sheeting, legends, borders, and symbols have been applied to the substrate, trim all excess material flush with the edge of the sign panel. Sheeting may overlap HDO plywood substrate sign edges by 3 mm. Finish the edges of HDO substrate panels with a second coat of paint applied in accordance with the recommendation of the paint manufacturer.

(h) Edge Film. When SHOWN ON THE DRAWINGS, apply edge tape over the entire top edge of the panel precisely as prescribed in writing by the manufacturer of the material being used.

(i) Maker's Mark. Install a decal showing the Contractor's identification or trademark and the date of manufacture on the back upper left-hand corner of each sign.

633.05 Delineator Posts & Housing. Drive delineator posts at locations and to the depth SHOWN ON THE DRAWINGS. Attach the delineator housing to the post, in accordance with the manufacturer's direction.

633.06 Sign Erection. Erect sign supports plumb and in accordance with the details SHOWN ON THE DRAWINGS. Make length of supports as SHOWN ON THE DRAWINGS, and in accordance with MUTCD, or as described in the SCHEDULE OF ITEMS.

Securely fasten the sign panels to the posts, as SHOWN ON THE DRAWINGS.

To reduce specular glare, erect the sign panel face in accordance with MUTCD, section 2A-26.

633.07 Sign Removal. Remove sign assemblies that are to be replaced, as SHOWN ON THE DRAWINGS. Ensure that sign replacement assemblies are replaced within 24 hours. Dispose of all removed sign material as SHOWN ON THE DRAWINGS. Remove signposts to a minimum of 75 mm below natural ground line. Backfill and compact remaining post holes with suitable material. When regulatory and warning signs are removed, immediately place replacement signs.

Measurement

633.08 Method. Use the method of measurement that is DESIGNATED IN THE SCHEDULE OF ITEMS.

Compute quantities of sign face area using the dimensions SHOWN ON THE DRAWINGS.

Make no deduction for rounded corners.

Compute the area for irregularly shaped signs, such as "Stop" signs, by multiplying the extreme width by the extreme height of the sign face.

For sign removal, treat an assembly of posts and signs as only one sign when these materials are integrally connected and standing at one location.

Payment

633.09 Basis. The accepted quantities will be paid for at the contract unit price for each PAY ITEM DESIGNATED IN THE SCHEDULE OF ITEMS.

Payment will be made under:

Pay Item		Pay Unit
633 (01)	Wood posts	Meter
633 (02)	Steel posts	Meter
633 (03)	Aluminum posts	Meter
633 (04)	Plastic posts	Each
633 (05)	Fiberglass-reinforced plastic posts	Each
633 (06)	Aluminum sign panels	Square Meter
633 (07)	Plywood sign panels	Square Meter
633 (08)	Steel sign panels	Square Meter
633 (09)	Plastic sign panels	Square Meter
633 (10)	Fiberglass-reinforced plastic sign panels	Square Meter
633 (11)	Delineators	Each
633 (12)	Sign	Each
633 (13)	Sign removal	Each
633 (14)	Sign and post(s), installation only	Each
633 (15)	Regulatory signs	Each
633 (16)	Warning signs and markers	Each
633 (17)	Object markers	Each

Section 634—Painted Traffic Markings

Description

634.01 Work. Apply permanent pavement markings and raised pavement markers on the completed pavement.

Pavement markings are designated as follows:

Type A	Conventional traffic paint with type 1 glass beads
Type B	Waterborne traffic paint with type 1 glass beads
Type C	Waterborne traffic paint with type 3 glass beads
Type D	Epoxy markings with type 1 glass beads
Type E	Epoxy markings with type 1 and type 4 glass beads
Type F	Polyester markings with type 1 glass beads
Type G	Polyester markings with type 1 and type 4 glass beads
Type H	Thermoplastic markings with type 1 glass beads
Type I	Thermoplastic markings with type 1 and type 5 glass beads
Type J	Preformed plastic markings
Type K	Nonreflectorized markings

Materials

634.02 Requirements. Furnish material that conforms to the MUTCD and the following subsections:

Conventional Traffic Paint 718.16
Epoxy Markings 718.18
Epoxy Resin Adhesives 718.26
Glass Beads 718.22
Polyester Markings 718.19
Preformed Plastic Markings 718.21
Raised Pavement Markers.. 718.23
Thermoplastic Markings 718.20
Waterborne Traffic Paint 718.17

Construction

634.03 Performance. Where existing and final pavement marking locations are identical, stake the limits of all existing pavement markings (no-passing zones, edge stripes, etc.) before any pavement work. Upon completion of the final surface course, establish line limits for the new pavement for approval before marking. Establish markings according to the MUTCD.

Remove loose particles, dirt, tar, grease, and other deleterious material from the surface to be marked. Where markings are placed on Portland cement concrete pavement less than 1 year old, clean the pavement of all residue and curing compounds. Remove temporary pavement markings the same day permanent pavement markings are applied. Apply markings to a clean, dry surface according to the MUTCD.

Furnish a written copy of the manufacturer's marking recommendations to the CO at least 7 days before starting pavement marking application. A field demonstration may be required to verify the adequacy of recommendations.

Ship marking material in appropriate containers plainly marked with the following information, as appropriate for the material being furnished:

(a) Manufacturer's name and address.

(b) Name of product.

(c) Lot/batch numbers.

(d) Color.

(e) Net weight and volume of contents.

(f) Date of manufacture.

(g) Date of expiration.

(h) Statement of contents, if mixing of components is required.

(i) Mixing proportions and instructions.

(j) Safety information.

Apply pavement markings in the direction of traffic according to the manufacturer's recommendations. Apply all markings to provide a clean-cut, uniform, and workmanlike appearance by day and night.

Make lines 100 mm wide. Make broken lines 3 m long, with 9-m gaps. Make dotted lines 0.5 m long, with 1.0-m gaps. Separate double lines with a 100-mm space.

Protect marked areas from traffic until the markings are dried to no-tracking condition. Remove all tracking marks, spilled marking material, markings in unauthorized areas, and defective markings.

634.04 Conventional Traffic Paint (Type A). Apply paint when the pavement and air temperatures are above 4 °C. Spray paint at a 0.38-mm minimum wet film thickness before glass beads, or at a rate of 2.6 m^2/L. Immediately apply glass beads on the paint at a minimum rate of 0.7 kg/L of paint.

Apply two coats with glass beads to all centerline stripes. Coats are applied in opposite directions on bituminous surface treatment or chip seal surfaces. Second coats are applied from 4 to 48 hours after the first coat.

634.05 Waterborne Traffic Paint (Types B & C). Apply paint when the pavement and air temperatures are above 10 °C. Spray paint at 0.38-mm minimum wet film thickness before glass beads, or at a rate of 2.6 m^2/L.

(a) Type B. Immediately apply type 1 glass beads on the paint at a minimum rate of 0.7 kg/L of paint.

(b) Type C. Immediately apply type 3 glass beads on the paint at a minimum rate of 1.4 kg/L of paint.

Apply two coats with glass beads to all centerline stripes. Coats are applied in opposite directions on bituminous surface treatment or chip seal surfaces. Second coats are applied from 4 to 48 hours after the first coat.

634.06 Epoxy Markings (Types D & E). Heat components A and B separately at 43 ± 17 °C and mix. Discard all material heated over 60 °C. Apply epoxy when the pavement and air temperatures are above 10 °C. Apply as a spray at 43 ± 17 °C (gun tip temperature) at a 0.38-mm minimum dry film thickness, or a rate of 2.6 m^2/L.

(a) Type D. Immediately apply type 1 glass beads on the epoxy at a minimum rate of 1.8 kg/L of epoxy.

(b) Type E. Use two bead dispensers. Immediately apply type 4 glass beads on the epoxy at a minimum rate of 1.4 kg/L of epoxy, immediately followed by an application of type 1 glass beads at a minimum rate of 1.4 kg/L.

634.07 Polyester Markings (Types F & G). Apply polyester when the pavement and air temperatures are above 10 °C. Spray at 53 ± 4 °C (gun tip temperature) at a 0.38-mm minimum dry film thickness, or at a rate of 2.6 m^2/L. Discard all material heated over 66 °C. Do not use fast-dry polyester markings on hot asphalt concrete pavements that are less than 1 year old.

(a) Type F. Immediately apply type 1 glass beads on the polyester at a minimum rate of 1.8 kg/L of polyester.

(b) Type G. Use two bead dispensers. Immediately apply type 4 glass beads on the polyester at a minimum rate of 1.4 kg/L of polyester, immediately followed by an application of type 1 glass beads at a minimum rate of 1.4 kg/L.

634.08 Thermoplastic Markings (Types H & I). On areas to be marked on Portland cement concrete pavements and old asphalt pavements, apply an epoxy resin primer/sealer according to the thermoplastic manufacturer's recommendations. Allow the primer/sealer to dry.

Apply thermoplastic when the pavement and air temperatures are above 10 °C. Spray or extrude the thermoplastic at 220 ± 3 °C. For centerlines and lane lines, spray or extrude a 2.3-mm minimum dry film thickness, or at a rate of 0.44 m^2/L. For edge lines, spray or extrude a 1.5-mm minimum dry film thickness, or at a rate of 0.66 m^2/L.

(a) Type H. Immediately apply type 1 glass beads on the thermoplastic at a minimum rate of 0.59 kg/m^2.

(b) Type I. Use two bead dispensers. Immediately apply type 5 glass beads on the thermoplastic at a minimum rate of 0.59 kg/m^2, immediately followed by an application of type 1 glass beads at a minimum rate of 0.59 kg/m^2.

Ensure that the minimum bond strength of the thermoplastic is 1.2 MPa on Portland cement concrete pavements.

634.09 Preformed Plastic Markings (Type J). Install to form a durable, weather-resistant bond to the pavement. Apply preformed plastic markings according to the manufacturer's recommendation.

Where applied during final compaction of asphalt pavement, apply preformed plastic when the pavement temperature is about 60 °C. Roll the marking onto the surface with a steel-wheeled roller. The finished pavement marking may extend approximately 0.25 mm above the final surface.

634.10 Nonreflectorized Markings (Type K). Apply conventional traffic paint, waterborne traffic paint, epoxy markings, polyester markings, or thermoplastic markings as described above, but with no glass beads added.

634.11 Raised Pavement Markers. Install raised pavement markers when the pavement and air temperatures are above 10 °C. Apply raised pavement markers with epoxy resin or asphalt adhesive.

Heat epoxy components A and B separately with indirect heat; mix, and apply at 21 ± 6 °C. Discard all material heated over 49 °C or stiffened by polymerization.

Heat and apply asphalt adhesives at 211 ± 7 °C. Discard all material heated over 232 °C.

Space and align the markers to within 13 mm of the required location. Do not place raised pavement markers over pavement joints.

Make the minimum bond strength 12 kPa, or a total tensile strength of 110 N.

Measurement

634.12 Method. Use the method of measurement that is DESIGNATED IN THE SCHEDULE OF ITEMS.

Measure pavement markings by the meter, kilometer, liter, or square meter.

When pavement markings are measured by the meter or kilometer, measure the length of line applied along the centerline of each 100-mm-wide line applied, regardless of color. Measure broken or dotted pavement lines from end to end of the line, including gaps. Measure solid pavement lines from end to end of each continuous line. For line widths other than 100 mm, the measured length of line is adjusted in the ratio of the required width to 100 mm.

When pavement markings are measured by the square meter, measure the number of square meters of symbol or letter marking based on the marking area shown in the contract or, if not shown, the area of each marking measured in place to the nearest square meter.

Measure raised pavement markers by the each.

Payment

634.13 Basis. The accepted quantities will be paid for at the contract unit price for each PAY ITEM DESIGNATED IN THE SCHEDULE OF ITEMS.

Payment will be made under:

Pay Item		Pay Unit
634 (01)	Pavement markings, type _____, color _____	Meter
634 (02)	Pavement markings, type _____, color _____	Kilometer
634 (03)	Pavement markings, type _____, color _____	Square Meter
634 (04)	Pavement markings, type _____, color _____	Liter
634 (05)	Raised pavement markers, type _____, color _____	Each

Section 637—Equipment Rental

Description

637.01 Work. Furnish and operate equipment for construction work ordered by the CO and not otherwise provided for under the contract.

Equipment

637.02 Requirements. The CO will order in writing rental equipment for use on the project. Submit the model number and serial number for each piece of equipment before use. Make equipment available for inspection and approval before use.

Furnish and operate equipment with such auxiliary attachments, oilers, and so forth, as are usually needed for efficient operation of the equipment. Keep the equipment in good repair and capable of operating 90 percent of the working time.

Obtain approval for the length of workday and workweek before beginning work. Keep daily records of the number of unit-hours of operation. Submit the records along with certified copies of the payroll.

Measurement

637.03 Method. Use the method of measurement that is DESIGNATED IN THE SCHEDULE OF ITEMS.

Measure rental equipment by the hour. When the equipment is operated part-time in any half shift and is operative and not used on other work during the half shift, measure the full half shift. Measure time in excess of 40 hours per week at the same rate as the first 40 hours.

Do not make deductions for nonoperating time for reasonable interruptions for minor repairs if the nonoperating time does not exceed one-half hour per workday. Do not measure nonoperating time in excess of the one-half hour per workday. Do not measure equipment dependent upon another piece of nonoperable equipment.

Do not measure standby time, or time for moving equipment to or from the project or between project worksites.

Measure quantities to include the actual hours, to the nearest half hour, that the equipment is in operation performing the required work. Each day, record the actual hours that the equipment is in operation on the required work.

Payment

637.04 Basis. The accepted quantities will be paid for at the contract unit price for each PAY ITEM DESIGNATED IN THE SCHEDULE OF ITEMS.

Payment will be made under:

Pay Item		Pay Unit
637 (01)	______________________________ *Type and size of equipment*	Hour

Section 640—Road Closure Devices

Description

640.01 Work. Furnish and install, or install only, road closure devices using fabricated gates and accessories, combination post and rail barriers, concrete barriers, earth mound barriers, and other devices as SHOWN ON THE DRAWINGS.

Materials

640.02 Requirements. Furnish materials to be used in fabricating gates and barriers as SHOWN ON THE DRAWINGS.

Furnish metal beam elements, steel posts, structural steel, and steel pipe that meet the requirements SHOWN ON THE DRAWINGS.

Ensure that all hardware is galvanized in accordance with AASHTO M 232 and meets the requirements of ASTM A 307. Furnish plain or cut washers that are American Standard Washers.

Furnish timber posts, rails, and lumber that meet the requirements of AASHTO M 168. Provide timber of the species and type, and rate of preservative treatment, that are SHOWN ON THE DRAWINGS.

Furnish concrete that meets the requirements of Subsection 602.03, method B or C, as SHOWN ON THE DRAWINGS.

Construct earth mound barriers as SHOWN ON THE DRAWINGS from excavated material adjacent to the barrier location, or from other locations as SHOWN ON THE DRAWINGS.

Construction

640.03 Performance. Place road closure devices at the location SHOWN ON THE DRAWINGS. Construct all devices to the dimensions SHOWN ON THE DRAWINGS.

In assembling gates, perform required welding in accordance with the best modern practice and the applicable requirements of AWS D1.1.

After assembly, clean nongalvanized steel pipe gates and paint them with one coat of zinc-rich primer and two coats of exterior enamel of the type and color SHOWN ON THE DRAWINGS or in the SPECIAL PROJECT SPECIFICATIONS.

Set all posts vertically and embed them to the depth SHOWN ON THE DRAWINGS. Place concrete for embedment against undisturbed earth within an excavation sized to achieve the embedment dimensions. Compact the backfill in 150-mm layers to finished grade.

Furnish and install all signs and/or reflective warning markers accessory to the road closure device, as SHOWN ON THE DRAWINGS.

Measurement

640.04 Method. Use the method of measurement that is DESIGNATED IN THE SCHEDULE OF ITEMS.

Payment

640.05 Basis. The accepted quantities will be paid for at the contract unit price for each PAY ITEM DESIGNATED IN THE SCHEDULE OF ITEMS.

Payment will be made under:

Pay Item		Pay Unit
640 (01)	Furnish and install road closure device, type ____________, size ____________	Each
640 (02)	Install road closure device, type ____________, size __________	Each
640 (03)	Furnish and install road closure barrier, type ___________, size ___________________	Each
640 (04)	Install road closure barrier, type ______________, size ___________	Each

DIVISION 700
Materials

Section 701—Cement, Grout, & Mortar

701.01 Cement

Ensure that cement meets requirements in the following specifications:

(a) **Portland Cement.** Ensure that Portland cement meets requirements specified in AASHTO M 85.

(b) **Blended Hydraulic Cements, Excluding Types S & SA.** Ensure that they meet requirements specified in AASHTO M 240.

(c) **Masonry Cement.** Ensure that masonry cement meets requirements specified in ASTM C 91.

Fly ash or pozzolan may be substituted for Portland cement, provided the proportions of cement and fly ash or pozzolan conform to the requirements specified in Section 552 or 602.

When blended cement (AASHTO M 240) is proposed for use, meet all requirements for fly-ash-modified concrete in the applicable sections.

Ensure that fly ash or pozzolan materials conform to the requirements specified in Subsection 725.04.

Use the product of only one manufacturing plant and only one brand of any one type of Portland cement on the project.

Store the cement and protect it against dampness. Reject cement that for any reason has become partially set or that contains lumps of caked cement. Do not use cement salvaged from discarded or used bags.

701.02 High-Strength Nonshrink Grout

Furnish grout that is packaged and ready for use with the addition of water at the construction site. Ensure that each bag is stamped to show the last date on which it may be used. Use grout that consists of a hydraulic cementitious system, graded and processed natural fine aggregate, and additional technical components such that the product meets the following conditions:

(a) It is free of inorganic accelerators, including chlorides.

(b) It is free of oxydizing catalysts.

(c) It is free of gas-producing agents.

(d) When mixed to 130 percent flow on flow table (ASTM C 230 at 10 drops), it does not reduce in linear dimension when tested in accordance with ASTM C 157. Take measurements at 72 hours and 7 days.

(e) It produces no bleeding for the first 2 hours after mixing when mixed to 130 percent flow on flow table (ASTM C 230 at 10 drops), as tested in accordance with ASTM C 232.

(f) It has a minimum strength as follows when tested in accordance with ASTM C 109:

(1) After 72 hours, 25 MPa.

(2) At 7 days, 40 MPa.

(3) After 28 days, 50 MPa.

Provide performance characteristics at 115 to 120 percent flow on flow table (ASTM C 230 at 10 drops).

(g) It must be designed, as stated by the manufacturer, to be mixed, placed, and cured at atmospheric temperatures of 5 °C to 30 °C. Submit products proposed for use for approval by the CO, and accompany them with manufacturer's submittals substantiating all requirements in this subsection, including graphs or charts showing the time, temperature, and humidity requirements for curing to achieve the specified grout strengths; and recommendations for storage, mixing, application, and curing procedures.

701.03 Low-Strength Grout

Furnish grout mixtures that conform to the following for the type or types SHOWN ON THE DRAWINGS:

(a) Hydraulic Cement Grout. Furnish a mixture of Portland cement, fine aggregate, water, expansive admixture, and/or fly ash such that the product meets the following requirements:

(1) 7-day compressive strength, AASHTO T 106 .... 4 MPa min.

(2) Flow, FLH T 502 or ASTM C 939, conforming to the following:

(a) Time of efflux[1] .... 16 to 26 seconds

[1]A more fluid mix having a flow cone time of efflux from 9 to 15 seconds may be used during initial injection.

Submit the following with the Certificate of Compliance:

- Mill certifications for the cement.
- Physical and chemical analysis for the pozzolans.
- Independent laboratory test results (1-day, 3-day, and 7-day strengths, flow cone times, shrinkage and expansion observed, and time of initial set).

(b) Plaster Mix (Grout). Ensure that plaster mix (grout) conforms to the following:

(1) Adhesive strength, 28-day, sheer bond adhesion testing method .... 2 MPa min.

(2) Freeze-thaw resistance, ASTM C 666, method B ...... No cracks or delamination after 300 cycles

(3) Accelerated weathering, 5,000-hour .... No visible defects

(4) Slat spray resistance, 300-hour .... No deterioration or loss of adhesion

(5) Absorption, ASTM C 67 .... 3.5% max.

(6) Flexural strength, ASTM C 348, 28-day .... 6.8 MPa min.

(7) Compressive strength, AASHTO T 106, 28-day .... 27.5 MPa min.

(c) Portland Cement Grout. Furnish one part Portland cement and three parts sand. Thoroughly mix with water to produce a thick, creamy consistency.

701.04 Mortar

Furnish mortar that is packaged and ready for use with the addition of water at the construction site. Ensure that each bag is stamped to show the latest date on which it may be used. Use mortar that consists of a cementitious system made up of:

(a) Natural aggregate, 10 mm in maximum size, that meets the requirements specified in ASTM C 33 except for grading. Accomplish grading by blending sieve sizes to obtain the optimum density.

(b) Metallic aggregate free from nonferrous material, soluble alkaline compounds, and visible rust.

(c) Water reducers, workability agents, air-entraining agents, and catalysts.

Blend the materials to minimize bleeding, increase workability, resist exposure to freeze-thaw cycles and deicing salts, and prevent shrinkage within and at the perimeter of the patch, keyway, or other area to be filled.

Ensure that the minimum compressive strength of the mortar, as tested by ASTM C 109 for a 75-mm slump, is:

- 24-hour .... 35 MPa
- 7-day .... 60 MPa
- 28-day .... 70 MPa

Ensure that the durability of the products when tested at 300 cycles, ASTM C 666, procedure A, is:

Submerged in:	DF (%)
Water	98
5% $CaCl_2$ solution	95
5% NaCl solution	85

Ensure that the scaling resistance has a rating of 3, Moderate Scaling, after 50 cycles when tested in accordance with ASTM C 672.

Provide certification from the manufacturer that the product is compatible for work that is 25 mm or more in depth and more than 25 mm in width; and where the mixing, placing, and curing temperatures may range from 5 °C to 30 °C.

Submit products proposed for use to the CO for approval, and accompany them with the manufacturer's submittals substantiating all requirements in this section, including (1) graphs or charts showing the time, temperature, humidity, and curing requirements to achieve mortar strengths equal to the adjacent concrete; and (2) complete recommendations for storage, mixing, application, and curing procedures.

701.05 Polymer Grout

Furnish a polymer binder and fine aggregate in the proportions recommended by the polymer manufacturer with a minimum compressive strength of 25 MPa in 4 hours.

Section 702—Bituminous Material

702.01 Asphalt Cement

Ensure that asphalt cement conforms to AASHTO M 20 or M 226, AASHTO MP 1, or applicable State department of transportation specifications for asphalt materials for the grade specified. When modified asphalt cement is used, test it in accordance with AASHTO PP 5.

702.02 Cutback Asphalt

Ensure that cutback asphalt conforms to the following specifications:

(a) Rapid-curing, AASHTO M 81.

(b) Medium-curing, AASHTO M 82.

(c) Slow-curing, ASTM D 2026.

702.03 Emulsified Asphalt

Ensure that anionic emulsified asphalt conforms to AASHTO M 140, and cationic emulsified asphalt to AASHTO M 208, except as specified below. When specified for tack coat, an equivalent anionic grade emulsion may be substituted for a cationic grade, and vice versa. Unless otherwise noted, test the emulsion in accordance with the procedures in AASHTO T 59.

(a) CRS–2 Emulsions. Ensure that CRS–2 emulsions conform to AASHTO M 208.

(b) CRS–1h & CRS–2h Emulsions. Ensure that CRS–1h and CRS–2h emulsions conform to the requirements of CRS–1 and CRS–2, respectively, except as follows:

(1) Penetration, 25 °C, 100 g, 5-second .... 40 to 90 mm

(c) CMS–2 & CMS–2h Emulsions. Ensure that CMS–2 and CMS–2h emulsions conform to AASHTO M 208; but in table 1, revise the percent of oil distillate as follows:

(1) Oil distillate, by volume of emulsion .... 5 to 12%

(d) Quick-Set Emulsions. Ensure that quick-set emulsions conform to the following:

(1) Viscosity, Saybolt Furol at 25 °C .... 20 to 100 seconds

(2) Residue by distillation .... 57% min.

(3) Sieve test .... 0.10% max.

(4) Tests on residue from distillation:

(a) Penetration, 25 °C, 100 g, 5-second, AASHTO T 49 .... 40 to 100 mm

(b) Solubility in trichloroethylene, AASHTO T 44 .... 97.5% min.

(c) Ductility, 25 °C, 50 mm per minute, AASHTO T 51 ... 40 mm min.

(e) CRS–2 Polymer-Modified Emulsions. Ensure that polymer-modified emulsions conform to the following:

(1) Viscosity, 50 °C, Saybolt Furol .... 100 to 400 seconds

(2) Storage stability test after 24 hours .... 1.0% max.

(3) Demulsibility .... 40% min.

(4) Particle charge .... Positive

(5) Sieve test .... 0.3% max.

(6) Residual by distillation .... 65% max.

(7) Oil distillate by volume of emulsion .... 3% max.

(8) Test on residue from distillation:

(a) Penetration, 25 °C, 100 g, 5-second .... 90 to 200 mm

(b) Solubility in trichloroethylene .... 97.5% min.

(c) Torsional recovery, CAL TRANS test no. 332 .... 18% min.
or toughness/tenacity N•m[1] .... 5.6/2.8 min.

[1] Benson Method of Toughness and Tenacity, Scott tester, mm-kg at 25 °C, 500 mm per minute pull. Tensionhead 22 mm diameter (ASTM D 4, proposed P 243).

Ensure that polymer is milled into the emulsion during the manufacturing process.

702.04 Application Temperatures

Apply asphalts within the temperature ranges shown in table 702-1.

Table 702-1.—Application temperatures, range (°C).

Type and Grade of Asphalt	Temperature Ranges (minimum–maximum)	
	Spraying Temperatures	Mixing Temperatures[b]
Cutback asphalt:		
MC–30	30–[a]	–
RC, MC, or SC–30	50–[a]	–
RC, MC, or SC–250	75–[a]	60–80[c]
RC, MC, or SC–800	95–[a]	75–100[c]
RC, MC, or SC–3000	110–[a]	80–115[c]
Emulsified asphalt:		
RS–1	20–60	–
RS–2	50–85	–
MS–1	20–70	20–70
MS–2, –2h	–	20–70
HFMS–1, –2, –2h, –2s	20–70	10–70
SS–1, –1h; CSS–1, –1h	20–70[d]	20–70
CRS–1, –1h	50–85	–
CRS–2, –2h, –2 modified	60–85	–
CMS–2, –2h	40–70	50–60
Asphalt cement, all grades	180 max.	180 max.

a. The maximum temperature at which fogging or foaming does not occur.
b. Temperature of mix immediately after discharge.
c. Temperature may be above flash point. Take precautions to prevent fire or explosion.
d. For fog seals and tack coats.

702.05 Recycling Agent

Use recycling agents that conform to ASTM D 4552, or use a preapproved petroleum product additive that restores aged asphalt to the required specifications.

702.06 Asphalt Mastic

Use asphalt mastic that conforms to AASHTO M 243.

702.07 Antistrip Additive

(a) Furnish commercially produced heat-stable liquid products that have the chemical and physical properties when added to an asphalt to prevent separation of the asphalt from aggregates.

(b) Furnish cement that conforms to Subsection 701.01 or fly ash that conforms to Subsection 725.04.

(c) Furnish hydrated lime conforming to Subsection 725.03.

702.08 Cold Asphalt Concrete

Provide an asphalt concrete mixture composed of crushed stone or gravel and asphalt cement mixed in an approved plant. Ensure that the gradation and quality of the aggregate and the grade and quality of asphalt binder conform to those normally used in the construction of highways by Federal or State agencies.

Do not use an aggregate asphalt mixture that strips. Use an asphalt grade that leaves the mix pliable and workable at a temperature of –10 °C.

Section 703—Aggregate

703.01 Fine Aggregate for Portland Cement Concrete

As fine aggregate, use sand that conforms to the requirements shown below.

(a) For structural concrete, fine aggregate is sand that conforms to AASHTO M 6, class B, but limit the material that passes the 75-µm sieve to 3.0 percent. Also meet the supplementary requirements of AASHTO M 6 for reactive aggregates. Use material that conforms to sand equivalent value, AASHTO T 176, alternate method number 2, 75 minimum.

(b) For structural concrete, lightweight fine aggregate is sand that conforms to AASHTO M 195, where applicable.

(c) For minor concrete structures, meet requirements specified in AASHTO M 6.

703.02 Coarse Aggregate for Portland Cement Concrete

Furnish coarse aggregate that conforms to the requirements listed below.

(a) For structural concrete, furnish coarse aggregate that conforms to AASHTO M 80, class A, but use aggregates with a percentage of wear that is not more than 40 percent, in accordance with AASHTO T 96. In concrete used in bridge decks or for paving, do not use aggregates known to polish, or carbonate aggregates containing by weight less than 25 percent insoluble residue, as determined by ASTM D 3042.

Ensure that the adherent coating on the aggregate does not exceed 1.0 percent when tested in accordance with FLH T 512.

(b) For structural concrete, furnish lightweight coarse aggregate that conforms to AASHTO M 195, when applicable.

(c) For minor concrete structures, aggregate must meet AASHTO M 80 and the class designations that are appropriate for end use and weathering exposure.

703.03 Granular Backfill

(a) Coarse Granular Backfill. Furnish backfill material that conforms to AASHTO M 80, class E, and AASHTO M 43, grading number 3, 4, 5, 7, 57, or 67.

Minor variations in the gradation and the deleterious substance content are subject to approval by the CO.

(b) Fine Granular Backfill. Furnish backfill material that conforms to AASHTO M 6. The soundness test is not required. Minor variations in the gradation and the deleterious substance content are subject to approval by the CO.

703.04 Sheathing Material

Furnish either fine aggregate meeting gradation requirements of AASHTO M 6, or coarse aggregate consisting of sound, durable particles of gravel, slag, or crushed stone, as specified in table 703-1.

Table 703-1.—Sheathing material gradation.

Sieve Designation	% by Weight Passing Standard Sieves (AASHTO T 11 and T 27)
75 mm	100
19 mm	50–90
4.75 mm	20–50
75 µm	0.0–2.0

703.05 Subbase, Base, & Surface Course Aggregate

(a) General. Furnish aggregates that consist of hard, durable particles or fragments of crushed stone, crushed slag, or crushed gravel meeting the appropriate gradation, as shown in table 703-2 or 703-3, and conforming to the following:

(1) Los Angeles abrasion, AASHTO T 96 .... 40% max.

(2) Sodium sulfate soundness loss (five cycles), AASHTO T 104 .... 12% max.

(3) Durability index (coarse), AASHTO T 210 .... 35 min.

(4) Durability index (fine), AASHTO T 210 .... 35 min.

(5) Fractured faces, FLH T 507 .... 50% min.

Furnish a material that is free from organic matter and lumps or balls of clay. Do not use material that breaks up when alternately frozen and thawed or wetted and dried.

Obtain the aggregate gradation by crushing, screening, and blending processes as necessary. Ensure that fine aggregate (material passing the 4.75-mm sieve) consists of natural or crushed sand and fine mineral particles.

Table 703-2.—Gradation TV ranges for subbase and base.

Sieve Size	% by Mass Passing Designated Sieve (AASHTO T 27 and T 11)				
	Grading Designation				
	A (Subbase)	B (Subbase)	C (Base)	D (Base)	E (Base)
63 mm	100	–	–	–	–
50 mm	97–100	100	100	–	–
37.5 mm	–	97–100	97–100	100	–
25 mm	65–79 (6)	–	–	97–100	100
19 mm	–	–	67–81 (6)	–	97–100
12.5 mm	45–59 (7)	–	–	–	–
9.5 mm	–	–	–	56–70 (7)	67–79 (6)
4.75 mm	28–42 (6)	40–60 (8)	33–47 (6)	39–53 (6)	47–59 (7)
425 µm	9–17 (4)	–	10–19 (4)	12–21 (4)	12–21 (4)
75 µm	4.0–8.0 (3)	0.0–12.0 (4)	4.0–8.0 (3)	4.0–8.0 (3)	4.0–8.0 (3)

Note: Allowable deviations (±) from TV are shown in parentheses.

Table 703-3.—Gradation TV ranges for surface courses.

Sieve Size	% by Mass Passing Designated Sieve (AASHTO T 27 and T 11)	
	Grading Designation	
	F	G
37.5 mm	100	–
25 mm	97–100	100
19 mm	76–89 (6)	97–100
9.5 mm	56–68 (6)	70–80 (6)
4.75 mm	43–53 (7)	51–63 (7)
1.18 mm	23–32 (6)	28–39 (6)
425 µm	15–23 (5)	19–27 (5)
75 µm	10–16 (4)	10–16 (4)

Note: Allowable deviations (±) from TV are shown in parentheses. If the plasticity index (PI) is greater than 0, the TV range for the 75-µm sieve size is 6–12 (± 4).

(b) Subbase & Base Aggregates. Furnish subbase or base aggregate that conforms to specifications in Subsection 703.05(a), and to the following:

(1) Liquid limit, AASHTO T 89 .... 25 max.

(2) Plastic limit, AASHTO T 90 .... Nonplastic

(c) Surface Course Aggregate. Furnish surface course aggregate that conforms to specifications in Subsection 703.05(a), and to the following:

(1) Liquid limit, AASHTO T 89 .... 35 max.

(2) Plasticity index, AASHTO T 90:

(a) If the percent passing the 75-µm sieve is less than 12% .... 2 to 9

(b) If the percent passing the 75-µm sieve is greater than 12% 0

Do not furnish material that contains asbestos fibers.

703.06 Crushed Aggregate

Furnish crushed hard, durable particles or fragments of stone or gravel meeting the size and quality requirements for crushed aggregate material normally used locally in the construction and maintenance of highways by Federal or State agencies.

Furnish crushed aggregate with a maximum size of 25 mm as determined by AASHTO T 27 and AASHTO T 11. Furnish crushed aggregate that is uniformly graded from coarse to fine and is free of organic matter and lumps or balls of clay.

703.07 Hot Asphalt Concrete Pavement Aggregate

Aggregate for hot asphalt concrete pavement consists of hard, durable particles or fragments of crushed stone, crushed slag, or crushed gravel.

Size, grade, and combine the aggregate fractions for the mixture in proportions such that the resulting composite blend conforms to the gradation shown in table 703-4 for the grading designated.

Furnish a blend that is reasonably free from organic or other deleterious material and does not contain more than 1.0 percent clay lumps and friable particles when tested in accordance with AASHTO T 112.

Local State department of transportation requirements for gradation and quality of hot asphalt concrete pavement may be substituted for the above requirements when DESIGNATED IN THE SCHEDULE OF ITEMS.

Table 703-4.—Aggregate gradation requirements for hot asphalt concrete pavement.

Sieve Size	% by Weight Passing Designated Sieve (AASHTO T 27 and T 11) Grading Designation A	B	D	F
50 mm	–	–	–	–
37.5 mm	100	–	–	–
25 mm	97–100	100	–	–
19 mm	–	97–100	100	–
12.5 mm	–	76–88	97–100	–
9.5 mm	53–70	–	–	100
4.75 mm	40–52	49–59	57–69	33–47
2.36 mm	25–39	36–45	41–49	7–13
600 µm	12–22	20–28	22–30	–
300 µm	8–16	13–21	13–21	–
75 µm	3–8	3–7	3–8	2–4

(a) Coarse Aggregate. Coarse aggregate (aggregate retained on the 4.75-mm sieve) consists of crushed stone, crushed slag, or crushed gravel that conforms to the following:

(1) Los Angeles abrasion, AASHTO T 96 .... 40% max.

(2) Sodium sulfate soundness loss (five cycles), AASHTO T 104 .... 12% max.

(3) Fractured faces, FLH T 507 .... 75% min.

(4) Durability index (coarse), AASHTO T 210 .... 35 min.

Do not use aggregates known to polish, or carbonate aggregates containing by weight less than 25 percent insoluble residue when tested in accordance with ASTM D 3042.

(b) Fine Aggregate. Fine aggregate (aggregate that passes a 4.75-mm sieve) consists of natural sand, stone screenings, slag screenings, or a combination thereof conforming to AASHTO M 29. Exclude the grading requirements and include the sodium sulfate soundness test and the following:

(1) Durability index (fine), AASHTO T 210 .... 35 min.

(2) Sand equivalent value, AASHTO T 176, alternate method number 2 .... 45 min.

(c) Lightweight Aggregate (Slag). Only use crushed slag that conforms to the quality requirements specified in AASHTO M 195. Other kinds or types of lightweight aggregates covered in AASHTO M 195 are not permitted.

703.08 Cold Asphalt Concrete Pavement Aggregate

Furnish aggregate for cold asphalt concrete pavement consisting of hard, durable particles or fragments of crushed stone, crushed slag, or crushed gravel.

Size, grade, and combine the aggregate fractions for the mixture in proportions such that the resulting composite blend conforms to the applicable gradation requirements shown in table 703-5 for dense-graded mixtures and table 703-6 for open-graded mixtures.

Ensure that the composite blend is reasonably free from organic or other deleterious material and contains less than 1.0 percent clay lumps and friable particles when tested in accordance with AASHTO T 112.

(a) Coarse Aggregate. Furnish coarse aggregate consisting of crushed stone, crushed slag, or crushed gravel that conforms to the following:

(1) Los Angeles abrasion, AASHTO T 96 .... 40% max.

(2) Sodium sulfate soundness loss (five cycles), AASHTO T 104 .... 12% max.

(3) Fractured faces, FLH T 507 .... 75% min.

(4) Durability index (coarse), AASHTO T 210 .... 35 min.

Do not use aggregates known to polish, or carbonate aggregates containing by weight less than 25 percent insoluble residue, in accordance with ASTM D 3042.

Table 703-5.—Aggregate gradation requirements and TV ranges for dense-graded cold bituminous pavement.

Sieve Size	% by Weight Passing Designated Sieve (AASHTO T 27 and T 11)				
	Grading Designation				
	DA	DB	DC	DD	DE
50 mm	100	–	–	–	–
37.5 mm	95–100	100	–	–	–
25 mm	–	95–100	100	–	–
19 mm	60–80 (7)	–	95–100	100	–
12.5 mm	–	60–80 (7)	–	95–100	100
9.5 mm	–	–	60–80 (7)	–	95–100
4.75 mm	20–55 (7)	25–60 (7)	35–65 (7)	45–70 (7)	60–80 (7)
2.36 mm	10–40 (6)	15–45 (6)	20–50 (6)	25–55 (6)	35–65 (6)
300 µm	2–16 (4)	3–18 (4)	3–20 (4)	5–20 (4)	6–25 (4)
75 µm	0–5 (3)	1–7 (3)	2–8 (3)	2–9 (3)	2–10 (3)

Note: Allowable deviations (±) from TV are shown in parentheses.

Table 703-6.—Aggregate gradation requirements and TV ranges for open-graded cold bituminous pavement.

Sieve Size	% by Weight Passing Designated Sieve (AASHTO T 27 and T 11)			
	Grading Designation			
	OA	OB	OC	OD
37.5 mm	100	–	–	–
25 mm	95–100	100	–	–
19 mm	–	95–100	100	–
12.5 mm	25–65 (7)	–	95–100	100
9.5 mm	–	20–55 (7)	35–40 (7)	85–100 (7)
4.75 mm	0–10 (5)	0–10 (5)	–	–
2.36 mm	0–5 (3)	0–5 (3)	3–7 (3)	–
1.18 mm	–	–	–	0–5 (3)
75 µm	0–2 (1)	0–2 (1)	0–1 (1)	0–2 (1)

Note: Allowable deviations (±) from TV are shown in parentheses.

(b) Fine Aggregate. Furnish fine aggregate consisting of natural sand, stone screenings, slag screenings, or a combination thereof conforming to AASHTO M 29. Exclude the grading requirements and include the sodium sulfate soundness test and the following:

(1) Durability index (fine), AASHTO T 210 .... 35 min.

(2) Sand equivalent value, AASHTO T 176, alternate method number 2 .... 35 min.

703.09 Asphalt Surface Treatment Aggregate

Furnish aggregate for single and multiple surface treatment courses consisting of hard, durable particles or fragments of crushed stone, crushed slag, or crushed gravel.

Size, grade, and combine the aggregate fractions to conform to specifications in table 703-7 for the gradation designated. Ensure that the composite blend is reasonably free from organic or other deleterious material and contains less than 1.0 percent clay lumps and friable particles when tested in accordance with AASHTO T 112.

Table 703-7.—Aggregate gradation requirements for single- and multiple-course surface treatments.

	% by Weight Passing Designated Sieve (AASHTO T 27 and T 11)					
	Grading Designation					
Sieve Size	A	B	C	D	E	F
37.5 mm	100	–	–	–	–	–
25 mm	90–100	100	–	–	–	–
19 mm	0–35	90–100	100	–	–	–
12.5 mm	0–8	0–35	90–100	100	–	–
9.5 mm	–	0–12	0–35	85–100	100	100
4.75 mm	–	–	0–12	0–35	85–100	85–100
2.36 mm	–	–	–	0–8	0–23	–
75 µm	0–2	0–2	0–2	0–2	0–2	0–10

Use only one type of aggregate on a project, and ensure that aggregates meet the following quality requirements:

(a) Los Angeles abrasion, AASHTO T 96 .... 40% max.

(b) Sodium sulfate soundness loss, AASHTO T 104 .... 12% max.

(c) Loose unit weight shoveling procedure, AASHTO T 19M .... 1,100 kg/m³ min.

(d) Coating and stripping of bitumen-aggregate mixtures, AASHTO T 182[1] .... 95% min.

(e) Fractured faces, FLH T 507 .... 75% min.

(f) Flakiness index, FLH 508 .... 30 max.

(g) Durability index (coarse), AASHTO T 210 .... 35 min.

(h) Durability index (fine), AASHTO T 210 .... 35 min.

(i) Adherent coating on the aggregate, FLH T 512 .... 0.5% max.

[1]An approved chemical additive may be used to meet this requirement.

Do not use lightweight aggregate as defined in AASHTO M 195.

703.10 Slurry Seal Aggregate

For slurry seals, furnish aggregate that is a natural or manufactured sand, slag, crushed fines, or other mineral aggregate conforming to table 703-8 and the following:

(a) Los Angeles abrasion, AASHTO T 96 .... 35% max.

(b) Sand equivalent value, AASHTO T 176, alternate method number 2 .... 45 min.

(c) Sand content by weight of total combined aggregate with < 1.25% water absorption .... 50% max.

(d) Sodium sulfate soundness loss, AASHTO T 104 .... 12% max.

(e) Fine durability index, AASHTO T 210 .... 60% max.

Ensure that aggregate gradation is as shown in table 703-8 for the type specified.

Table 703-8.—Slurry seal aggregate gradation requirements and application rates.[a]

Sieve Size	% by Weight Passing Designated Sieve (AASHTO T 27 and T 11)		
	Type of Slurry Seal		
	I	II	III
9.5 mm	–	100	100
4.75 mm	100	90–100	70–90
2.36 mm	90–100	65–90	45–70
1.18 mm	65–90	45–70	28–50
600 µm	40–65	30–50	19–34
300 µm	25–42	18–30	12–25
150 µm	15–30	10–21	7–18
75 µm	10–20	5–15	5–15
Application rate (kg/m^2)	3.3–5.5	5.5–8.2	≥ 8.2

a. Based on the dry weight of the aggregate.

703.11 Choker Aggregate

Furnish aggregate for choker consisting of hard durable particles or fragments of crushed gravel or crushed stone meeting the gradation shown in table 703-9. Furnish a material that is free from organic matter and clay balls and has a minimum sand equivalent value of 75, as determined by AASHTO T 176, referee method.

Table 703-9.—Choker aggregate gradation.

Sieve Size	% by Weight Passing Designated Sieve (AASHTO T 27 and T 11)
9.5 mm	100
4.75 mm	70–100
75 µm	0.0–5.0

703.12 Blotter

Furnish aggregate for blotter material consisting of sound, durable particles of gravel or crushed stone with a gradation such that all particles will pass a sieve with 9.5-mm square openings. Furnish material that is free from organic matter and has a liquid limit, established by AASHTO T 89, of less than 25.

703.13 Aggregate for Lean Concrete Backfill

Furnish hard, clean, durable, nonplastic, nonorganic, nonreactive aggregate.

703.14 Superpave Asphalt Concrete Pavement Aggregate

Furnish hard, durable particles, or fragments of crushed stone, crushed slag, or crushed gravel conforming to the following:

(a) Los Angeles abrasion, AASHTO T 96 .... 35% max.

(b) Sodium sulfate soundness loss, AASHOT T 104 (five cycles) .... 12% max.

(c) Durability index, AASHTO T 210 (coarse and fine) .... 35 min.

(d) Fractured faces, FLH T 507 .... 55 min.

(e) Sand equivalent value, AASHTO T 176, alternate method number 2 .... 40 min.

(f) Size, grade, and combine the aggregate fractions for the mixture in proportions such that the resulting composite blend is located between the control points for the appropriate nominal maximum size of aggregate shown in table 703-10, 703-11, or 703-12, and figure 703-1, 703-2, or 703-3. The nominal maximum size is one sieve size greater than the first sieve to retain more than 10 percent of the combined aggregate. Use the appropriate table and figure in accordance with the nominal maximum size of aggregate. The gradation should not pass through the restricted zone when plotted. Test in accordance with AASHTO T 11 and T 27.

Table 703-10.—Superpave requirements for 12.5-mm nominal size aggregate.

Sieve Size (mm)	Control Points		0.45 Chart Max. Density	Restricted Zone		TV's	Allowable Deviation[b]
				Min. Boundary	Max. Boundary		
19.00	–	100.0	100.0	–	–	–	–
12.50	100.0	90.0	82.8	–	–	–	–
9.50	–	–	73.2	–	–	–	–
4.75	–	–	53.6	–	–	–[a]	6
2.36	58.0	28.0	39.1	39.1	39.1	–[a]	6
1.18	–	–	28.6	25.6	31.6	–	–
0.60	–	–	21.1	19.1	23.1	–[a]	4
0.30	–	–	15.5	15.1	15.1	–[a]	3
0.15	–	–	11.3	–	–	–	–
0.075	10.0	2.0	8.3	–	–	–[a]	2

a. Establish TV's as part of the job-mix formula. Establish aggregate gradation TV's to the nearest 0.1 percent.
b. Plus or minus from established TV's.

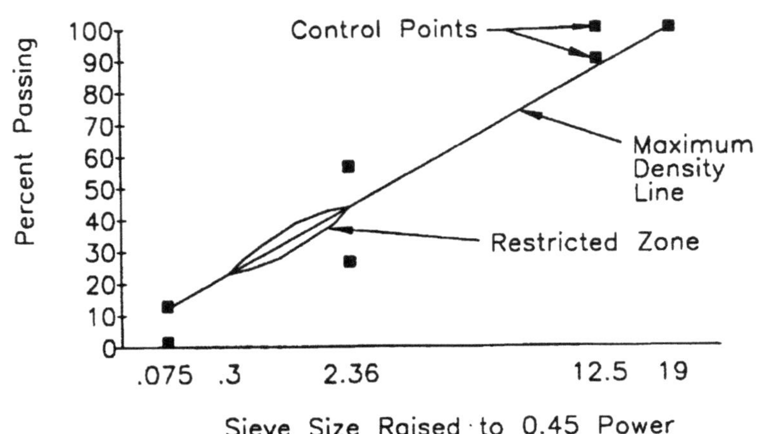

Figure 703-1.—Gradation chart for 12.5-mm nominal size aggregate.

Table 703-11.—Superpave requirements for 19-mm nominal size aggregate.

Sieve Size (mm)	Control Points		0.45 Chart Max. Density	Restricted Zone Min. Boundary	Restricted Zone Max. Boundary	TV's	Allowable Deviation[b]
25.00	–	100.0	100.0	–	–	–	–
19.00	100.0	90.0	88.4	–	–	–	–
12.50	–	–	73.2	–	–	–	–
9.50	–	–	64.7	–	–	–	–
4.75	–	–	47.4	–	–	–[a]	6
2.36	49.0	23.0	34.6	34.6	34.6	–[a]	6
1.18	–	–	25.3	22.3	28.3	–	–
0.60	–	–	18.7	16.7	20.7	–[a]	4
0.30	–	–	13.7	13.7	13.7	–[a]	3
0.15	–	–	10.0	–	–	–	–
0.075	8.0	2.0	7.3	–	–	–[a]	2

a. Establish TV's as part of the job-mix formula. Establish aggregate gradation TV's to the nearest 0.1 percent.
b. Plus or minus from established TV's.

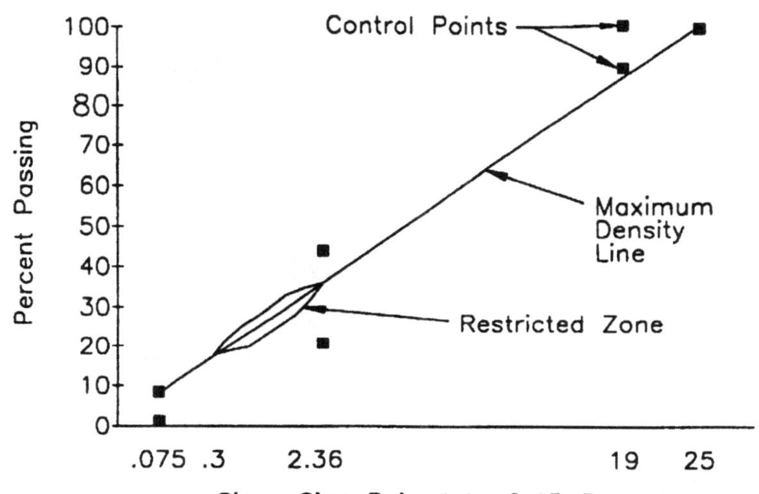

Figure 703-2.—Gradation chart for 19-mm nominal size aggregate.

Table 703-12.—Superpave requirements for 25-mm nominal size aggregate.

Sieve Size (mm)	Control Points		0.45 Chart Max. Density	Restricted Zone		TV's	Allowable Deviation[b]
				Min. Boundary	Max. Boundary		
37.5	–	100.0	100.0	–	–	–	–
25.0	100.0	90.0	83.3	–	–	–	–
19.00	–	–	73.6	–	–	–	–
12.50	–	–	61.0	–	–	–	–
9.50	–	–	53.9	–	–	–	–
4.75	–	–	39.5	39.5	39.5	–[a]	6
2.36	45.0	19.0	28.8	26.8	30.8	–[a]	6
1.18	–	–	21.1	18.1	24.1	–	–
0.60	–	–	15.6	13.6	17.6	–[a]	4
0.30	–	–	11.4	11.4	11.4	–[a]	3
0.15	–	–	8.3	–	–	–	–
0.075	7.0	1.0	6.1	–	–	–[a]	2

a. Establish TV's as part of the job-mix formula. Establish aggregate gradation TV's to the nearest 0.1 percent.

b. Plus or minus from established TV's.

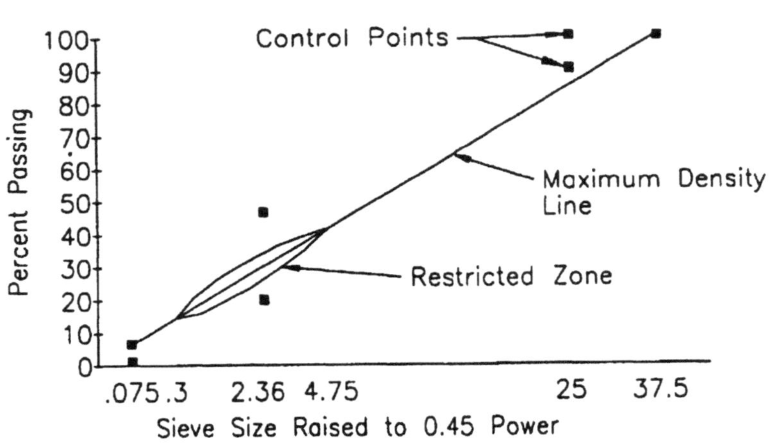

Figure 703-3.—Gradation chart for 25-mm nominal size aggregate.

Section 704—Soil

704.01 Foundation Fill

Furnish granular material free of excess moisture, frozen lumps, roots, sod, and other deleterious material and conforming to the following:

(a) Material passing 50-mm sieve 100%

(b) Soil classification, AASHTO M 145 .... A-1-a

(c) In wet environments, material passing 75-µm sieve ... 6% max.

704.02 Bedding

Furnish material that conforms to the following for the class specified:

(a) Class A Bedding. Furnish concrete in accordance with specifications in Section 602.

(b) Class B Bedding. Furnish approved sand or selected sandy soil free of excess moisture, muck, frozen lumps, roots, sod, and other deleterious material, and conforming to the following:

(1) Material passing 9.5-mm sieve 100%

(2) Material passing 75-µm sieve, AASHTO T 27 and T 11 .... 10% max.

(c) Class C Bedding. Furnish approved sand or fine granular material free of excess moisture, muck, frozen lumps, roots, sod, and other deleterious material. Remove all rock particles and hard earth clods larger than 38 mm.

704.03 Backfill Material

Furnish granular material or fine compatible soil free of excess moisture, muck, frozen lumps, roots, sod, and other deleterious material. Remove all rock particles and hard earth clods larger than 75 mm in the longest dimension.

704.04 Structural Backfill

Furnish free-draining granular material free of excess moisture, muck, frozen lumps, roots, sod, and other deleterious material. Remove all rock particles and hard earth clods larger than 75 mm in the longest dimension. Ensure that material conforms to the following:

(a) Material passing 75-µm sieve, AASHTO T 27 and T 11 .... 15% max.

(b) Liquid limit, AASHTO T 89 .... 30 max.

704.05 Topping

Furnish a granular material free of excess moisture, muck, frozen lumps, roots, sod, and other deleterious material. Remove all rock particles larger than 100 mm in the longest dimension. Ensure that material conforms to AASHTO M 145, table 2, soil classification A-1 or A-3.

704.06 Unclassified Borrow

Furnish granular material free of excess moisture, muck, frozen lumps, roots, sod, and other deleterious material. Remove all rock fragments and boulders greater than 600 mm in the longest dimension. Ensure that material conforms to AASHTO M 145, table 2, soil classification A-1, A-3, or A-2-4.

704.07 Select Borrow

Furnish crushed, partially crushed, or natural material free of excess moisture, muck, frozen lumps, roots, sod, and other deleterious material. Ensure that material conforms to the following:

(a) Gradation .... Table 704-1

(b) Liquid limit, AASHTO T 89 .... 30 max.

Table 704-1.—Select borrow gradation.

Sieve Size	% by Weight Passing Designated Sieve (AASHTO T 27 and T 11)
75 mm	100
25 mm	70–100
4.75 mm	30–70
150 µm	0–15

704.08 Select Topping

Furnish crushed, partially crushed, or natural material free of excess moisture, muck, frozen lumps, roots, sod, and other deleterious material. Ensure that material conforms to the following:

(a) Gradation, uniform coarse to fine .... Table 704-2

(b) Liquid limit, AASHTO T 89 .... 30 max.

Table 704-2.—Select topping gradation.

Sieve Size	% by Weight Passing Designated Sieve (AASHTO T 27 and T 11)
75 mm	100
75 μm	0–15

704.09 Bed Course

Furnish porous, free-draining granular material free of excess moisture, muck, frozen lumps, roots, sod, and other deleterious material. Ensure that material conforms to the following:

(a) Gradation, uniform coarse to fine .... Table 704-3

(b) Liquid limit, AASHTO T 89 .... 30 max.

Table 704-3.—Bed course gradation.

Sieve Size	% by Weight Passing Designated Sieve (AASHTO T 27 and T 11)
12.5 mm	100
75.0 μm	0–10

704.10 Select Granular Backfill

Furnish sound, durable granular material free from organic matter or other deleterious material (such as shale or other soft particles with poor durability). Ensure that material conforms to the specifications below.

(a) Quality Requirements. Furnish material that meets the following quality requirements:

(1) Gradation .... Table 704-4

(2) Shear angle of internal friction, AASHTO T 236[1] ...... 34° min.

(3) Sodium sulfate soundness loss (five cycles), AASHTO T 104 .... 15% max.

(4) Los Angeles abrasion, AASHTO T 96 .... 50% max.

(5) Liquid limit, AASHTO T 89 .... 30 max.

[1] Compact samples for AASHTO T 236 to 95 percent of the maximum density determined in accordance with AASHTO T 99, method C or D, and corrected for oversized material as set forth in AASHTO T 99.

Table 704-4.—Select granular backfill gradation.

Sieve Size	% by Weight Passing Designated Sieve (AASHTO T 27 and T 11)
100 mm	100
75 mm	75–100
75 µm	0–15

(b) Electrochemical Requirements. Furnish material that meets the following electrochemical requirements:

(1) Resistivity, AASHTO T 288, any method .... 3,000 Ω × centimeter min.

(2) pH, AASHTO T 289, any method .... 5.0 to 10.0

(3) Sulfate content, AASHTO T 290, any method[1] .... 1,000 ppm max.

(4) Chloride content, AASHTO T 291, any method[1] .... 200 ppm max.

[1] Tests for sulfate and chloride content are not required when pH is between 6.0 and 8.0 and the resistivity is greater than 5,000 Ω × centimeter.

704.11 Special Grout Backfill

Furnish lean grout slurry composed of three parts Portland cement and eight parts fine aggregate by volume. Fly ash may be substituted for two of the three parts Portland cement. Ensure that material conforms to the following:

(a) Water/cement ratio 1.5

(b) Portland cement 701.01

(c) Fly ash .... 725.04, type C

(d) Fine aggregate 703.01

(e) Water 725.01

704.12 Crib Wall Backfill

Furnish material in accordance with Subsection 704.10, but conform to the following:

(a) Gradation .... Table 704-5

(b) Unit weight .... 1,900 kg/m^3 min.

Table 704-5.—Crib wall backfill gradation.

Sieve Size	% by Weight Passing Designated Sieve (AASHTO T 27 and T 11)
75 mm	100
4.75 mm	25–70
300 µm	5–20
75 µm	0–5

Section 705—Rock

705.01 Gabion Rock

Furnish hard, durable rock that is resistant to weathering and reasonably free of organic and spoil material. Ensure that rock conforms to the following specifications:

(a) Coarse durability index, AASHTO T 210 .... 52 min.

(b) Unit weight of a filled basket .... 1,600 kg/m^3 min.

(c) Gradation:

(1) Baskets 0.3 m or greater in the vertical dimension:

(a) Max. dimension .... 200 mm

(b) Min. dimension .... 100 mm

(2) Baskets less than 0.3 m in the vertical dimension:

(a) Max. dimension .... 150 mm

(b) Min. dimension .... 75 mm

705.02 Riprap Rock

Furnish hard, durable, angular rock free of organic and spoil material and resistant to weathering and water action. Do not use rounded rock, boulders, shale, or rock with shale seams. Furnish rock that conforms to the following:

(a) Apparent specific gravity, AASHTO T 85 .... 2.50 min.

(b) Absorption, AASHTO T 85 .... 4.2% max.

(c) Coarse durability index, AASHTO T 210 .... 52 min.

(d) Gradation for the class specified .... Table 705-1

Table 705-1.—Gradation requirements for riprap.

Class	% of Rock by Mass	Mass (kg)	Approximate Cubic Dimension[b,c] (mm)
1	20	10–15	150–200
	30	5–10	125–150
	40	0.5–5	50–125
	10[a]	0–0.5	0–50
2	20	25–50	200–250
	30	10–25	150–200
	40	1–10	75–150
	10[a]	0–1	0–75
3	20	100–150	350–400
	30	50–100	250–350
	40	5–50	125–250
	10[a]	0–5	0–125
4	20	250–350	450–500
	30	100–250	350–450
	40	10–100	150–350
	10[a]	0–10	0–150
5	20	700–1,000	650–700
	30	350–700	500–650
	40	25–350	200–500
	10[a]	0–25	0–200
6	20	850–1,600	700–850
	30	500–850	550–700
	40	50–500	250–550
	10[a]	0–50	0–250

a. Furnish spalls and rock fragments graded to provide a stable compact mass.
b. The volume of a rock with these cubic dimensions will have a mass approximately equal to the specified rock mass.
c. Furnish stone with breadth and thickness at least one-third its length.

705.03 Rock for Masonry Structures

Furnish sound, durable rock that is native to the vicinity of the work or is similar in texture and color to the native rock and has been proven satisfactory for the intended use.

Furnish dimensioned masonry rock free of reeds, rifts, seams, laminations, and minerals that may cause discoloration or deterioration from weathering.

(a) **Sizes & Shapes.** Do not use rock with depressions or projections that might weaken it or prevent it from being properly bedded.

When no dimensions are shown on the plans, furnish the rocks in the sizes and with the face areas necessary to produce the general characteristics and appearance indicated on the plans.

Unless otherwise specified, furnish rock fragments with the following minimum dimensions:

(1) Min. thickness .... 125 mm

(2) Min. width .... 300 mm or 1-1/2 times the thickness, whichever is greater

(3) Min. length .... 1-1/2 times the width

When headers are required, furnish headers with lengths no less than the width of bed of the widest adjacent stretcher plus 300 mm.

Ensure that at least 50 percent of the total volume of masonry consists of rock with a volume of at least 0.03 m^3.

(b) **Dressing.** Dress the rocks to remove any thin or weak portions. Dress face rocks to provide bed and joint lines with a maximum variation from true line as follows:

(1) Cement rubble masonry .... 40 mm

(2) Class B masonry .... 20 mm

(3) Class A masonry .... 5 mm

(4) Dimensioned masonry .... Reasonably true

(c) **Bed Surfaces.** Bed surfaces of face rock normal to the faces of the rocks for 75 mm. Beyond that point, do not permit the departure from normal to exceed 25 mm in 300 mm for dimensioned masonry, and 50 mm in 300 mm for all other classes.

(d) **Joint Surfaces.** For dimensioned masonry, dress face rock joint surfaces normal to the bed surface. In all classes of masonry except dimensioned masonry, ensure that the joint surfaces of face rocks form an angle with the bed surfaces of not less than 45°.

Dress face rock joint surfaces normal to the bed surfaces and to the exposed faces of the rock for at least 50 mm. Beyond that point, do not permit the departure from normal to exceed 25 mm in 300 mm.

Do not round corners at the meeting of the bed and joint lines in excess of the following radii:

(1) Cement rubble masonry .... 40 mm

(2) Class B masonry .... 25 mm

(3) Class A masonry .... No rounding

(4) Dimensioned masonry .... No rounding

(e) **Arch Ring Rock Joint Surfaces.** Dress ring rock joint surfaces radial to the arch or normal to the front face to a depth of 75 mm. Beyond that point, the departure from the radial or normal may not exceed 20 mm in 300 mm.

Dress the back surface adjacent to the arch barrel concrete parallel to the front face and normal to the intrados to a depth of 150 mm. When concrete is placed after the masonry is constructed, vary adjacent ring stones at least 150 mm in depth.

(f) **Finish for Exposed Faces.** Remove all drill or quarry marks from exposed faces. Pitch face stones to the line along all beds and joints. Finish the exposed faces as SHOWN ON THE DRAWINGS. The following symbols are used to represent the type of surface or dressing specified:

(1) Fine Pointed (F.P.). The point depressions are approximately 10 mm apart with surface variations not to exceed 3 mm from the pitch line.

(2) Medium Pointed (M.P.). The point depressions are approximately 15 mm apart with surface variations not to exceed 5 mm from the pitch line.

(3) Coarse Pointed (C.P.). The point depressions are approximately 30 mm apart with surface variations not to exceed 10 mm from the pitch line.

(4) Split or Seam Faced (S.). The surface presents a smooth appearance that is free from tool marks, with no depressions below the pitch line, and no projection exceeding 20 mm beyond the pitch line.

(5) Rock Faced (R.F.). The face is an irregular projecting surface without indications of tool marks, with no concave surfaces below the pitch line, and with projections beyond the pitch line. Do not permit the projections to exceed the maximum specified.

For example, where "40 R.F." is specified, do not permit projections beyond the pitch line to exceed 40 mm. Where a "variable rock face" is specified, uniformly distribute stones of the same height of projection.

705.04 Rock for Mechanically Placed Embankments

Furnish hard, durable rock that is angular in shape, resistant to weathering, and graded in a well-balanced range of sizes. Furnish material that conforms to table 705-2.

Table 705-2.—Gradation for mechanically placed rock.

% of Rock Fragments by Mass	Mass (kg)	Equivalent Cubic Dimension (mm)
50	> 900	> 700
50	40–900	250–700

Rock placed below the high water mark of live streams shall have breadth and thickness at least one-third its length and conform to the following:

(a) Apparent specific gravity, AASHTO T 85 .... 2.50 min.

(b) Absorption, AASHTO T 85 .... 4.2% max.

(c) Coarse durability index, AASHTO T 210 .... 52 min.

705.05 Rock for Hand-Placed Embankments

Furnish hard, durable rock that is angular in shape, resistant to weathering, and graded in a well-balanced range of sizes. Furnish material that conforms to table 705-3.

Table 705-3.—Gradation for hand-placed rock.

% of Rock Fragments by Mass	Mass (kg)	Equivalent Cubic Dimension (mm)
75	> 75	> 300
25	40–75	250–300

Rock placed below the high water mark of live streams shall have breadth and thickness at least one-third its length and conform to the following:

(a) Apparent specific gravity, AASHTO T 85 .... 2.50 min.

(b) Absorption, AASHTO T 85 .... 4.2% max.

(c) Coarse durability index, AASHTO T 210 .... 52 min.

705.06 Stone Curbing

(a) Stone Curb, Type I. Ensure that stone conforms to the size and shape specified, and to the following:

(1) Furnish quarried limestone, sandstone, or granite from an approved source. Use one type of stone throughout the project. Do not use stone with visible drill marks on the exposed faces.

(2) Saw or point the top surface of all vertical stone curb to an approximate true plane with no depression or projection on the top surface of more than 6 mm. Pitch the front and back arris lines straight and true. Do not permit any projection or depression on the back surface to exceed a batter of 25 mm horizontal to 75 mm vertical.

(3) Saw, point, or smooth quarry split the front exposed face of the vertical stone curb and form to an approximately true plane. Do not permit the remaining face distance to have any projections or depressions greater than 25 mm from the plane of the exposed face.

(4) Square the ends of vertical stone curb with the top back and face and finish so that when the sections are placed end to end, no space more than 13 mm shows in the joint for the full width of the top surface and for the entire exposed front face. Do not permit the remainder of the end to break back more than 100 mm from the plane of the joint. Cut the joints of circular or curved stone curb on radial lines.

(5) Ensure that the minimum length of any segment of vertical stone curb is 1.2 m, except where a depressed or modified section of curb is required for driveways, crossings, closures, and so forth, where the length may vary.

(b) Stone Curb, Type II. Ensure that slope stone curb conforms to the requirements for type I stone curb, except as follows:

(1) On a horizontal top surface, limit the maximum allowable projection or depression to 13 mm. On other exposed faces, limit the maximum allowable projection or depression to 25 mm.

(2) For unexposed surfaces, limit the maximum allowable projection or depression from a true plane on a 0.5-m length to 75 mm.

(3) On exposed faces between adjacent segments of slope stone curb, limit the maximum allowable space that shows to 19 mm. Ensure that the minimum length of any segment of slope stone curb is 0.5 m.

Section 706—Concrete & Plastic Pipe

706.01 Nonreinforced Concrete Pipe

Furnish pipe that conforms to AASHTO M 86M for the diameters and strength classes specified.

706.02 Reinforced Concrete Pipe

Furnish pipe that conforms to AASHTO M 170M for the diameters and strength classes specified. Ensure that precast reinforced concrete end sections conform to the cited specifications, to the extent to which they apply.

706.03 Perforated Concrete Pipe

Furnish pipe that conforms to AASHTO M 175M, type 1 or type 2, and to AASHTO M 86M for the diameters and strength classes specified.

706.04 Reinforced Arch-Shaped Concrete Pipe

Furnish pipe that conforms to AASHTO M 206M for the diameters and strength classes specified.

706.05 Reinforced Elliptical-Shaped Concrete Pipe

Furnish pipe that conforms to AASHTO M 207M for the diameters, placement design (horizontal or vertical), and strength classes specified.

706.06 Reinforced D-Load Concrete Pipe

Furnish pipe that conforms to AASHTO M 242M for the diameters specified.

706.07 Precast Reinforced Concrete Box Sections

Furnish sections that conform to AASHTO M 259M or M 273M, as applicable, for the dimensions and loading conditions specified.

706.08 Plastic Pipe

Furnish perforated and nonperforated plastic pipe that conforms as shown below for the sizes and types SHOWN ON THE DRAWINGS. Ensure that joints specified as watertight conform to ASTM D 3212.

(a) Smooth Wall Polyethylene Pipe. Furnish 300- to 1,050-mm-diameter pipe conforming to ASTM F 714 and minimum cell class, ASTM D 3350, 335434C.

(b) Corrugated Polyethylene Pipe. Furnish 300- to 900-mm-diameter pipe conforming to AASHTO M 294 and minimum cell class, ASTM D 3350, 315412C or 324420C.

(c) Profile Wall (Ribbed) Polyethylene Pipe. Furnish 450- to 1,200-mm-diameter pipe conforming to ASTM F 894 and minimum cell class, ASTM D 3350, 334433C or 335434C.

(d) Corrugated Polyethylene Drainage Tubing. Furnish 75- to 250-mm-diameter tubing conforming to AASHTO M 252.

(e) Smooth Wall PVC Pipe. Furnish 100 to 375-mm-diameter pipe conforming to AASHTO M 278 and minimum cell class, ASTM D 1784, 12454C or 12364C. For sanitary sewer conditions, conform to ASTM D 3034.

(f) Profile Wall (Ribbed) PVC Pipe. Furnish 100- to 1,200-mm-diameter pipe conforming to AASHTO M 304M and minimum cell class, ASTM D 1784, 12454C or 12364C. For sanitary sewer conditions, conform to ASTM F 794 or F 949.

(g) ABS Pipe. Furnish pipe conforming to AASHTO M 264. When perforated pipe is specified, ensure that perforations conform to AASHTO M 278.

Section 707—Metal Pipe

707.01 Ductile Iron Culvert Pipe

Furnish pipe that conforms to ASTM A 716 for the sizes specified.

707.02 Metallic-Coated Corrugated Steel Pipe

Furnish pipe, special sections (such as elbows, branch connections, and prefabricated flared end sections), and coupling bands that conform to AASHTO M 36M and AASHTO M 218, M 274, or M 289 for the dimensions and thicknesses specified.

Fabricate underdrain pipe from a minimum of 1.32-mm steel sheets. Use any class of perforation specified in AASHTO M 36M.

707.03 Aluminum-Alloy Corrugated Pipe

Furnish pipe, special sections (such as elbows, branch connections, and prefabricated flared end sections), and coupling bands that conform to AASHTO M 196M for the sectional dimensions and thicknesses specified.

Fabricate underdrain pipe from a minimum of 1.22-mm aluminum sheets. Use any class of perforation.

707.04 Asphalt-Coated Pipe

Furnish pipe, special sections (such as elbows, branch connections, and prefabricated flared end sections), and coupling bands that conform to Subsections 707.02, 707.03, 707.08, 707.09, and 707.13, as applicable for the kinds of pipes to be coated. Coat the pipe with bituminous material conforming to AASHTO M 190 for the type of coating specified.

Coat special sections (such as elbows, branch connections, and end sections) and coupling bands in accordance with AASHTO M 190. Coat flared end sections with a type A bituminous coating conforming to AASHTO M 190, or with a field-applied asphalt mastic coating conforming to AASHTO M 243.

707.05 Steel Structural-Plate Structures

Furnish structures and assembly fasteners for connecting plates that conform to AASHTO M 167M for the sizes and types specified.

707.06 Aluminum-Alloy Structural-Plate Structures

Furnish structures and assembly fasteners for connecting plates that conform to AASHTO M 219M for the sizes and types specified.

707.07 Asphalt-Coated Structural-Plate Structures

Furnish structures that conform to either Subsection 707.05 or Subsection 707.06, as applicable. Apply a bituminous coating at the place of fabrication conforming to AASHTO M 190 for a type A coating, or apply an onsite asphalt mastic coating conforming to AASHTO M 243, as specified.

If asphalt coating is applied to the plates before field erection, identify each plate's nominal metal thickness by appropriately painting the data on the inside surface of the plate after coating. Other methods of plate identification may be used if approved.

707.08 Polymer-Coated Steel Pipe

Furnish pipe, special sections (such as elbows and branch connections), and coupling bands that conform to AASHTO M 245M and M 246M. Furnish the pipe with a 250/250 polymer coating.

707.09 Fiber-Bonded Bituminous-Coated Steel Pipe

Furnish pipe, special sections (such as elbows, branch connections and prefabricated flared end sections), and coupling bands that conform to Subsection 707.02, but use a zinc metallic coating impregnated with an aramid fiber composite conforming to ASTM A 885.

After fabrication, coat the pipe sections with an asphalt material in accordance with AASHTO M 190 for the type of coating specified.

Coat coupling bands with a bituminous material in accordance with AASHTO M 190, type A. Coupling bands do not require fiber bonding.

707.10 Slotted Drain Pipe

Furnish pipe that conforms to AASHTO M 36M and AASHTO M 218, M 274, or M 289 for the dimensions and thicknesses specified. Fabricate the pipe with either angle or grate slots and as detailed on the plans.

Ensure that slot angles for the angle slot drain conform to ASTM A 570 M, grade 36, and that grate assemblies for the grate slot drain conform to ASTM A 570. Galvanize slot angles and grate slot assemblies in accordance with Subsection 725.12.

707.11 Metallic-Coated Spiral Rib Pipe

Furnish pipe, special sections (such as elbows and branch connections), and coupling bands that conform to AASHTO M 36M, type IR and IIR, and AASHTO M 218, M 274, or M 289 for the dimensions and thicknesses specified.

707.12 Aluminum-Alloy Spiral Rib Pipe

Furnish pipe, special sections (such as elbows and branch connections), and coupling bands that conform to AASHTO M 196M, type IR and IIR, for the dimensions and thicknesses specified.

707.13 Concrete-Lined Corrugated Steel Pipe

Furnish pipe, special sections (such as elbows and branch connections), and coupling bands that conform to Subsection 707.02 for the dimensions and thicknesses specified.

Fully line the pipe and special sections with concrete, in accordance with ASTM A 849, class C.

707.14 Invert-Paved Corrugated Steel Pipe

Furnish pipe, special sections (such as elbows and branch connections), and coupling bands that conform to Subsection 707.02 for the dimensions and thicknesses specified.

Pave the invert of the pipe and special sections with concrete or asphalt material, in accordance with ASTM A 849, class C or B, as specified.

707.15 Repair of Damaged Coatings

Repair damaged coatings in accordance with AASHTO M 36M and ASTM A 849.

Section 708—Paint

708.01 Paint, General

Furnish a contrasting color for each coat of paint. For the finish coat color, conform to FSS 595 B. If requested by the CO, provide color chips from the paint supplier.

(a) Packaging. Furnish paint in strong, substantial containers plainly marked with the following:

(1) Trade name or trademark.

(2) Paint type, color, formulation, lot number, and date of manufacture.

(3) Net weight.

(4) Volume, including the percent of solids and the percent of volatile organic compound (VOC).

(5) Storage requirements.

(6) Mixing and equipment cleanup instructions.

(7) Name and address of the manufacturer.

(b) VOC Content. Conform to the following VOC limits for both shop and field paintings:

(1) Clear (unpigmented) coatings .... 520 g/L

(2) Other coatings .... 350 g/L

(c) Lead Content. Furnish paint with a maximum lead content of 0.06 percent by weight in the dried film.

(d) Other Properties. Furnish paint that:

(1) Does not show excessive settling in a freshly opened full can.

(2) Easily redisperses with a paddle to a smooth, homogeneous state free of curdling, livering, caking, color separation, lumps, and skins.

(3) Does not skin within 48 hours in a closed container that is three-fourths full.

(4) Brushes on easily.

(5) Possesses good leveling properties.

(6) Shows no running or sagging tendencies when applied to smooth steel vertical surfaces.

(7) Dries to a smooth uniform finish, free from roughness, grit, unevenness, and other surface imperfections.

(8) Shows no streaking or separation when flowed on clean glass.

(9) Shows no thickening, curdling, gelling, or hard caking after 6 months storage in a full, tightly covered container at a temperature of 20 °C.

708.02 Paint for Timber Structures

(a) Primer. Conform to FSS TT–P–25, TT–P–96D, or TT–P–001984.

(b) Paint. Conform to FSS TT–P–102, class A; TT–P–96D; TT–P–102F; or TT–P–19D.

708.03 Paint for Concrete & Masonry Block Structures

Conform to FS TT–P–19. Color tint with universal or all-purpose concentrates.

708.04 Paint for Steel Structures

(a) Inorganic Zinc Primer. Conform to AASHTO M 300, type II.

(b) Vinyl Wash Primer. Conform to MIL–P–15328 or SSPC number 27.

(c) Aliphatic Urethane Coating. Conform to U.S. Product Standard C–644, type I.

(d) Acrylic Latex Coating. Conform to SSPC number 24.

708.05 Penetrating Stain

Conform to the following:

(a) Weatherometer on base material, ASTM G 23 .... 1,000 hours

(b) Acrylic dispersion .... 73.4% of nonvolatile vehicle

(c) Viscosity .... 58 ± 2 Kerb units

(d) Solids volatile content 40.3%

Store stain in accordance with the manufacturer's recommendations.

Section 709—Reinforcing Steel & Wire Rope

709.01 Reinforcing Steel

(a) General. Furnish the following information with each shipment of steel to the project:

(1) Name and location of the steel rolling mill.

(2) Manufacturing process.

(3) Heat number(s).

(4) Size(s).

(5) Specifications.

(6) Copies of mill test analyses for chemical and physical tests.

(7) Consignee and destination of shipment.

(b) Reinforcing Bars. Furnish deformed, grade 400 bars conforming to AASHTO M 31M, M 42M, or M 53M.

(c) Epoxy-Coated Reinforcing Bars. Conform to AASHTO M 284M.

Inspect the reinforcing bars after the near white blast cleaning. Reject all bars with steel slivers or scabs. Selective sorting and rejection at the fabricator's shop may avoid unnecessary delays and subsequent rejection of bars during the precoating inspection at the coating applicator's shop.

Coat epoxy-coated reinforcing steel in a plant certified by CRSI as a fusion-bonded epoxy applicator.

(d) Tie Bar. Furnish deformed, grade 400 bars conforming to AASHTO M 31M or M 42M, except do not use AASHTO M 42M steel for tie bars bent and restraightened during construction.

(e) Hook Bolts. Furnish plain, grade 400 bars conforming to AASHTO M 31M or M 42M with M14 rolled threads or M16 cut threads. Furnish a threaded sleeve nut capable of sustaining a minimum axial load of 67 kN.

(f) **Dowel Bars.** Conform to AASHTO M 254, type A or B. Use plain round bars, free from burring or other deformation restricting free movement in the concrete. Paint half the length of each dowel bar with one coat of tar paint. When the paint dries and immediately before placing the dowels, lubricate the painted end to prevent concrete from bonding to the painted end.

For expansion joints, furnish a dowel cap that snugly covers 50 mm ± 5 mm of the dowel, has a closed end, and has a suitable stop to hold the closed end 25 mm from the end of the dowel bar.

Lubricants for type B dowels may be rapid-curing cutback asphalt, medium-setting emulsified asphalt, or a flaked graphite and vehicle. Lubricants are not required for type A coated dowel bars.

Furnish dowel assemblies that hold dowel bars within 6 mm tolerance vertically and horizontally during concrete placement and permit unrestricted movement of the pavement slab.

Use wire conforming to AASHTO M 32M for dowel assemblies. Coat dowel assemblies with the same material as the dowel bar. Recoat or repair damaged coatings equivalent to the manufacturer's original coating.

(g) **Deformed Steel Wire.** Conform to AASHTO M 225M.

(h) **Welded Steel Wire Fabric.** Conform to AASHTO M 55M.

(i) **Cold-Drawn Steel Wire.** Conform to AASHTO M 32M.

(j) **Welded Deformed Steel Wire Fabric.** Conform to AASHTO M 221M.

(k) **Fabricated Deformed Steel Bar or Rod Mats.** Conform to AASHTO M 54M.

(l) **Low-Alloy Steel Deformed Bars.** Conform to ASTM A 706M.

709.02 Wire Rope or Wire Cable

Conform to AASHTO M 30 for size and strength class specified.

709.03 Prestressing Steel

Fabricate from one of the following:

(a) Stress-relieved wire strand, AASHTO M 204M, type BA or WA.

(b) Stress-relieved seven-wire strand, AASHTO M 203M, grade 270.

(c) High-strength steel bars, AASHTO M 275M, type II.

Protect all prestressing steel against physical damage, rust, or corrosion at all times. Do not use damaged prestressing steel.

Package prestressing steel to protect it from physical damage and corrosion during shipping and storage. Place a corrosion inhibitor in the package. Use a corrosion inhibitor that will have no deleterious effect on the steel, concrete, or bond strength of steel to concrete. Immediately replace or restore damaged packaging.

Mark the shipping package with a statement that the package contains high-strength prestressing steel and a warning to use care in handling. Identify the type, kind, and amount of corrosion inhibitor used, including the date when placed, safety regulations, and instructions for use. For identification purposes, assign a lot number and tag to all wire, strand, anchorage assemblies, or bars shipped to the site.

Submit representative samples from members fabricated offsite. In the case of wire or strand, take the sample from the same master roll.

Section 710—Fence & Guardrail

710.01 Barbed Wire

Furnish galvanized barbed wire of the coating class specified in conformance with AASHTO M 280, and aluminum-coated steel barbed wire that conforms to AASHTO M 305, type I.

710.02 Woven Wire

Furnish galvanized woven wire fence fabric that conforms to AASHTO M 279 for the design number, grade, and coating specified, and aluminum-coated woven wire fence fabric that conforms to ASTM A 584.

710.03 Chain Link Fence

Furnish chain link fabric, posts, rails, ties, bands, bars, rods, and other fittings and hardware that conform to AASHTO M 181 for the kind of metal, coating, size of wire, and mesh specified.

Furnish coiled spring steel tension wire that is 4.5 mm, conforms to ASTM A 641M, and has hard temper with a class 3 galvanized coating or an aluminized coating with a minimum coating weight of 120 g/m^2 of aluminum. Use the same coating on the coiled spring steel tension wire as used on the rest of the chain link fence.

710.04 Fence Posts

(a) Wood. Furnish wood posts with the details and dimensions SHOWN ON THE DRAWINGS. Furnish wood posts of sound, seasoned wood, peeled and with ends cut as SHOWN ON THE DRAWINGS. Furnish posts that are straight and have all knots trimmed flush with the surface. Where treated posts are called for, provide the kind and type of treatment that meets the requirements SHOWN ON THE DRAWINGS. The requirements for peeling may be omitted for Redcedar posts or bracing.

(b) Concrete. Ensure that all dimension timber and lumber required for fences or gates is sound, straight, and reasonably free from knots, splits, shakes, and other defects. Furnish the species and grades SHOWN ON THE DRAWINGS, and dress and finish on four sides.

Furnish concrete posts made of concrete that meets the requirements specified in Subsection 602.03, method A or B. Furnish steel reinforcement, as SHOWN ON THE DRAWINGS, that meets the requirements specified in Section 709.

(c) Steel. Furnish steel posts for line-type fencing that are manufactured in accordance with AASTM A 702 and galvanized in accordance with AASHTO M 111 (ASTM A 123), but ensure that tubular steel posts are galvanized in accordance with ASTM A 120. Furnish fittings, hardware, and other appurtenances that are galvanized in accordance with ASTM A 120 by current standard practice, and are of standard commercial grade. Furnish weathering steel posts that meet the requirements of AASHTO M 222.

(d) Aluminum. Furnish aluminum alloy posts that meet the requirements of AASHTO M 181.

710.05 Fence Gates

Furnish frame gates used with chain link fences that conform to the applicable requirements of AASHTO M 181 for the types and sizes specified. Ensure that the fabric in the gate conforms to fabric in the chain link fence.

Furnish frame and wire gates used with woven wire and barbed wire fences that conform to the dimensions and material SHOWN ON THE DRAWINGS.

710.06 Metal Beam Rail

(a) Galvanized Steel Rail. Furnish W-beam or thrie-beam rail elements fabricated from corrugated sheet steel that conform to AASHTO M 180 for the designated shape, class, type, and weight of coating specified.

(b) Corrosion-Resistant Steel Rail. Furnish W-beam and thrie-beam rail elements and associated weathering steel hardware that conform to the following:

(1) Shapes and plates .... AASHTO M 222M

(2) Rail elements .... ASTM A 606, type 4

(3) Fasteners .... AASHTO M 164M, type 3

710.07 Box Beam Rail

Furnish steel box beam rail elements that conform to the applicable standards contained in the AASHTO–Associated General Contractors of America (AGC)–ARTBA "Guide to Standardized Highway Barrier Hardware," 1995 edition.

710.08 Steel-Backed Timber Rail

Furnish timber that conforms to AASHTO M 168. Fabricate the 150 × 250-mm timber rail and the 100 × 225-mm blockouts from dry, well-seasoned, and dressed rough-sawn Douglas Fir, Southern Pine, or other species with a stress grade of at least 10 MPa.

Treat the timber rail and blockout elements with CCA, ACZA, or ACA preservative treatment conforming to AWPA C14, but ensure that the minimum retention is 9.6 kg/m^3.

Fabricate the steel backing elements from 9.5-mm structural steel conforming to AASHTO M 222M. Furnish fastener hardware that conforms to ASTM A 325M, type 3.

710.09 Guardrail Posts

(a) Box Beam Post. Furnish guardrail posts for metal beam guardrail that conform to the applicable standards contained in the AASHTO–AGC–ARTBA "Guide to Standardized Highway Barrier Hardware," 1995 edition.

(b) Steel-Backed Timber Post. Furnish 250 × 300-mm guardrail posts for steel-backed timber rail that conform to specifications in Subsection 710.08. Use the post lengths SHOWN ON THE DRAWINGS.

(c) Wood Post. Do not use a wooden guardrail post that has a through check, shake, or end split in the same plane as the bolt hole, or in a plane parallel to the bolt hole, and that extends from the top of the post to within 75 mm of the bolt hole.

710.10 Guardrail Hardware

Furnish guardrail hardware for use with galvanized steel beam rail that conforms to the standards contained in the AASHTO–AGC–ARTBA "Guide to Standardized Highway Barrier Hardware," 1995 edition. Ensure that guardrail hardware for corrosion-resistant steel conforms to the requirements specified in Subsection 710.06.

Except for material covered in Subsection 710.06, make all angles, channels, wide flanges, and plates not contained in the above standard conform to ASTM A 36, but make the structural tubing for the short steel post conform to ASTM A 500 or A 513, grade 1008. Galvanize soil plates and structural tubing in accordance with ASTM A 123. Do not punch, drill, cut, or weld the metal after galvanizing.

Manufacture reflector tabs from 4-mm aluminum or galvanized steel sheets. Use adhesive that resists peeling with a force of 0.89 kg/cm of width. Use mildew-resistant adhesive that has no staining effect on the reflective sheeting.

710.11 Temporary Plastic Fence

Furnish plastic noncorrosive fence fabricated from high-density polyethylene (HDPE) and ultraviolet-stabilized for outdoor weathering. Furnish material that conforms to the following:

(a) Height .... 1,200 mm min.

(b) Mesh openings .... 80 to 85 mm

(c) Color .... International orange

(d) Mass .... 0.25 kg/m min.

710.12 Crash Cushion Barrels

Furnish 900-mm-diameter barrels made of HDPE structural foam or equal material. Furnish lids of the same material as the barrels, but of a thinner gauge. Furnish appropriate height cores made of polystyrene or equivalent material.

710.13 Timber Rails

Furnish timber rail that is cut from dry, well-seasoned, and dressed timber stock that meets the requirements of AASHTO M 168 for the grade and species SHOWN ON THE DRAWINGS.

Provide preservative treatment that meets the requirements specified in Subsection 716.03, or as SHOWN ON THE DRAWINGS.

Furnish rustic rails that are straight, sound, and free of injurious defects, and are cut from live trees not less than 30 days, and not more than 1 year, before use. Ensure that they are stripped of bark before seasoning or stored under water. Immediately before the logs are used in the work, trim all knots and projections smooth and, if logs are water cured, peel all bark.

Section 711—Concrete Curing Material & Admixtures

711.01 Curing Material

Conform to the following:

(a) Burlap cloth .... AASHTO M 182

(b) Waterproof paper .... AASHTO M 171

(c) Polyethylene film .... AASHTO M 171

(d) Liquid membrane-forming compounds .... AASHTO M 148

711.02 Air-Entraining Admixtures

Conform to AASHTO M 154.

For structural concrete, furnish air-entraining admixtures classified as vinsol resin or neutralized vinsol resin.

711.03 Chemical Admixtures

Furnish water-reducing, set-retarding, and set-accelerating additives, or combinations thereof, that conform to AASHTO M 194. Do not combine chemical admixtures together in a mixture unless they are compatible. Furnish supporting documentation of compatibility from the manufacturers. Do not use chloride accelerators.

711.04 Latex Modifier

Furnish a homogeneous, nontoxic, film-forming polymeric emulsion with stabilizers added at the point of manufacture. Conform to the following:

(a) Color .... White

(b) Styrene butadiene polymer type .... 68 ± 4% styrene
32 ± 4% butadiene

(c) Chlorides 0%

(d) Polymer particle size .... 0.15 to 0.25 μm avg.

(e) Emulsion stabilizers .... Anionic and nonionic surfactant

(f) Solids .... 46.5 to 49.0%

(g) Mass .... 1.00 to 1.02 kg/L

(h) pH .... 9 to 13

(i) Shelf life .... 2 years min.

Section 712—Joint Material

712.01 Sealants, Fillers, Seals, & Sleeves

Conform to the following:

(a) Joint Sealants & Crack Fillers. Furnish a commercial certification identifying the batch and/or lot number, material, quantity of batch, date and time of manufacture, and name and address of the manufacturer. Conform to the following:

(1) Concrete joint sealer, hot-poured elastic type .... ASTM D 1190

(2) Joint sealants, hot-poured, for concrete and asphalt pavement .... ASTM D 3405

(3) Crack filler, hot-applied, for asphalt concrete and Portland cement concrete pavements .... ASTM D 5078

(4) For proprietary asphalt-rubber products, furnish the following:

(a) Source and grade of asphalt cement.

(b) Total granulated rubber content and mass, as a percent of the asphalt-rubber mixture.

(c) Granulated rubber type(s) and content of each type (if blend).

(1) Mass as a percent of combined rubber.

(2) Gradation of granulated rubber.

(d) Type of asphalt modifier, if any.

(e) Quantity of asphalt modifier and mass as a percent of asphalt cement.

(f) Other additives.

(g) Heating and application temperatures.

(h) Manufacturer's recommended application procedures.

(b) Preformed Expansion Joint Fillers. Furnish in a single piece for the depth and width required for the joint:

(1) Preformed expansion joint filler for concrete (bituminous type) .... AASHTO M 33

(2) Preformed sponge rubber expansion joint fillers for concrete paving and structural construction .... AASHTO M 153

(3) Preformed cork expansion joint fillers for concrete paving and structural construction[1] .... AASHTO M 153

(4) Preformed expansion joint fillers for concrete paving and structural construction (nonextruding and resilient bituminous types) .... AASHTO M 213

[1] Do not use in concrete structures.

(c) **Preformed Joint Seals & Sleeves.** Furnish material in accordance with the following:

*(1) **Paving Applications.*** Furnish a polychloroprene elastomeric seal conforming to AASHTO M 220. Use a lubricant adhesive with a minimum solids content of 22 percent by weight, in accordance with ASTM D 2369, and a maximum peel strength of 10 MPa, in accordance with ASTM D 903. Use within 9 months of manufacture.

*(2) **Manhole, Inlet, & Drainage Applications.*** Furnish a multisectional neoprene rubber and ethylene propylene dimonomer rubber seal with a minimum thickness of 1.5 mm. Before shipping, coat the rubber with a nonhardening butyl rubber sealant to produce a watertight seal when installed. Properties and values are shown in table 712-1.

Table 712-1.—Preformed joint seals.

Physical Properties	ASTM Test Method	EPDM	Neoprene	Butyl Mastic
Tensile (MPa)	D 412	10	12	–
Elongation (%)	D 412	440	230	280
Tear resistance (N/mm)	D 624 (Die B)	40	20	–
Rebound (%, 5 min)	C 972 (mod.)	–	–	11
Rebound (%, 2 h)	C 972 (mod.)	–	–	12

(d) **Foam Filler.** Furnish an expanded polystyrene filler having a compressive strength of not less than 70 kPa.

(e) **Cold-Poured Sealer.** Furnish a one-part, low-modulus silicone rubber-base joint-sealing compound conforming to FSS TT–S–1543, class A, with an ultimate elongation of 1,200 percent.

(f) Low-Modulus Silicone Joint Sealant. Furnish a one-part silicone formulation conforming to the following:

(1) Flow, MIL–S–8802 .... 8 mm max.

(2) Extrusion rate, MIL–S–8802 .... 75 to 250 g/min

(3) Tack-free time, MIL–S–8802 .... 20 to 75 minutes

(4) Specific gravity, ASTM D 792, method A .... 1.010 to 1.515

(5) Durometer hardness, shore A, ASTM D 2240 .... 10 to 25

(6) Tensile stress at 150% elongation, ASTM D 412 .... 520 kPa max.

(7) Elongation, ASTM D 412 .... 500% min.

(8) Peel (adhesion), MIL–S–8802 .... ≥ 9 kg with ≥ 75% cohesive failure

(9) Age from manufacturing .... 6 months max.

(g) Backer Rod. Furnish a closed-cell polyethylene conforming to ASTM D 3204, type 1. Use a compatible sealant as recommended by the manufacturer of the rod. Select size as shown in table 712-2.

Table 712-2.—Backer rod sizes.

Joint Width	Rod Diameter
8 mm	9 mm
9 mm	13 mm
13 mm	16 mm
16 mm	19 mm
19 mm	25 mm
25 mm	32 mm
32 mm	38 mm
38 mm	50 mm

712.02 Joint Mortar

Furnish Portland cement and fine aggregate conforming to Subsections 701.01 and 703.01, respectively. Mix one part Portland cement and two parts approved sand, with water as necessary to obtain a usable consistency. Use the mortar within 30 minutes after mixing.

712.03 Watertight Gaskets

For ring gaskets for rigid pipe, conform to AASHTO M 198, type A or B. For ring gaskets for flexible metal pipe, conform to ASTM C 361M. For continuous flat gaskets for flexible metal pipe with bands or bands with projections, conform to ASTM D 1056, grade SCE 41, and use a gasket with a thickness 13 mm greater than the nominal depth of the pipe corrugations. For continuous flat gaskets for flexible metal pipe with corrugated bands, conform to ASTM D 1056, grade SCE 43, and use a 9-mm-thick gasket.

712.04 Oakum

Fabricate oakum from a thoroughly corded and finished hemp (*Cannabis sativa*) line, Benares Sunn fiber, or a combination thereof that is reasonably free from lumps, dirt, and extraneous matter.

712.05 Mortar for Masonry Beds & Joints

(a) Composition. Mix one part masonry cement, Portland cement, or air-entraining Portland cement with two parts fine aggregate by volume. Lime or fly ash may be added in an amount not to exceed 10 percent of the Portland cement by weight. In lieu of air-entraining cement, Portland cement may be used with an air-entraining admixture, in accordance with the applicable provisions of Subsections 552.06 and 552.08.

(b) Material. Conform to the following:

(1) Masonry cement/Portland cement 701.01

(2) Fine aggregate .... 703.01 or AASHTO M 45

(3) Hydrated lime 725.03

(4) Fly ash 725.04

(5) Water 725.01

(6) Air-entraining admixtures 711.02

(c) Comprehensive Strength. Use mortar with a minimum 28-day comprehensive strength of 14 MPa when tested according to AASHTO T 22 and T 23, except that samples shall consist of cylinders with a length-to-diameter ratio of 2 to 1.

712.06 Copper Water Stops or Flashings

Furnish sheet copper for water stops or flashings that conform to AASHTO M 138M, copper USN number C11000. The resistivity test is not required.

712.07 Rubber Water Stops

Furnish molded or extruded rubber with a uniform cross section that is free from porosity or other defects. If approved, an equivalent standard shape may be furnished.

Fabricate rubber water stops from a compound of natural rubber, synthetic rubber, or a blend of the two, together with other compatible material. Do not use any reclaimed material. Furnish a certification from the producer showing the composition of the material. Conform to the following:

(a) Hardness (shore), 3021[1] .... 60 to 70

(b) Compression set, 3311[1] .... 30% max.

(c) Tensile strength, 4111[1] .... 17 MPa min.

(d) Elongation at breaking, ASTM D 412 .... 450% min.

(e) Tensile stress, 300% elongation, 4131[1] .... 6 MPa min.

(f) Water absorption by weight, 6631[1] .... 5% max.

(g) Tensile strength after aging, 7111[1] .... 80% of original, min.

[1] Federal Test Method Standard number 601.

712.08 Plastic Water Stops

Fabricate from a homogeneous, elastomeric, plastic compound of basic PVC and other material. Form to a uniform cross section that is free from porosity and other defects. If approved, an equivalent standard shape may be furnished. Conform to the following:

(a) Tensile strength, ASTM D 638M .... 9.6 MPa min.

(b) Elongation at breaking, ASTM D 638M .... 250% min.

(c) Hardness, ASTM D 2240 .... 60 to 75 shore

(d) Specific gravity, 5011[1] .... Manufacturer's value ± 0.02

(e) Resistance to alkali,[2] ASTM D 543:

(1) Mass change .... – 0.10 to + 0.25%

(2) Hardness change .... ± 5 shore max.

(3) Tensile strength change .... 15% max.

(f) Water absorption (48 hours), ASTM D 570 .... 0.50% max.

(g) Cold bending[3] .... No cracking

(h) Volatile loss, ASTM D 1203 .... Not more than manufacturer's value

[1] Federal Test Method Standard number 406.

[2] Use a 10 percent solution of NaOH for a 7-day test period.

[3] Subject a 25 × 150-mm strip that is 3 mm thick to a temperature of 29 °C for a period of 2 hours. After the 2 hours, immediately bend the sample 180° around a 3-mm-diameter rod. Apply sufficient force to maintain contact with the rod during bending. Examine the sample for evidence of cracking. Test and report results for at least three individual samples from each lot.

Furnish the manufacturer's test results for the above properties with the product certification. If directed, furnish samples in lengths adequate for performing the specified tests.

Section 713—Roadside Improvement Material

713.01 Topsoil

(a) Furnished Topsoil. Furnish fertile, friable, free-draining, sandy loam soil free of subsoil, refuse, stumps, roots, brush, weeds, rocks or stones larger than 25 mm, and other substances detrimental to the development of vegetative growth. Demonstrate that the soil will sustain healthy crops of grass, shrubs, or other plant growth. Furnish material that conforms to the following:

(1) Texture:

(a) Organic matter, AASHTO T 267 .... 3 to 10%

(b) Sand, AASHTO T 88 .... 20 to 70%

(c) Silt, AASHTO T 88 .... 10 to 60%

(d) Clay, AASHTO T 88 .... 5 to 30%

(2) pH, AASHTO T 289 .... 6 to 8

(b) Conserved Topsoil. Conserve natural humus-bearing soils from the overlying portions of the roadway excavation and embankment areas, in accordance with Subsection 203.06(e).

713.02 Agricultural Limestone

Furnish calcic or dolomitic ground limestone conforming to the standards of the Association of Official Analytical Chemists International, applicable State and Federal regulations, and the following:

(a) Purity (calcium and magnesium) carbonates .... 75% min.

(b) Gradation .... Table 713-1

Granulated slag or other approved natural sources of lime may be used, provided that the application rate is adjusted to equal the total neutralizing power of the specified ground limestone.

Table 713-1.—Agricultural limestone gradation.

Sieve Size	Minimum % by Weight Passing Designated Sieve (AASHTO T 27)
2 mm	90
425 µm	50

713.03 Fertilizer

Furnish standard commercial-grade dry formulated fertilizer conforming to the standards of the Association of Official Analytical Chemists International, applicable State and Federal regulations, and required minimum percentages of available nutrients.

Supply the fertilizer in new, clean, sealed, and properly labeled containers with name, weight, and guaranteed analysis of contents clearly marked.

A liquid form of fertilizer containing the minimum percentage of available nutrients may be used.

713.04 Seed

Furnish seed that conforms to FSS JJJ–S–181 for seed testing and quality, and is in conformance with the State Seed Acts. If seed species in the specified seed mix are not listed in FSS JJJ–S–181, furnish certified weed-free seed. Do not use wet, moldy, or otherwise contaminated or damaged seed.

Provide seeds as follows:

(a) Furnish each seed type in a separate standard sealed container. Clearly label each container with the following:

   (1) Name and type of seed.

   (2) Lot number.

   (3) Net weight.

   (4) Percent of purity, germination, and hard seed.

   (5) Percent of maximum weed seed content.

   Inoculate legume seed with approved cultures, in accordance with the manufacturer's instructions.

(b) Furnish a product certification for each kind or type of seed, certifying that the seed was tested by a recognized laboratory within 6 months of the date of delivery. Include the following:

(1) Name and address of testing laboratory.

(2) Date of test.

(3) Seed identification.

(4) Test results showing the percentages of purity, germination, and weed content.

(5) Certified weed-free seed.

713.05 Mulch

(a) Straw. Obtain straw for mulching from oats, wheat, rye, or other grain crops that are free from weeds, mold, and other objectionable material. Furnish straw mulch in an air-dry condition suitable for placing with mulch blower equipment.

(b) Hay. Obtain hay from herbaceous mowing. Ensure that it is free from weeds, mold, and other objectionable material. Furnish hay in an air-dry condition suitable for placing with mulch-blower equipment.

(c) Wood Fiber. Furnish processed wood fiber from wood chips that is:

(1) Colored with a green dye noninjurious to plant growth.

(2) Readily dispersible in water.

(3) Nontoxic to seed or other plant material.

(4) Free of growth or germination inhibiting substances.

(5) Certified weed-free seed.

(6) Air dried to an equilibrium moisture content of 12 ± 3 percent.

(7) Packaged in new, labeled containers.

(8) Packaged in a condition appropriate for mixing in a homogeneous slurry suitable for application with power-spray equipment.

(d) Grass Straw Cellulose Fiber. Furnish processed grass straw fiber that is:

(1) Colored with a green dye noninjurious to plant growth.

(2) Readily dispersible in water.

(3) Nontoxic to seed or other plant material.

(4) Free of growth- or germination-inhibiting substances.

(5) Certified weed-free seed.

(6) Air dried to a moisture content of 10 ± 0.2 percent.

(7) Air dried to a uniform weight of ± 5 percent.

(8) Packaged in new containers labeled with the manufacturer's name and air-dry weight.

(9) Packaged in a condition appropriate for mixing in a homogeneous slurry suitable for application with power-spray equipment

(e) Sawdust. Obtain sawdust from wood that has not been subjected to conditions that would cause the sawdust to lose its value or usefulness as mulch. Ensure that sawdust contains no toxic substances and has been naturally aged for at least 5 years.

(f) Peat Moss. Furnish a granulated sphagnum peat moss that is air dried, in conformance with State and Federal regulations, and meets the following requirements:

(1) Sticks, stones, and mineral matter 0%

(2) Partially decomposed stems and leaves of sphagnum .... 75% min.

(3) Color .... Brown

(4) Texture .... Porous fibrous to spongy fibrous

(5) pH .... 3.5 to 7.5

(g) Mature Compost. Furnish partially decomposed organic materials, such as leaves, grass, shrubs, and yard trimmings, cured for 4 to 8 weeks. Maturity is indicated by temperature stability and soil-like odor. Furnish friable, dark brown, weed- and pathogen-free mature compost with the following properties:

(1) Carbon/nitrogen ratio .... 25:1 to 35:1

(2) Carbon/phosphorus ratio .... 120:1 to 240:1

(3) pH .... 6.0 to 7.8

(4) Water content .... 40% max.

(5) Particle size:

(a) Seeding and sodding .... 12 mm max.

(b) Erosion control .... 25 mm max.

(6) Organic material .... 50% min.

(7) Manmade inserts (plastic, glass, and metal) .... 2% max.

(h) Straw for Hydroseeding. Use clean agricultural straw. Mill fibers to 25 mm or less in length. Dry the fibers to 10 percent moisture for compaction. Bale in heat-sealed plastic bags.

(i) Bonded Fiber Matrix Hydramulch. Furnish a mixture of long wood fibers and bonding agent that, when hydraulically applied and dried, produces a matrix that:

(1) Does not dissolve or disperse when wetted.

(2) Holds at least 1,000 g of water per 100 g of dry matrix.

(3) Has no germination- or growth-inhibiting factors.

(4) Forms no water-insensitive crust.

(5) Contains material that is 100 percent biodegradable.

713.06 Plant Material

Conform to "American Standard for Nursery Stock."

(a) Quality of Plant Material. Furnish plants that are excellent representatives of their normal species or varieties. Ensure that all plants are nursery grown stock that has been transplanted or root-trimmed two or more times, in accordance with the kind and size of plants. Furnish plants with a normal developed branch system that is free from disfiguring knots, sun-scald, injuries, abrasions of the bark, dead or dry wood, broken terminal growth, and other objectionable disfigurements.

Furnish trees that have reasonably straight stems and are well branched and symmetrical, in accordance with their natural habits of growth.

(b) Plant Names. For scientific and common plant names, conform to "Standardized Plant Names," as adopted by the American Joint Committee on Horticultural Nomenclature. Legibly tag and identify all plants by name and size.

(c) Grading Standards. For grading of plants, conform to "American Standard for Nursery Stock," as approved by ANSI.

(d) Nursery Inspection and Plant Quarantine. Furnish plants that are essentially free from plant diseases and insect pests.

Comply with all nursery inspection and plant quarantine regulations of the States of origin and destination, and with Federal regulations governing interstate movement of nursery stock. Provide a valid copy of the certificate of inspection with each package, box, bale, and carload shipped or otherwise delivered.

(e) Balled & Burlapped Plants. Obtain the plants from the original and undisturbed soil in which the plants were grown. Dig balled and burlapped plants to retain as many fibrous roots as possible. Wrap, transport, and handle the plants so the soil ball and small and fibrous roots remain intact.

713.07 Erosion Control Mats, Roving, & Geocell

(a) Erosion Control Mats. Erosion control mats are designated as types 1, 2, 3, 4, and 5, described below.

(1) Type 1—Erosion Control Mats. Type 1 mats are designated as follows:

(a) Straw Erosion Control Mat. Furnish a mat consisting of clean agricultural straw, in accordance with Subsection 713.05(a), that is attached to a photodegradable polypropylene netting by sewing with cotton thread. Ensure that material conforms to the specifications shown in table 713-2.

Table 713-2.—Straw erosion control mat.

Material	Specification Minimums
Straw[a]	240 g/m^2
Netting	Photodegradable netting on one side. 5- to 20-mm square mesh[b] with a 1.5 kg/100 m^2 weight.

a. Moisture content shall not exceed 20 percent.
b. Dimensions are approximate and may vary to meet manufacturer's standards.

(b) Burlap. Furnish burlap fabric in a standard weave with a weight of 145 ± 20 g/m².

(c) Jute Mesh. Furnish jute mesh with a uniform open plain weave fabricated from jute yarn that does not vary in thickness by more than one-half its normal diameter, and that conforms to the following:

(1) Mesh size .... 25 × 25 mm max.

(2) Mesh weight, ASTM D 1776 .... 0.5 kg/m² ± 5%

(d) Woven Paper or Sisal Mesh Netting. Furnish mesh netting of woven paper or woven sisal twisted yard conforming to the following:

(1) Mesh openings .... 3 to 6 mm

(2) Shrinkage after wetting .... 20% max.

(2) Type 2—Erosion Control Mats. Type 2 mats are designated as follows:

(a) Straw and Coconut Mat. Furnish mat consisting of undyed, untreated, biodegradable, jute, coconut coir, synthetic polypropylene fibers, or other approved yarn woven into a plain weave mesh with 16- to 25-mm square openings. Ensure that material conforms to the specifications shown in table 713-3.

Table 713-3.—Straw and coconut mat.

Material	Specification Minimums
Straw[a] 70%	240 g/m²
Coconut 30%	240 g/m²
Netting	Photodegradable netting on one side. 16- to 25-mm square mesh[b] with a 1.5 kg/100 m² weight.

a. Moisture content shall not exceed 20 percent.
b. Dimensions are approximate and may vary to meet manufacturer's standards.

(b) Excelsior Blanket. Furnish a blanket of uniform thickness consisting of curled wood excelsior secured on the top side to biodegradable, photodegradable extruded plastic mesh. Make the blanket smolder resistant without the use of chemical additives. Conform to the following:

(1) Excelsior fibers ≥ 200 mm length .... 80% min.

(2) Mesh size .... 25 × 50 mm

(3) Blanket mass/area .... 0.53 ± 0.05 kg/m²

(c) Mulch Blanket. Furnish a 3- to 13-mm-thick blanket consisting of organic, biodegradable mulch such as straw, curled wood cellulose, coconut coir, or other material evenly distributed on one side of a photodegradable polypropylene mesh with a minimum weight of 0.27 kg/m².

(3) Type 3—Coconut Mats. Furnish coconut mat consisting of undyed, untreated, biodegradable jute, coconut coir, synthetic polypropylene fibers, or other approved yarn woven into a plain weave mesh with approximately 16- to 25-mm square openings. Ensure that material conforms to the specifications shown in table 713-4.

Table 713-4.—Coconut erosion control mat.

Material	Specification Minimums
Coconut[a]	240 g/m²
Netting	Photodegradable netting on one side. 16- to 25-mm square mesh[b] with a 1.5 kg/100 m² weight.

a. Do not permit moisture content to exceed 20 percent.
b. Dimensions are approximate and may vary to meet manufacturer's standards.

(4) Type 4—Synthetic Erosion Control Mats and Meshes. Type 4 erosion control mats are designated as follows:

(a) Synthetic Erosion Control Mat. Furnish a machine-produced flexible mat consisting of polyolefin monofilament fibers positioned between two biaxially oriented nets, and mechanically bound together by parallel stitching with polyolefin thread to form a three-dimensional weblike weave that is highly resistant to environmental and chemical deterioration. Ensure that material conforms to the specifications shown in table 713-5.

(b) Synthetic Polypropylene Mesh. Furnish a flexible woven geotextile mesh fabricated from polypropylene fibers that have been spun in one direction. Ensure that material conforms to the specifications shown in table 713-6.

(c) Synthetic Mulch Control Netting. Furnish a uniformly extruded, rectangular, plastic mesh netting with 50 × 50-mm nominal mesh openings and weighing at least 8 g/m².

Table 713-5.—Synthetic erosion control mat.

Property	Specification	Test Method
Color	Green	Visual
Thickness	6 mm (min.)	ASTM D 1777
Strength[a]	1,590 × 525 N/m (min.)	ASTM D 5035
Elongation[a]	50% (max.)	ASTM D 5035
Porosity[b]	85% (min.)	Calculated
Resiliency[c]	80%	ASTM D 1777
Ultraviolet stability[d]	80%	ASTM D 4355

a. Values for both machine and cross-machine directions under dry or saturated conditions. Machine direction specimen for 50-mm strip test includes one machine-direction polyolefin stitch line centered within its width and extending the full length of the specimen.

b. Calculation based upon weight, thickness, and specific gravity.

c. The percent of original thickness retained after three cycles of a 690-kPa load for 60 seconds, followed by 60 seconds without load. Thickness measured 30 minutes after load removed.

d. Tensile strength retained after 1,000 hours in a Xenon ARC weatherometer.

Table 713-6.—Synthetic polypropylene erosion control mesh.

Property	Specification	Test Method
Color	Beige	Visual
Weight	59 g/m^2 (min.)	ASTM D 5261
Tensile strength	6,700 × 3,700 N/m	ASTM D 5035
Elongation at break	40% (max.)	ASTM D 5035
Mullen burst strength	515 kPa (min.)	ASTM D 3786

(d) Organic Mulch Control Netting. Furnish a leno weave mesh netting fabricated from 12.7-kg biodegradable cellulose fiber yarn with five twists per 25 mm. Make the size of the mesh grid 13 to 25 mm square. Finish the selvedge to prevent raveling or fraying.

(5) ***Type 5—Turf Reinforcement Mats.*** Furnish a web of mechanically or melt bonded polymer netting, monofilaments, or fibers that are entangled to form a strong and dimensionally stable mat. Bonding methods include polymer welding, thermal or polymer fusion, and the placement of fibers between two high-strength, biaxially oriented nets mechanically bound together by parallel stitching with polyolefin thread. Ensure that the mat is resistant to biological, chemical, and ultraviolet degradation. Ensure that material conforms to the specifications shown in table 713-7.

(b) Roving. Roving types are described below.

(1) Fiberglass Roving. Form fiberglass roving from continuous fibers drawn from molten glass, coated with a chrome-complex sizing compound, collected into strands and lightly bound together into roving without the use of clay, starch, or other similar deleterious substances. Wind the roving into a cylindrical package approximately 300 mm high so the roving can be continuously fed from the center of the package through an ejector driven by compressed air and expanded into a mat of glass fibers on the soil surface. Ensure that the material contains no petroleum solvents or other agents known to be toxic to plant or animal life, and that it conforms to the specifications shown in table 713-8.

Table 713-7.—Synthetic polypropylene erosion control mat.

Property	Specification	Test Method
Color	Black	Visual
Thickness	13 mm (min.)	ASTM D 1777
Tensile strength[a]	1,370 × 790 N/m (min.)	ASTM D 5035
Elongation[a]	50% (max.)	ASTM D 5035
Porosity[b]	90% (min.)	Calculated
Resiliency[c]	80%	ASTM D 1777
Ultraviolet stability[d]	80%	ASTM D 4355

a. Values for both machine and cross-machine directions under dry or saturated conditions using 50-mm strip method.
b. Calculation based upon weight, thickness, and specific gravity.
c. The percent of original thickness retained after three cycles of a 690-kPa load for 60 seconds, followed by 60 seconds without load. Thickness measured 30 minutes after load removed.
d. Tensile strength retained after 1,000 hours in a Xenon ARC weatherometer.

Table 713-8.—Fiberglass roving.

Property	Specification	Test Method
Strands per rove	56–64	End count
Fibers per strands	184–234	End count
Fiber diameter (trade designation G)	0.009–0.013 mm	ASTM D 578
m/kg of rove	340–600 m/kg	ASTM D 578
km/kg of strand	2.62–2.82	ASTM D 578
Organic content, % max.	1.65	ASTM D 578

(2) Polypropylene Roving. Form polypropylene roving from continuous strands of fibrillated polypropylene yarn. Wind the roving into a cylindrical package so the roving can be continuously fed from the outside of the package through an ejector driven by compressed air, and can be expanded into a mat of polypropylene strands. Ensure that the material contains no agents that are toxic to plant or animal life, and that it conforms to the specifications shown in table 713-9.

Table 713-9.—Polypropylene roving.

Property	Specification	Test Method
Tensile strength	15.6 N	ASTM D 2256
Elongation at break	15.5%	ASTM D 2256
Mass of strand	360 denier	ASTM D 1907
Strands per rove	24	Measured
Ultraviolet stability	50% retained after 200 h	ASTM D 4355

(c) Geocell (Cellular Confinement System). Furnish a flexible honeycomb three-dimensional structure fabricated from HDPE that has been ultraviolet-stabilized with carbon black and/or hindered anime light stabilizers.

713.08 Miscellaneous Planting Material

(a) Stakes for Bracing and Anchoring. Fabricate stakes for bracing and anchoring trees from rough cypress, cedar, locust, or other approved wood that is essentially free from knots, rot, cross grain, and other defects that would impair the strength of the stake.

Ensure that stakes are a minimum 50 × 50 mm square in cross section, and of adequate length. Furnish stakes that conform to basic requirements of the American Lumber Standards Committee (ALSC).

Furnish anchor stakes that conform to the same size and quality as bracing stakes. The diameter and length of deadman will be SHOWN ON THE DRAWINGS.

(b) Hose. Furnish 25-mm-diameter garden or steam hose (rubber and fabric) to be used with wire for bracing and anchoring trees.

(c) Wire. Use 3.8-mm-diameter soft annealed galvanized steel wire for bracing and anchoring trees.

(d) Wrapping Material. Use 100-mm-wide rolls of waterproof paper (triple lamination 30–30–30) or 150-mm-wide rolls of burlap for wrapping trees.

(e) Twine. Use two-ply twine for trees 75 mm and less in diameter and three-ply twine for trees more than 75 mm in diameter for tying wrapping material to the trees.

(f) Antidesiccant. If approved, use a commercially available antidesiccant emulsion that will provide a film over plant surfaces permeable enough to permit transpiration.

(g) Tree Wound Dressing. Use commercially available products that have an asphalt base and contain a fungicide. Furnish a material that is antiseptic, waterproof, adhesive, and elastic. Do not use material that would be harmful to living tree tissue, such as kerosene, coal tar, creosote, and so forth.

713.09 Sprigs

Furnish healthy living stems (stolons or rhizomes) and attached roots of the perennial turf-forming grasses as SHOWN ON THE DRAWINGS. Obtain sprigs from approved heavy and thickly matted sources in the locality of the work. Remove all Johnson grass and other objectionable grasses, weeds, and other detrimental material.

713.10 Sod

Furnish living vigorous sod of the type of grass and thickness as SHOWN ON THE DRAWINGS. Furnish grass with a dense root system contained in suitable sod and reasonably free from noxious weeds and grasses. When the sod is cut, its top growth shall not be more than 75 mm in height.

713.11 Pegs for Sod

Fabricate square or round pegs from sound wood. Ensure that pegs conform to the following:

(a) Length .... 200 mm min.

(b) Approximate cross-sectional area .... 600 mm^2

713.12 Stabilizing Emulsion Tackifiers

Furnish a commercially available product containing no solvents or other diluting agents toxic to plant life. Furnish material that conforms to one of the following:

(a) Emulsified asphalt, grades SS–1, SS–1h, CSS–1, or CSS–1h.

(b) Nonasphalt emulsions with a water-soluble natural vegetable gum, blended with gelling and hardening agents or a water-soluble blend of hydrophilic polymers, viscosifiers, sticking agents, and gums.

(c) Polyvinyl acetate using emulsion resins and containing 60 ± 1 percent total solids by weight.

713.13 Bales

(a) Straw Bales. Tie the bales with either a commercial-quality baling wire or string. Ensure that straw and bales conform to the following:

(1) Straw .... 713.05(a)

(2) Approximate length .... 1 m

(3) Shape .... Rectangular

(4) Approximate mass .... 30 kg

(b) Wood Excelsior Bales. Furnish bales of curled wood excelsior. Tie the bales with either commercial bailing wire, plastic, or string. Ensure that bales conform to the following:

(1) Approximate dimensions .... 400 × 450 × 900 mm

(2) Approximate mass .... 33 kg

713.14 Sandbags

Use clean, silt-free material for sand filler. Furnish material that conforms to the following:

(a) Bag material .... Canvas or burlap

(b) Volume per bag .... 0.01 m^3 min.

713.15 Erosion Control Culvert Pipe

Furnish culvert pipe fabricated from corrugated metal, plastic, or concrete for use in diverting live streams through work areas. Provide for AASHTO M 18 loading on temporary culvert pipe placed beneath the traveled way.

713.16 Silt Fence

Furnish silt fence consisting of a combination of the following materials, constructed as specified:

(a) Posts. Furnish 75-mm-diameter wood or 1.86-kg/m steel fence posts.

(b) Supports. Furnish 2.03-mm steel wire with a mesh spacing of 150 × 150 mm or a prefabricated polymeric mesh of equivalent strength.

(c) **Geotextile.** Furnish geotextile conforming to Subsection 714.01 and table 714-5, as applicable.

(d) **Height.** Ensure that minimum height above the ground is 760 mm, and that minimum embedment depth is 150 mm.

If approved, variations from the above may be permitted to accommodate premanufactured fences.

Section 714—Geotextile, Geocomposite Drain Material, & Geogrids

714.01 Geotextiles

Use long-chain synthetic polymers composed by weight of at least 95 percent polyolefins or polyesters to manufacture geotextile or the threads used to sew geotextiles. Form the geotextiles, including selvedges, into a stable network such that the filaments or yarns retain their dimensional stability relative to each other.

(a) Physical Requirements. For the specified type, see the following tables:

(1) Subsurface drainage, type I (A–F) .... Table 714-1

(2) Separation, type II (A–C) .... Table 714-2

(3) Stabilization, type III (A–B) .... Table 714-3

(4) Permanent erosion control, type IV (A–F) .... Table 714-4

(5) Temporary silt fence, type V (A–C) .... Table 714-5

(6) Paving fabric, type VI .... Table 714-6

All property values in these specifications, with the exception of apparent opening size (AOS), represent minimum average roll values in the weakest principal direction (i.e., ensure that average test results of any roll in a lot sampled for conformance or quality assurance testing shall meet or exceed the specified values). Values for AOS represent maximum average roll values.

Elevate and protect rolls with a waterproof cover if stored outdoors. When using a geotextile for a permanent installation, limit the geotextile exposure to ultraviolet radiation to less than 10 days.

(b) Evaluation Procedures. Furnish a product certification, including the name of the manufacturer, product name, style number, chemical composition of the filaments or yarn, and other pertinent information to fully describe the geotextile.

When samples are required, remove a 1-m-long full-width sample from beyond the first outer wrap of the roll. Label the sample with the lot and batch number, date of sampling, project number, item number, manufacturer name, and product name.

Table 714-1.—Physical requirements for subsurface drainage geotextile.

Property	Test Method	Units	Specifications[a]					
			Type I–A	Type I–B	Type I–C	Type I–D	Type I–E	Type I–F
Grab strength	ASTM D 4632	N	1,100/700	1,100/700	1,100/700	800/500	800/500	800/500
Sewn seam strength	ASTM D 4632	N	990/630	990/630	990/630	720/450	720/450	720/450
Tear strength	ASTM D 4533	N	400[c]/250	400[c]/250	400[c]/250	300/175	300/175	300/175
Puncture strength	ASTM D 4833	N	400/250	400/250	400/250	300/175	300/175	300/175
Burst strength	ASTM D 3786	kPa	2,700/1,300	2,700/1,300	2,700/1,300	2,100/950	2,100/950	2,100/950
Permittivity	ASTM D 4491	s^{-1}	0.5	0.2	0.1	0.5	0.2	0.1
Apparent opening size	ASTM D 4751	mm	0.45[b]	0.25[b]	0.22[b]	0.45[b]	0.25[b]	0.22[b]
Ultraviolet stability	ASTM D 4355	%	50[d]	50[d]	50[d]	50[d]	50[d]	50[d]

a. The first values in a column apply to geotextiles that break at < 50 percent elongation (ASTM D 4632). The second values in a column apply to geotextiles that break at ≥ 50 percent elongation (ASTM D 4632).

b. Maximum average roll value.

c. The minimum average roll tear strength for woven monofilament geotextile is 245 N.

d. After 500 hours of exposure.

Table 714-2.—Physical requirements for separation geotextile.

Property	Test Method	Units	Specifications[a] Type II–A	Type II–B	Type II–C
Grab strength	ASTM D 4632	N	1,400/900	1,100/700	800/500
Sewn seam strength	ASTM D 4632	N	1,260/810	990/630	720/450
Tear strength	ASTM D 4533	N	500/350	400[c]/250	300/180
Puncture strength	ASTM D 4833	N	500/350	400/250	300/180
Burst strength	ASTM D 3786	kPa	3,500/ 1,700	2,750/ 1,300	2,100/950
Permittivity	ASTM D 4491	s^{-1}	0.02	0.02	0.02
Apparent opening size	ASTM D 4751	mm	0.60[b]	0.60[b]	0.60[b]
Ultraviolet stability	ASTM D 4355	%	50[d]	50[d]	50[d]

a. The first values in a column apply to geotextiles that break at < 50 percent elongation (ASTM D 4632). The second values in a column apply to geotextiles that break at ≥ 50 percent elongation (ASTM D 4632).

b. Maximum average roll value.

c. The minimum average tear strength for woven monofilament geotextile is 245 N.

d. After 500 hours of exposure.

Table 714-3.—Physical requirements for stabilization geotextile.

Property	Test Method	Units	Specifications[a] Type III–A	Type III–B
Grab strength	ASTM D 4632	N	1,400/900	1,100/700
Sewn seam strength	ASTM D 4632	N	1,260/810	990/630
Tear strength	ASTM D 4533	N	500/350	400[c]/250
Puncture strength	ASTM D 4833	N	500/350	400/250
Burst strength	ASTM D 3786	kPa	3,500/1,700	2,750/1,300
Permittivity	ASTM D 4491	s^{-1}	0.05	0.05
Apparent opening size	ASTM D 4751	mm	0.43[b]	0.43[b]
Ultraviolet stability	ASTM D 4355	%	50[d]	50[d]

a. The first values in a column apply to geotextiles that break at < 50 percent elongation (ASTM D 4632). The second values in a column apply to geotextiles that break at ≥ 50 percent elongation (ASTM D 4632).

b. Maximum average roll value.

c. The minimum average tear strength for woven monofilament geotextile is 245 N.

d. After 500 hours of exposure.

Table 714-4.—Physical requirements for permanent erosion control geotextile.

Property	Test Method	Units	Specifications[a]					
			Type IV–A	Type IV–B	Type IV–C	Type IV–D	Type IV–E	Type IV–F
Grab strength	ASTM D 4632	N	1,400/900	1,400/900	1,400/900	1,100/700	1,100/700	1,100/700
Sewn seam strength	ASTM D 4632	N	1,260/810	1,260/810	1,260/810	990/630	990/630	990/630
Tear strength	ASTM D 4533	N	500/350	500/350	500/350	400[c]/250	400[c]/250	400[c]/250
Puncture strength	ASTM D 4833	N	500/350	500/350	500/350	400/250	400/250	400/250
Burst strength	ASTM D 3786	kPa	3,500/1,700	3,500/1,700	3,500/1,700	2,750/1,300	2,750/1,300	2,750/1,300
Permittivity	ASTM D 4491	s^{-1}	0.7	0.2	0.1	0.7	0.2	0.1
Apparent opening size	ASTM D 4751	mm	0.43[b]	0.25[b]	0.22[b]	0.43[b]	0.25[b]	0.22[b]
Ultraviolet stability	ASTM D 4355	%	50[d]	50[d]	50[d]	50[d]	50[d]	50[d]

a. The first values in a column apply to geotextiles that break at < 50 percent elongation (ASTM D 4632). The second values in a column apply to geotextiles that break at ≥ 50 percent elongation (ASTM D 4632).

b. Maximum average roll value.

c. The minimum average roll tear strength for woven monofilament geotextile is 245 N.

d. After 500 hours of exposure.

Table 714-5.—Physical requirements for temporary silt fence.

Property	Test Method	Units	Specifications		
			Type V–A	Type V–B[b]	Type V–C[c]
Maximum post spacing	–	m	1.2	1.2	2
Grab strength:					
Machine direction	ASTM D 4632	N	400	550	550
Cross direction		N	400	450	450
Permittivity	ASTM D 4491	s^{-1}	0.05	0.05	0.05
Apparent opening size	ASTM D 4751	mm	0.60[a]	0.60[a]	0.60[a]
Ultraviolet stability	ASTM D 4355	%	70[d]	70[d]	70[d]

a. Maximum average roll value.
b. Elongation at break ≥ 50 percent elongation (ASTM D 4632).
c. Elongation at break < 50 percent elongation (ASTM D 4632).
d. After 500 hours of exposure.

Table 714-6.—Physical requirements for paving fabric.

Property	Test Method	Units	Specifications, Type VI
Grab strength	ASTM D 4632	N	500
Ultimate elongation	ASTM D 4632	N	50% at break
Asphalt retention	Texas DOT item 3099	L/m^2	0.90
Melting point	ASTM D 276	°C	150

In addition, when geotextile joints are sewn, submit the seam assembly description and a sample of the sewn material. In the description, include the seam type, seam allowance, stitch type, sewing thread tex ticket number(s) and type(s), stitch density, and stitch gage. If the production seams are sewn in both the machine and cross-machine directions, provide sample sewn seams that are oriented in both the machine and cross-machine directions. Furnish a sewn sample that has a minimum 2 m of sewn seam and is at least 1.5 m in width. Sew the sample seams with the same equipment and procedures that are used to sew the production seams. Ensure that seams sewn onsite conform to the manufacturer's recommendations and are approved before installation.

714.02 Geocomposite Drains

Geocomposite drains consist of a polymeric drainage core with a geotextile conforming to Subsection 714.01(a)(1) attached to or encapsulating the core. Ensure that the geocomposite drain includes all necessary fittings and material to splice one sheet, panel, or roll to the next and to connect the geocomposite drain to the collector and outlet piping.

Fabricate the drainage core in sheet, panel, or roll form of adequate strength to resist installation stresses and long-term loading conditions. Furnish core material that consists of long chain synthetic polymers composed by weight of at least 85 percent polypropylene, polyester, polyamide, PVC, polyolifin, or polystyrene. Build the core up in thickness by means of columns, cones, nubs, cusps, meshes, stiff filaments or other configurations.

Ensure that geocomposite drains have a minimum compressive strength of 275 kPa when tested in accordance with ASTM D 1621, procedure A. Ensure that all splices, fittings, and connections have sufficient strength to maintain the integrity of the system during construction handling and permanent loading, and do not impede flow or damage the core.

Identify, ship, and store the geocomposite drains in accordance with AASHTO M 288. Elevate and protect sheets, panels, and rolls with a waterproof and ultra-violet-resistant cover if stored outdoors.

When using a geocomposite drain for a permanent installation, limit the geocomposite exposure to ultraviolet radiation to less than 10 days.

When samples are required, provide a 1-m-square sample from products supplied as sheets or panels, or a 1-m-length full-roll-width sample from products supplied in rolls. Label the sample with the lot and batch number, date of sampling, project number, item number, manufacturer's name, and product name.

(a) Geocomposite Underdrains. Ensure that the horizontal and vertical flow of water within the core interconnects at all times for the full height of the core, and water can pass from one side of the core to the other. Ensure that the drainage core with the geotextile in place provides a minimum flow rate of 0.1 L/s/m of width when tested in accordance with ASTM D 4716 under the following test conditions:

(1) A specimen 300 mm long.

(2) An applied load of 69 kPa.

(3) A gradient of 0.1.

(4) A 100-hour seating period.

(5) A closed-cell foam rubber between platens and geocomposite.

Firmly attach the geotextile to the core so folding, wrinkling, and other movement cannot occur either during handling or after placement. Achieve bonding using nonwater-soluble adhesive, heat sealing, or another method recommended by the manufacturer. Do not use adhesive on areas of the geotextile fabric where flow is intended to occur.

If heat sealing is used, do not weaken the geotextile below the required strength values. Extend the geotextile below the bottom of the core far enough to completely encapsulate the collector pipe.

(b) Geocomposite Sheet Drains. Ensure that the horizontal and vertical flow of water within the sheet drain interconnects at all times for the full height of the core. Ensure that the drainage core with the geotextile in place provides a minimum flow rate of 0.1 liters per second per meter of width when tested in accordance with ASTM D 4716 under the following test conditions:

(1) A specimen 300 mm long.

(2) An applied load of 69 kPa.

(3) A gradient of 0.1.

(4) A 100-hour seating period.

(5) A closed-cell foam rubber between platens and geocomposite.

If core construction separates the flow channel into two or more sections, only the flow rate on the inflow face is considered in determining the core's acceptability.

Firmly attach the geotextile to the core so folding, wrinkling, and other movement cannot occur either during handling or after placement. Achieve bonding using nonwater-soluble adhesive, heat sealing, or another method recommended by the manufacturer. Do not use adhesive on areas of the geotextile fabric where flow is intended to occur.

If heat sealing is used, do not weaken the geotextile below the required strength values. Extend the geotextile below the bottom of the core far enough to completely encapsulate a the collector pipe.

(c) Geocomposite Pavement Edge Drains. Ensure that the geotextile tightly encapsulates the geocomposite edge drain, and that the edge drains permit inflow from both sides. Ensure that the drain core with the geotextile in place provides a minimum flow rate of 3 liters per second per meter of width when tested in accordance with ASTM D 4716 under the following test conditions:

(1) A specimen 300 mm long.

(2) An applied load of 69 kPa.

(3) A gradient of 0.1.

(4) A 100-hour seating period.

(5) A closed-cell foam rubber between platens and geocomposite.

If the geocomposite polymer core separates the flow channel into two or more parts, consider only the tested flow rate of the channel facing the pavement.

Firmly attach the geotextile to the core so folding, wrinkling, and other movement cannot occur during handling or after placement. Achieve bonding using nonwater-soluble adhesive, heat sealing, or another method recommended by the manufacturer. Do not use adhesive on areas of the geotextile fabric where flow is intended to occur.

If heat sealing is used, do not weaken the geotextile below the required strength values. Extend the geotextile below the bottom of the core far enough to completely encapsulate the collector pipe.

Furnish nonperforated plastic pipe conforming to Subsection 706.08 for all pipe and pipe fittings used for an outlet to the edge drain.

Furnish solvent cement for the outlet pipe and fittings in accordance with ASTM D 2564. Ensure that the material composition of the outlet fittings is compatible for direct solvent welding to PVC.

714.03 Geogrids

Furnish geogrids consisting of polymeric materials such as polypropylene, polyethylene, or polyester formed into a stable network of bars or straps fixed at their junctions such that the bars retain their relative position to each other. Ensure that the geogrid is treated to resist ultraviolet degradation, and that it conforms to the physical strength requirements shown in table 714-7 in accordance with ASTM D 4595.

Table 714-7.—Physical strength requirements for geogrids.

Category	Minimum Strength at 5% Strain (kN/m)	Minimum Ultimate Strength at Breakage (kN/m)
1	9	13
2	13	21
3	17	29
4	28	61
5	53	98
6	70	125

Furnish the CO with a certificate signed by a legally authorized official from the company that manufactured the geogrid. Ensure that the certificate attests that the geogrid meets the chemical, physical, material, and manufacturing requirements stated in the specification. When requested by the CO, furnish a sample of the geogrid from each lot for verification testing.

During shipment and storage, wrap the geogrid in a heavy-duty protective covering. Protect the geogrid from mud, soil, dust, debris, and sunlight prior to installation.

Ensure that the geogrid meets the minimum average roll values for the wide-width strip tensile strength tests performed in accordance with ASTM D 4595 for the category SHOWN ON THE DRAWINGS. Provide test results to the CO prior to incorporating the geogrid into the work.

Ensure that the aperture size for all geogrids is from 22 to 75 mm. Square and rectangular openings are permitted. Strengths shown in table 714-7 are for both the machine and cross directions.

Section 715—Piling

715.01 Untreated Timber Piles

Conform to ASTM D 25. Fabricate the piles from the following species for the sizes and dimensions as SHOWN ON THE DRAWINGS:

(a) Douglas Fir

(b) Larch

(c) Norway Pine

(d) Red Oak

(e) Southern Yellow Pine

Install steel straps along the length of the pile at not more than 3-m centers. In addition, place a strap at 75, 150, and 300 mm from the tip and two additional straps within 600 mm of the butt. Use 32-mm-wide × 0.8-mm-thick steel strapping material fabricated from cold-rolled, heat-treated, high-tensile steel with a minimum tensile strength of 22 kN.

Hold straps in place with clips that are secured by crimping twice in the clip length with a notch-type sealer. Fabricate the clips from 57 × 0.9-mm-thick steel. The clip joint shall develop at least 75 percent of the strap tensile strength. Straps shall encircle the pile once and shall be tightened by hand-operated or power-assisted tensioning tools.

Furnish one copy each of the supplier's certification of species and the certification that the piling meets the requirements specified in ASTM D 25.

715.02 Treated Timber Piles

Conform to Subsection 715.01, except furnish only Douglas Fir or Southern Yellow Pine piles for use in saltwater. Treat the piles with preservative in accordance with AASHTO M 133 for the types and quantities of preservatives as SHOWN ON THE DRAWINGS.

Use the pressure method procedure prescribed in AWPA standard C1. Apply the treatment to the piles after all millwork is completed.

Ensure that the treating plant imprints legible symbols or legends on the end of all piles, identifying the name of the treating company and type and year of treatment in accordance with AWPA standards M1 and M6.

Furnish one copy of the following Certificates of Compliance to the CO upon delivery of the piling to the jobsite:

(a) Supplier certification of species, and certification that the piling meets the requirements specified in ASTM D 25.

(b) Certificate of Conformance to AASHTO M 133, including type of treatment, retention (Assay method), and penetration from an ALSC-accredited agency.

Have the compliance certification made by a qualified testing and inspection agency.

715.03 Concrete Piles

Fabricate piles from class A (AE) concrete conforming to Section 552. Furnish billet steel and rail steel reinforcement bars conforming to Subsection 709.01. For prestressing reinforcement steel, conform to Subsection 709.03.

Construct precast concrete piles in accordance with Section 552. Construct prestressed concrete piles according to Section 553. When lifting anchors are used, maintain at least a 25-mm clearance from the pile reinforcing steel or prestressing steel.

Use metal, plywood, or dressed lumber forms that are watertight, rigid, and true-to-line. Use a 25-mm chamfer strip in all corners of the forms.

Cast piles separately or, if alternate piles are cast in a tier, cast the intermediate piles at least 4 days after the adjacent piles have been poured. Separate piles cast in tiers with tar paper or other suitable separating material. Place concrete in each tier in a continuous operation that prevents the formation of stone pockets, honeycombs, and other defects. Leave forms in place for at least 24 hours.

When the forms are removed, make the pile surface true, smooth, even, and free from honeycombs and voids. Make piles straight so that a line stretched from butt to tip on any face will not be more than 25 mm from the face of the pile at any point.

Remove lifting anchors to a depth of at least 25 mm below the concrete surface, and fill the resulting hole with concrete. Finish the surface of each pile with a class 1 ordinary surface finish, according to Subsection 552.18. Cure the piles in accordance with Sections 552 and 553, as applicable.

If concrete test cylinders are made and tested in accordance with Section 552, do not move piles until the tests indicate a compressive strength of at least 80 percent of the design 28-day compressive strength. Do not transport or drive piles until the tests indicate that the minimum design 28-day compressive strength has been attained.

If concrete test cylinders are not made, do not move piles until they have cured for at least 14 days at a minimum temperature of 15 °C, or 21 days at a minimum temperature of 4 °C. Do not transport or drive piles until cured for at least 21 days at a minimum of 15 °C, or 28 days at a minimum of 4 °C. When high-early-strength cement is used, do not move, transport, or drive piles until cured for at least 7 days.

715.04 Steel Shells

Furnish either cylindrical or tapered pile shells of spiral welded, straight-seam welded, or seamless tube steel material. Use only one type of pile shell throughout a structure. Conform to the following minimum shell wall thickness:

- Outside cylinder diameter < 350 mm .... 6 mm
- Outside cylinder diameter ≥ 350 mm .... 10 mm
- Tapered or fluted .... 4.5 mm

(a) Shells Driven Without a Mandrel. For tapered or step-tapered cast-in-place concrete piles, furnish shells having a minimum 300-mm diameter at cutoff and a minimum 200-mm diameter at tip. For constant-diameter cast-in-place concrete piles, furnish shells having a minimum nominal diameter of 270 mm.

Fabricate the shells from not less than 4.5-mm plate stock conforming to AASHTO M 183M. Shells may be either spirally welded or longitudinally welded and either tapered or constant in section. Seal the tips as SHOWN ON THE DRAWINGS.

(b) Shells Driven With a Mandrel. Furnish shells of sufficient strength and thickness to withstand driving without injury and to resist harmful distortion and/or buckling due to soil pressure after being driven and the mandrel is removed. Butt and tip dimensions will be SHOWN ON THE DRAWINGS.

715.05 Steel Pipes

Conform to the following:

(a) Steel pipe to be filled with concrete .... ASTM A 252, grade 2

(b) Closure plates for closed end piles .... AASHTO M 183M

(c) Reinforced conical points for pipe closure at the tip ... AASHTO M 103M

(d) Unfilled tubular steel piles for welded and seamless steel pipe piles with chemical properties conforming to ASTM A 53, grade B .... ASTM A 252, grade 2

715.06 Steel H-Piles

Furnish steel H-piles from rolled steel sections of the weight and shape SHOWN ON THE DRAWINGS. Fabricate the H-piles from structural steel conforming to AASHTO M 183M, except do not use steel manufactured by the acid Bessemer treatment process.

For copper-bearing structural steel, furnish steel that contains not less than 0.20 percent or more than 0.35 percent copper.

715.07 Sheet Piles

Furnish steel sheet piles conforming to AASHTO M 202M or AASHTO M 223M. Make the joints practically watertight when the piles are in place.

715.08 Pile Shoes

Furnish shoes for timber piles that are prefabricated from cast steel conforming to ASTM A 27M.

715.09 Splices

Manufacture splices for H-piles or pipe piles from structural steel conforming to AASHTO M 183M.

Section 716—Material for Timber Structures

716.01 Untreated Structural Timber & Lumber

Furnish structural timber and lumber that conform to AASHTO M 168 and the applicable standards of the West Coast Lumber Inspection Bureau, Southern Pine Inspection Bureau, or another nationally recognized timber association. Ensure that all structural timber and lumber are seasoned and dried at the time of fabrication. Material that has become twisted, curved, or otherwise distorted prior to assembly into the final structure may be cause for rejection.

Do not use boxed-heart pieces of Douglas Fir or Redwood in stringer, floor beams, caps, posts, sills, curbs, rails, rail posts, and rail post blocks. Boxed-heart pieces are defined as timber so sawed that at any point in the length of a sawed pieced the pith lies entirely inside the four faces.

Legibly mark, stamp, or brand all pieces, identifying the inspection service, grade designation, species, and identity of the inspector. Furnish timber and lumber that conform to the species, design values, and nominal dimensions SHOWN ON THE DRAWINGS. Furnish an inspection certification as to the species and grade from an agency accredited by ALSC.

716.02 Hardware & Structural Steel

Furnish machine and carriage bolts that meet the requirements of ASTM A 307, drift pins and dowels that meet the requirements of ASTM A 575, and galvanized hardware that meets the requirements of AASHTO M 232.

Ensure that all structural steel shapes, rods, glued laminated deck panel dowels, and plates are structural steel that meets the requirements of AASHTO M 183. Ensure that galvanizing meets the requirements of AASHTO M 111.

Furnish bolts with square or hexagonal heads, nuts or dome-heated bolts as SHOWN ON THE DRAWINGS, and nails that are cut or round nails of standard form Use cut, round, or boat spikes, as specified. Use washers that are malleable iron castings, and plain or cut washers that are American Standard Plain Washers.

Use ring or shear plate timber connectors conforming to AASHTO's "Standard Specifications for Highway Bridges," division II, article 16.2.6, Timber Connectors.

716.03 Treated Structural Timber & Lumber

Furnish wood in accordance with Subsection 716.01. Treat the wood and mark each piece of treated timber in accordance with AASHTO M 133. Use the type of treatment and minimum net retention of preservative that are SHOWN ON THE DRAWINGS. Completely and accurately fabricate all treated timber before it is treated. Except for Southern Pine, incise all surfaces greater than 50 mm in width, including glued laminated members, before treatment. Treat glued laminated timbers in accordance with AWPA C28. Furnish inspection certification of treatment from an agency accredited by ALSC.

Use the assay method to determine retention of preservatives in all lumber and timbers.

Ensure that treatment meets the requirements in the current edition of the WWPI's "Best Management Practices for the Use of Treated Wood in Aquatic Environments."

716.04 Structural Glued Laminated Timber

Furnish structural glued laminated timber that meets the requirements specified in the current edition of AITC 117. Use the combination symbol, protection, quality marks, certificates, and preservation treatment that are SHOWN ON THE DRAWINGS. Ensure that manufacture, marking, and quality control of structural glued laminated timber are in conformance with ANSI/AITC A190.1, Structural Glued Laminated Timber.

Ensure that members are manufactured as industrial-appearance grade for wet use conditions, using a phenol-resorcinol resin type of adhesive throughout. Use only single- or multiple-piece laminations with boded edge joints.

Ensure that caulking compound used to seal deck panel joints meets the requirements of FSS TT–S–001543 (com.) and is brown or bronze in color.

716.05 Substitution for Solid Sawn Structural Timber & Lumber

Comparable glued laminated material may be substituted for solid sawn material. Ensure that all substitutions have approximately equal dimension and will provide equal or greater bonding and shear strength per member. Before fabrication, submit drawings that show revised details, including any changes in dimensions, elevation, and bolt length.

Section 717—Structural Metal

717.01 Structural Steel

Furnish structural carbon steel in accordance with AASHTO M 270M and as shown below.

(a) Structural Carbon Steel. For primary bridge members, furnish structural carbon steel that conforms to AASHTO M 270M, grade 250T. For fracture-critical bridge members, furnish structural carbon steel that conforms to AASHTO M 270M, grade 250F. For other shapes, plates, and bars, furnish structural carbon steel that conforms to AASHTO M 270M, grade 250.

(b) High-Strength Low-Alloy (HSLA) Structural Steel. For other shapes, plates, and bars, furnish HSLA steel that conforms to AASHTO M 270M, grade 345 or 345W.

For primary bridge members, furnish HSLA steel that conforms to AASHTO M 270M, grade 345T or 345WT. Ensure that fracture-critical bridge members conform to AASHTO M 270M, grade 345F or 345WF.

For welded members, furnish HSLA steel that conforms to AASHTO M 270M, grade 345T or 345WT. Ensure that fracture-critical welded members conform to AASHTO M 270M, grade 345F or 345WF.

(c) High-Strength Quenched & Tempered Steel. For other shapes, plates, and bars, ensure that all quenched and tempered steel provided conforms to AASHTO M 270M, grade 485W, 690, or 690W. For primary bridge members, furnish quenched and tempered steel that conforms to AASHTO M 270M, grade 485WT, 690T, or 690WT. Furnish fracture-critical bridge members that conform to AASHTO M 270M, grade 485WF, 690F, or 690 WF.

(d) Bolts & Nuts. Conform to ASTM A 307.

(e) High-Strength Bolts, Nuts, & Washers. Conform to either AASHTO M 164M or AASHTO M 253M, as specified. Furnish circular, clipped, and beveled hardened steel washers that conform to AASHTO M 293 (ASTM F 436).

(f) Load-Indicating Washers. Furnish load-indicating washers that conform to ASTM F 959, type 325 or 490. Use type 325 with AASHTO M 164 bolts, and type 490 with AASHTO M 253 bolts.

(g) Steel Anchor Bolts. Furnish steel anchor bolts that conform to AASHTO M 314 and are of the grade and dimensions SHOWN ON THE DRAWINGS. Ensure that the exposed portion of the bolt is zinc coated by hot dip of mechanical deposition.

717.02 Steel Forgings

Conform to AASHTO M 102, classes C, D, F, and G.

717.03 Pins & Rollers

Furnish pins and rollers that are more than 225 mm in diameter from annealed carbon-steel forgings that conform to AASHTO M 102, class C.

Furnish pins and rollers that are 225 mm or less in diameter either from annealed carbon-steel forgings that conform to AASHTO M 102, class C, or from cold-finished carbon-steel shafting that conforms to AASHTO M 169, grade 1016 to 1030, inclusive, with a minimum Rockwell Scale B hardness of 85. The hardness requirement may be waived if the steel develops a tensile strength of 480 MPa and a yield point of 250 MPa.

Furnish pin threads that conform to the ANSI B1.1 Coarse Thread Series, class 2A. Thread pin ends with a diameter of 35 mm or more with six threads in 25 mm.

717.04 Castings

Furnish castings that conform to the following:

(a) Steel Castings. Furnish steel castings that conform to AASHTO M 192M, class 485.

(b) Chromium Alloy Steel Castings. Furnish chromium alloy steel castings that conform to AASHTO M 163M, grade CA–15.

(c) Gray Iron Castings. Furnish gray iron castings that conform to AASHTO M 105, class number 30B, unless otherwise specified. Furnish iron castings that are free from pouring faults, sponginess, cracks, blow holes, and other defects in position affecting their strength and value for the service intended. Boldly fillet the castings at angles and make the arrises sharp and perfect. Sand blast all castings, or otherwise effectively remove the scale, and sand to present a smooth, clean, and uniform surface.

(d) Malleable Iron Castings. Furnish malleable iron castings that conform to ASTM A 47, grade number 35018, unless otherwise specified. Ensure that workmanship, finishing, and cleaning conform to Subsection 717.04(c).

717.05 Welded Stud Shear Connectors

Furnish shear connector studs that conform to AASHTO M 169 for standard-quality, cold-finished, carbon steel bars. Provide the connectors conforming to AASHTO's "Standard Specifications for Highway Bridges," division II, article 11.3.3, Welded Stud Shear Connectors.

717.06 Steel Pipe

Furnish galvanized steel pipe conforming to ASTM A 3, type F, standard weight class, and plain ends for the designation SHOWN ON THE DRAWINGS.

717.07 Galvanized Coatings

When galvanizing is specified, galvanize structural steel shapes, plates, bars, and their products in accordance with AASHTO M 111. Galvanize hardware in accordance with AASHTO M 232.

717.08 Sheet Lead

Furnish sheet lead that conforms to ASTM B 29 for common desilverized lead. Furnish the sheets in a uniform thickness of 6 mm ± 1 mm, and make them free from cracks, seams, slivers, scale, and other defects.

717.09 Steel Grid Floors

Furnish steel grid floors that conform to AASHTO M 270M, grade 250 or 345W. Ensure that steel furnished in accordance with AASHTO M 270M, grade 250, has a minimum copper content of 0.2 percent unless galvanized. Galvanize steel grid floors unless painting is specified.

717.10 Elastomeric Bearing Pads

Furnish elastomeric bearing pads that conform to AASHTO M 251.

717.11 TFE Surfaces for Bearings

(a) TFE Resin. Furnish virgin TFE resin material conforming to ASTM D 1457. Ensure that specific gravity is 2.13 to 2.19 and the melting point is 328 °C ± 1 °C.

(b) Filler Material. Furnish filler material consisting of milled glass fibers, carbon, or other approved inert material.

(c) Adhesive Material. Furnish epoxy resin adhesive conforming to FSS MMM–A–134, FEP film, or an approved equivalent.

(d) Unfilled TFE Sheet. Furnish unfilled TFE sheet from TFE resin conforming to the following:

(1) Min. tensile strength, ASTM D 1457 .... 19 MPa

(2) Min. elongation, ASTM D 1457 .... 200%

(e) Filled TFE Sheet. Furnish filled TFE sheet from TFE resin uniformly blended with inert filler material. For filled TFE sheets containing glass fiber or carbon, conform to specifications in table 717-1.

Table 717-1.—TFE sheeting.

Property	ASTM Method	15% Glass Fibers	25% Carbon
Mechanical:			
Minimum tensile strength	D 1457	14 MPa	9 MPa
Minimum elongation	D 1457	150%	75%
Physical:			
Minimum specific gravity	D 792	2.20	2.10
Melting point	D 1457	327 °C ± 10 °C	327 °C ± 10 °C

(f) Fabric Containing TFE Fibers. Furnish fabric from oriental multifilament TFE fluorocarbon and other fibers. Use TFE fibers that conform to the following:

(1) Min. tensile strength, ASTM D 2256 .... 165 MPa

(2) Min. elongation, ASTM D 2256 .... 75%

(g) Interlocked Bronze & Filled TFE Components. Furnish interlocked bronze and filled TFE components that consist of a phosphor bronze plate conforming to ASTM B 100 with an 0.25-mm-thick porous bronze surface layer conforming to ASTM B 103M, into which a TFE compound is impregnated. Overlay the surface with compounded TFE not less than 25 µm thick.

(h) TFE Metal Composite. Furnish virgin TFE molded on each side and completely through a 33-mm perforated stainless steel sheet conforming to ASTM A 240, type 304.

(i) Surface Treatment. For epoxy bonding, factory treat one side of the TFE sheet with a sodium naphthalene or sodium ammonia process.

(j) Stainless Steel Mating Surface. Furnish stainless steel mating surfaces that are at least 0.91 mm thick, conform to ASTM A 240, type 304, and have a surface finish less than 0.5 µm root mean square. Polish or roll stainless steel mating surfaces as necessary to provide the specified friction properties.

717.12 Structural Aluminum Alloy

Furnish structural aluminum material that conforms to the requirements SHOWN ON THE DRAWINGS and to "Specifications for Aluminum Structures," published by the Aluminum Association, Inc. (AA). For aluminum expansion joint material, furnish aluminum extrusion alloy 6061–T6.

717.13 Aluminum Alloy for Bridge Rail

Furnish aluminum alloys that conform to the applicable specifications of table 717-3, as specified.

717.14 Aluminum Bolt Heads & Nuts

Furnish aluminum bolt heads and nuts that conform to American Standard heavy hexagon ANSI B18.2. Ensure that threads conform to American Standard coarse series, class 2 fit, ANSI specification B1.1.

717.15 Aluminum Welding Wire

Furnish aluminum welding wire that conforms to the specifications in table 717-2.

Table 717-2.—Aluminum welding wire.

Alloys Series	Specification	Wire
3xxx and 6xxx	AWS 5.10	ER 4043
3xxx, 5xxx, and 6xxx	AWS 5.10	ER 5356
5xxx and 6xxx	AWS 5.10	ER 5556 or 5183

717.16 Elastomeric Compression Joint Seals

Furnish elastomeric compression joint seals that conform to AASHTO M 220.

717.17 Dowels

Furnish dowels that conform to the requirements of AASHTO M 31 (ASTM A 615) for grades 40 and 60, or AASHTO M 227 (ASTM A 663) for grades 70, 75, and 80.

Table 717-3.—Aluminum alloys for bridge railing systems (ASTM and AA alloy designation).

Railing Component	Sheet and Plate ASTM B 209	Drawn Formless Tubes ASTM B 210	Bars, Rods, and Wire ASTM B 211	Extruded Bars, Rods, Shapes, and Tubes ASTM B 221	Pipe ASTM B 241	Standard Structural Shapes ASTM B 308	Rivet Cold and Heading Wires and Rods ASTM B 316	Sand Castings ASTM B 26	Permanent Mold Castings ASTM B 108
Posts and post bases, structural:									
Wrought				6061–T6	6061–T6	6061–T6			
Cast					6063–T6				A444.0–T4
Posts, ornamental:									
Wrought				6063–T6	6063–T6			356.0–T6	A356.0–T6
Cast								356.0–T6	A356.0–T6
Rails and sleeves, structural:									
Wrought		6061–T6 6063–T6		6061–T6 6063–T6 6351–T5	6061–T6 6063–T6	6061–T6			
Bolts and screws, misc.:[a,b]									
Aluminum, wrought Stainless steel Galvanized steel Aluminized steel			2024–T4[c] 6061–T6[d]						
Nuts,[e] wrought:			2024–T4	6061–T6[f]					
6 mm and over[b]			6061–T6						
5 mm and under			6262–T9						

Note: "F" temper applies to products that acquire some temper from fabricating processes.

a. Use compatible stainless or coated steel nuts and washers. Do not use aluminum for anchor bolts.

b. Coat alloy 2024–T4 with a 5-µm minimum thickness anodic coating with a dichromate or boiling water seal.

c. Use alloy 2024–T4 for stress-carrying bolts and minor bolts.

d. Use alloy 6061–T6 as an alternate material for minor bolts.

e. Use with aluminum bolts and screws. Do not use aluminum for anchor bolt nuts and washers.

f. ASTM B 211 is an acceptable alternate.

Table 717-3.—Aluminum alloys for bridge railing systems (ASTM and AA alloy designation) (cont.).

Railing Component	Sheet and Plate ASTM B 209	Drawn Formless Tubes ASTM B 210	Bars, Rods, and Wire ASTM B 211	Extruded Bars, Rods, Shapes, and Tubes ASTM B 221	Pipe ASTM B 241	Standard Structural Shapes ASTM B 308	Rivet Cold and Heading Wires and Rods ASTM B 316	Sand Castings ASTM B 26	Permanent Mold Castings ASTM B 108
Washers, flat:[b]									
Wrought	Alclad 2024–T4 Alclad 2024–T3[c]								
Washers, springlock:[b]									
Wrought			7075–T6						
Rivets:									
Wrought				6061–T6			6061–T6[a,d] 6061–T4[a,e]		
Shims:									
Wrought	1100–0			6063–F[a]				443.0–F	
Cast									
Weld filler:									
Wrought						5356			
End caps:									
Wrought	6061–T6			6061–T6				356.0–T6[a]	
Cast								356.0–F 443.0–F	

Note: "F" temper applies to products that acquire some temper from fabricating processes.

a. Chemical, composition only.

b. Use with aluminum bolts and screws. Do not use aluminum for anchor bolt nuts and washers.

c. Use T3 temper for thicknesses less than 6 mm, and use T4 temper for thicknesses 6 mm and greater.

d. Use for cold-driven rivets.

e. Use for rivets driven at 530 °C to 565 °C.

Section 718—Traffic Signing & Marking Material

718.01 Retroreflective Sheeting

Furnish retroreflective sheeting material that conforms to AASHTO M 268, except that the minimum coefficients of retroreflection for brown type I sheeting shown in AASHTO M 268, table 1, are amended as follows:

(a) 2.0 cd/lx/m^2 at 0.2° observation angle and – 4° entrance angle.

(b) 1.0 cd/lx/m^2 at 0.2° observation angle and + 30° entrance angle, and at 0.5° observation angle and – 4° entrance angle.

(c) 0.5 cd/lx/m^2 at 0.5° observation angle and + 30° entrance angle.

Furnish retroreflective sheeting material that conforms to AASHTO M 268 supplemental requirement S1, if specified. Furnish reboundable retroreflective sheeting that conforms to AASHTO M 268, including supplemental requirement S2.

Furnish sheeting that is either heat activated or pressure sensitive (class 1), engineer grade, unless otherwise specified. Ensure that colors are as specified in the MUTCD and as SHOWN ON THE DRAWINGS.

Ensure that no more than 12 months elapse from the date of manufacture to the date of application.

When an adhesive is used, use backing class 1, 2, or 3, in accordance with AASHTO M 268.

718.02 Test Procedures

Use test procedures in accordance with AASHTO M 268, except that subsection S1.3.3 is amended as follows:

The stock cultures of *Aspergillus niger*, American Type Culture Collection number 6275, may be kept for not more than 4 months in a refrigerator at a temperature from 3 °C to 10 °C. Use subcultures incubated at 28 °C to 30 °C for 10 to 14 days in preparing the inoculum.

718.03 Plywood Panels

Fabricate the panels from HDO plywood, two sides Douglas Fir, exterior type, conforming to U.S. Product Standard 1 (current edition), with a B-grade veneer or better on both faces. Ensure that surfacing overlay material is high-density 90–90 resin-impregnated fiber, permanently fused to the base panel under heat and pressure. Furnish material that is suitable for sign manufacturing and compatible with reflective-sheeting adhesive. Do not permit any marks, blemishes, or damage of any kind. Overlay color may be either black or buff, unless specified otherwise. Ensure that each panel edge-brand includes the following: HDO B–B G1 EXT APA PS 1.

Use a minimum 13-mm-thick plywood for signs less than 600 mm in the longest dimension. Use a minimum 15-mm-thick plywood for all other signs cut from a single sheet. Use a minimum 19-mm-thick plywood for all signs requiring joining.

Abrade, clean, and degrease the face of the plywood panel in accordance with methods recommended by the manufacturer of the retroreflective sheeting. Treat the edges of the plywood panel with an approved edge sealant.

718.04 Steel Panels

Fabricate the panels from 2-mm continuous-coat galvanized sheet steel blanks conforming to ASTM A 525. Mill phosphatize the zinc coating (designation G 90) to a thickness of 1.1 ± 0.5 g/m^2 of surface area.

The finished plate shall be free of twist or buckles and the background substantially a plane surface. Clean, degrease, or otherwise prepare the panels in accordance with methods recommended by the sheeting manufacturer.

718.05 Aluminum Panels

Furnish sheets and plates that conform to ASTM B 209, alloy 6061–T6 or 5052–H38.

Fabricate temporary panels and permanent panels that are 750 × 750 mm or smaller from 2-mm-thick aluminum sheets. Fabricate larger permanent panels from 3-mm-thick aluminum sheets.

Furnish blanks that are free from laminations, blisters, open seams, pits, holes, and other defects that may affect their appearance or use. Ensure that thickness is uniform and the blank commercially flat. Perform shearing, cutting, and punching before preparing the blanks for application of reflective material.

Clean, degrease, and chromate the blanks, or otherwise properly prepare the panels in accordance with methods recommended by the sheeting manufacturer.

718.06 Plastic Panels

(a) Plastic. Fabricate the panels from sheets of lightweight, flexible, high-impact, and ultraviolet-chemical-resistant polycarbonate material or approved equivalent that will accept adhesives, coatings, and retroreflective sheeting material, as recommended for such material.

Fabricate panels that are 600 × 600 mm or smaller from 2-mm-thick plastic blanks. Fabricate larger panels from 3-mm-thick plastic blanks.

Furnish panels that are flat and free of buckles, warps, and other defects. Where multiple panels adjoin, ensure that the gap between adjacent panels is no greater than 16 mm. Ensure that signs larger than 600 × 600 mm have reinforcement stiffeners attached on the back for rigidity and for mounting on the supports.

(b) Fiberglass-Reinforced Plastic. Fabricate fiberglass-reinforced plastic signs from fiberglass-reinforced thermoset polyester acrylic modified laminate sheets. Furnish sign panel that is ultraviolet stabilized for outdoor weathering ability. Ensure that sign panel accepts adhesives, coatings, and retroreflective sheeting material, as recommended.

Furnish sign panel free of visible cracks, pinholes, foreign inclusions, and surface wrinkles that would affect implied performance, alter the specific dimensions of the panel, or otherwise affect the sign panel's serviceability.

Wipe sign panel surface clean with a slightly dampened cloth before applying reflective sheeting.

Furnish fiberglass-reinforced plastic that complies with the recommendations of the Fiberglass Reinforced Panel Council publication "Recommended Traffic Control Sign Panel Specification." Unless otherwise SHOWN ON THE DRAWINGS, furnish material that is brown, matching FSS 595a, color number 20059.

718.07 Extruded Aluminum Panels

Fabricate the panels from aluminum alloy 6063–T6 conforming to the requirements of ASTM B 221. Ensure that panel thickness and fabrication conform to Subsection 718.05. The maximum allowable deviation from flat on the face is 4 mm/m.

718.08 Paint

Furnish premium-grade exterior silicone alkyd enamel paint with a color to match the color of the HDO plywood substrate.

718.09 Silk Screen Inks

Furnish inks that are compatible with the sheeting, as determined by the manufacturer of the ink and the manufacturer of the sheeting. Furnish color as specified in the MUTCD and as SHOWN ON THE DRAWINGS.

718.10 Edge Film

Furnish edge film that is a pressure-sensitive, premium-quality, clear, ultraviolet-resistant, 75-mm-wide vinyl film.

718.11 Signposts

Fabricate traffic signposts from wood, steel, plastic, aluminum, or fiberglass-reinforced plastic, as specified.

(a) **Wood Posts.** Fabricate wood posts from dry number 1 structural-grade Douglas Fir, Southern or Ponderosa Pine, Hemlock, Spruce, or Western Larch conforming to AASHTO M 168. Treat the posts with waterborne preservative ACA, ACZA, or CCA, in accordance with AWPA standard C14, but ensure that the minimum preservative retention is 6 kg/m^3.

(b) **Steel Posts.** Fabricate steel posts from billet or rail steel conforming to ASTM A 499. Drill or punch 10-mm holes in the posts along the centerline of the web before galvanizing. Begin punching or drilling 25 mm from the top of the post and proceed on 25-mm centers for the entire length of the post. Galvanize the posts in accordance with ASTM A 123.

(c) **Aluminum Posts.** Fabricate aluminum posts in approved standard shapes and thicknesses using aluminum alloy 6061–T6, 6351–T5, 6063–T6, or 6005–T5 conforming to ASTM B 221.

(d) **Plastic Posts.** Fabricate flexible posts made with high-impact-resistant, ultraviolet, chemical-resistant polycarbonate material or approved equivalent.

(e) **Fiberglass-Reinforced Plastic Posts.** Fabricate fiberglass-reinforced plastic posts from fiberglass-reinforced thermoset polymers. Post to be ultraviolet stabilized for outdoor weathering ability.

718.12 Object Marker & Delineator Posts

Fabricate object marker and delineator posts from wood, steel, aluminum, plastic, or fiberglass-reinforced plastic.

(a) Wood Posts. Furnish 100 × 100-mm wood posts that conform to Subsection 718.11.

(b) Steel Posts. Furnish flanged U-channel steel posts that weigh not less than 3 kg/m and conform to ASTM A 36. Galvanize the posts in accordance with ASTM A 123.

(c) Aluminum Posts. Furnish standard-shaped 3-mm-thick aluminum posts conforming to ASTM B 221, alloy 6061–T6.

(d) Plastic Posts. Furnish flexible delineator posts made with high-impact-resistant, ultraviolet-chemical-resistant polycarbonate or approved equivalent.

(e) Fiberglass-Reinforced Plastic Posts. Fabricate fiberglass-reinforced plastic posts from fiberglass-reinforced thermoset polymers. Post to be ultraviolet stabilized for outdoor weathering ability.

718.13 Hardware

Use galvanized steel or aluminum alloy for fittings such as lag screws, washers, clip angles, wood screws, shear plates, U-bolts, clamps, bolts, nuts, and other fasteners.

Furnish high-strength steel bolts, nuts, and washers conforming to specifications in Subsection 717.01. Galvanize steel hardware in accordance with ASTM A 153.

Furnish aluminum alloy bolts, nuts, and washers conforming to specifications in Subsections 717.13 and 717.14, as applicable.

Furnish oversize bolt heads and oversize neoprene or nylon washers for plastic sign panels.

718.14 Letters, Numerals, Arrows, Symbols, & Borders

Furnish colors as SHOWN ON THE DRAWING and in accordance with Subsection 718.01 and the FHWA's "Standard Alphabets for Highway Signs," current edition.

Perform all silk-screening operations precisely as prescribed in writing by the manufacturers of the ink and the sheeting to which they are applied.

Furnish silk screen inks that are color matched to eliminate any visual difference between silk-screened material and applied material of the same color on the same sign.

Form letters, numerals, and other units to provide a continuous stroke width with smooth edges. Make the surface flat and free of warp, blisters, wrinkles, burrs, and splinters. Ensure that units of the sign message conform to the following:

(a) Type L–1 (Screen Process). Apply letters, numerals, arrows, symbols, and borders on the retroreflective sheeting or opaque background of the sign by direct or reverse screen process. Apply messages and borders that are of a darker color than the background to the paint or the retroreflective sheeting by direct process. Produce messages and borders that are of a lighter color than sign background by the reverse screen process.

Use opaque or transparent colors, inks, and paints in the screen process of the type and quality recommended by the retroreflective sheeting manufacturer.

Perform the screening in a manner that results in a uniform color and tone, with sharply defined edges of legends and borders and without blemishes on the sign background that will affect intended use.

Air dry or bake the signs after screening in accordance with manufacturer's recommendations to provide a smooth, hard finish. Any signs with blisters or other blemishes will be rejected.

(b) Type L–3 (Direct-Applied Characters). Cut letters, numerals, symbols, borders, and other features of the sign message from the type and color of the retroreflective sheeting specified and apply them to the sign background's retroreflective sheeting in accordance with the retroreflective sheeting manufacturer's instructions. Ensure that the retroreflective sheeting has a minimum coefficient of retroreflection (R_A) in accordance with ASTM D 4956.

718.15 Delineator & Object Marker Retroreflectors

Retroreflector units for delineators and object markers are either type 1 (acrylic plastic lens) or type 2 (retroreflective sheeting). Furnish the units ready for mounting.

(a) Type 1 (Acrylic Plastic Lens). The retroreflector unit has an acrylic plastic lens with a minimum area of 4,500 mm^2, prismatic optical elements, and a smooth, clear, transparent face. Fabricate the back from similar material and fuse it to the lens around the entire perimeter to form a homogenous unit. Retroreflection is provided by the lens prismatic optical elements. Permanently seal the units against the intrusion of dust, water, and air.

Ensure that the coefficient of (retroreflective) luminous intensity of each retroreflector unit equals or exceeds the minimum values shown in table 718-1, regardless of the orientation angle.

Mount the retroreflector unit in a housing fabricated from 1.6-mm aluminum alloy 3003–H–14 or similar material, or from cold-rolled, hot dip galvanized steel with a thickness of 1.6 mm. Provide antitheft attachment hardware.

Table 718-1.—Minimum coefficient of (retroreflective) luminous intensity (R_I) (cd/lx).

Observation Angle (°)	Entrance Angle (°)	White[a]	Yellow	Red
0.1	0	10.7	6.5	2.8
0.1	20	4.2	2.3	1.1

a. Crystal, clear, or colorless are acceptable color designations.

(b) Type 2 (Retroreflective Sheeting). The retroreflector unit is composed of a fungus-resistant type III, type IV, or type V retroreflective sheeting material with a class 1 or class 2 adhesive backing conforming to AASHTO M 268.

Attach type 2 retroreflective units to an aluminum or plastic support panel (target plate) of the size and dimension specified. Type 1 units require a sealed optical system complete with housing and assembly hardware and do not require attachment to a support panel unless otherwise specified.

Furnish post-mounting antitheft hardware consisting of bolts, nuts, washers, fastening plates, brackets, and so forth, as required.

718.16 Conventional Traffic Paint

Furnish an alkyd resin ready-mixed paint for use on asphalt and Portland cement concrete pavements conforming to FSS TT–P–115F.

718.17 Waterborne Traffic Paint

Furnish an acrylic water-based ready-mixed paint for use on asphalt and Portland cement concrete pavements conforming to the following:

(a) Composition. Furnish a paint composed of resin solids of 100 percent acrylic polymer with the exact formulation determined by the manufacturer. Conform to the following:

(1) Pigment, by mass, ASTM D 3723 .... 45 to 55%

(2) Nonvolatile vehicle, by mass, FTMS 141, method 6121 .... 40% min.

(3) Lead, chromium, cadmium, or barium 0%

(4) Volatile organic matter .... 250 g/L max.

(5) Mass of paint, ASTM D 1475 .... 1.44 kg/L min.

(b) Viscosity. Conform to ASTM D 562, 75 to 95 Krebs units.

(c) **Drying Time.** Conform to the following:

(1) Paint shall dry to a no-pickup condition, according to ASTM D 711, in a maximum of 10 minutes.

(2) Paint having 0.7 kg/L type 1 waterproofed glass beads shall dry to a no-tracking condition under traffic in a maximum of 90 seconds when applied at 0.38 mm ± 0.03 mm wet film thickness at 54 °C, or in a maximum of 10 minutes when applied at ambient temperatures.

(d) **Flexibility.** Conform to FSS TT–P–1952B. No cracking or flaking.

(e) **Dry Opacity.** Conform to FTMS 141, contrast ratio at 0.25 mm, 0.96 minimum.

(f) **Color.** Conform to the following:

(1) White .... FHWA standard highway white

(2) Yellow .... FHWA standard highway yellow

(g) **Daylight Reflectance.** Conform to the following, without glass beads:

(1) White, FTMS 141, method 6121 .... 84% relative to magnesium oxide standard

(2) Yellow, FTMS 141, method 6121 .... 55% relative to magnesium oxide standard

(h) **Bleeding Ratio.** Conform to FSS TT–P–1952B, 0.96 minimum.

(i) **Scrub Resistance.** Conform to ASTM D 2486, 300 cycles minimum.

(j) **Freeze-Thaw Stability.** Conform to FSS TT–P–1952B:

(1) Change in consistency .... ± 5 Krebs units max.

(2) Decrease in scrub resistance .... –10% max.

(k) **Storage Stability.** During a 12-month storage period, conform to the following:

(1) No excessive setting, caking, or increase in viscosity.

(2) Readily stirred to a consistency for use in the striping equipment.

718.18 Epoxy Markings

Formulate a two-component, 100-percent-solids type system for hot-spray application conforming to the following:

(a) Pigments. Furnish component A (percent by weight) as follows:

(1) White:

(a) Titanium dioxide (TiO_2), ASTM D 476, type II 16.5% min. at 100% purity) .... 18% min.

(b) Epoxy resin .... 75 to 82%

(2) Yellow:

(a) Chrome yellow ($PbCrO_4$), ASTM D 211, type III (20% min. at 100% purity) .... 23% min.

(b) Epoxy resin .... 70 to 77%

(b) Epoxy Content. For component A (weight per epoxy equivalent), ASTM D 1652, meet manufacturer's TV ± 50.

(c) Amine Value. For component B, ASTM D 2074, meet manufacturer's TV ± 50.

(d) Toxicity. Do not permit toxic or injurious fumes at application temperature.

(e) Color. Furnish 0.38-mm thickness (cured), as follows:

(1) White .... FHWA standard highway white

(2) Yellow .... FHWA standard highway yellow

(f) Directional Reflectance. Furnish directional reflectance without glass beads as follows:

(1) White, FSS 141, method 6121 .... 84% relative to magnesium oxide standard

(2) Yellow, FSS 141, method 6121 .... 55% relative to magnesium oxide standard

(g) Drying Time. Furnish 0.38-mm film thickness with beads as follows:

(1) Laboratory at 22 °C, ASTM D 711 .... 30 min max. to no-pickup condition

(2) Field at 25 °C, viewed from 15 m .... 10 min max. to no-tracking condition

(h) Abrasion Resistance. Ensure that wear index with a CS–17 wheel under a 1,000-g load for 1,000 cycles, ASTM C 501, is a maximum of 82.

(i) Hardness. Ensure that shore D hardness with 72- to 96-hour cure at 22 °C, ASTM D 2240, is 75 to 100.

(j) Storage. When stored for up to 12 months, individual epoxy components do not require mixing before use.

718.19 Polyester Markings

Formulate a two-component system conforming to the following:

(a) Directional reflectance (without glass beads):

(1) White, FSS 141, method 6121 .... 80% relative to magnesium oxide standard

(2) Yellow, FSS 141, method 6121 .... 55% relative to magnesium oxide standard

(b) Color:

(1) White .... FHWA standard highway white

(2) Yellow .... FHWA standard highway yellow

(c) Viscosity, uncatalyzed polyester at – 4 °C, ASTM D 562 .... 70 to 90 Krebs

(d) Bleeding, ASTM D 969 .... 6 min.

(e) Drying time in field, viewed from 15 m .... 45 min max. to no-tracking condition

718.20 Thermoplastic Markings

Conform to AASHTO M 249.

718.21 Preformed Plastic Markings

Furnish thermoplastic material consisting of a mixture of polymeric material, pigments, and glass beads homogeneously distributed throughout. Embed additional glass beads into the retroreflective surface. Provide a precoated adhesive system or liquid contact cement to make the marking material capable of being attached to asphalt and Portland cement pavements.

Ensure that the marking material molds itself to pavement contours by the action of traffic at normal pavement temperatures. Furnish marking material that can be used for patching worn areas of previously applied markings of similar composition under normal conditions of use.

Furnish material that conforms to ASTM D 4505, type I, V, VI, or VII, grade A, B, C, D, or E.

Ensure that the minimum thickness without adhesive is 1.5 mm. Ensure that a matrix with a raised pattern cross-sectional area has a minimum thickness of 0.5 mm between pattern configurations.

718.22 Glass Beads

Furnish glass beads for dropping or spraying on pavement markings that conform to AASHTO M 247 for the type specified. Treat glass beads with an adherence coating, as recommended by manufacturer.

AASHTO M 247, table 1, Gradation of Glass Beads, is supplemented by the gradations of glass beads shown in table 718-2.

Ensure that type 3, 4, and 5 glass beads also conform to the following:

(a) Treat beads with a reactive adherence coating, as recommended by the manufacturer.

(b) Ensure that roundness, FLH T 520, conforms to 70 percent minimum for each sieve size.

(c) Ensure that the refractive index, AASHTO M 247, is from 1.50 to 1.55.

Table 718-2.—Gradation of glass beads.

Sieve Size	Percent by Weight Passing Designated Sieve (ASTM D 1214) — Grading Designation: Type 3	Type 4	Type 5
2.36 mm	–	–	100
2.0 mm	–	100	95–100
1.7 mm	100	95–100	80–95
1.4 mm	95–100	80–95	10–40
1.18 mm	80–95	10–40	0–5
1.0 mm	10–40	0–5	0–2
850 μm	0–5	0–2	–
710 μm	0–2	–	–

718.23 Raised Pavement Markers

Furnish prismatic retroreflector-type markers consisting of a methyl methacrylate, polycarbonate, or suitably compounded ABS shell fitted with retroreflective lenses. Make the exterior surface of the shell smooth.

Use a retroreflector with a minimum coefficient of (retroreflected) luminous intensity conforming to table 718-3.

Table 718-3.—Minimum coefficient of (retroreflected) luminous intensity (R_I) (mcd/lx).

Observation Angle (°)	Entrance Angle (°)	White[a]	Yellow	Red
0.2	0	279	167	70
0.2	20	112	67	28

a. Crystal, clear, or colorless are acceptable color designations.

Make the base of the marker flat, patterned, or textured and free from gloss or substances that may reduce its bond to the adhesive. Do not permit deviation from a flat surface to exceed 1 mm.

718.24 Temporary Pavement Markings

(a) Preformed Retroreflective Tape. Furnish 100-mm-wide tape conforming to ASTM D 4592, type I (removable).

(b) Raised Pavement Markers. Furnish an L-shaped polyurethane marker body with retroreflective tape on both faces of the vertical section, capable of retroreflecting light from opposite directions, and with an adhesive on the base.

Provide a minimum coefficient of retroreflection of 1,200 cd/lx/m^2 at 0.1° observation angle and – 4° entrance angle.

Fabricate the marker body from polyurethane with a 1.5-mm minimum thickness. Fabricate the vertical leg about 50 mm high by about 100 mm wide. Fabricate the base for the marker body about 30 mm wide.

Factory apply a 3-mm minimum thickness and 19-mm-wide pressure-sensitive adhesive to the marker base and protect it with release paper.

If approved, variations in design and dimensions will be permitted in order to meet manufacturer's standards.

718.25 Temporary Traffic Control Devices

Furnish traffic control devices (barricades, cones, tubular markers, vertical panels, drums, portable barriers, warning lights, advance warning arrow panels, traffic control signals, and so forth) whose designs and configurations conform to the MUTCD.

Use suitable commercial-grade material for the fabrication of the temporary traffic control devices. Construct the devices from material that is capable of withstanding anticipated weather and traffic conditions and is suitable for the intended use. Do not use units that have been used on other projects without approval.

When interpreting the requirements in the applicable MUTCD sections, replace the word "should" with the word "shall."

718.26 Epoxy Resin Adhesives

Furnish epoxy resin adhesives for bonding traffic markers to hardened Portland cement and asphalt concrete that conform to AASHTO M 237.

Section 720—Structural Wall & Stabilized Embankment Material

720.01 Mechanically Stabilized Earth Wall Material

(a) Concrete Face Panels. Fabricate the panels in accordance with Section 552, except for the following:

(1) Furnish Portland cement concrete that conforms to class A(AE) and has a minimum 30 MPa 28-day compressive strength.

(2) In addition to meeting the requirements specified in Subsection 562.11 for removal of forms and falsework, fully support the units until the concrete reaches a minimum compressive strength of 7 MPa. The units may be shipped and/or installed after the concrete reaches a minimum compressive strength of 24 MPa.

(3) Finish the front face of the concrete panel surface with a class 1 finish, in accordance with Subsection 552.18. Give the rear face a uniform surface finish. Screed the rear face of the panel to eliminate open pockets of aggregate and surface distortions in excess of 6 mm. Cast the panels on a flat area. Do not attach galvanized connecting devices or fasteners to the face panel reinforcement steel.

(4) Clearly scribe on an unexposed face of each panel the date of manufacture, production lot number, and piece mark.

(5) Handle, store, and ship all units in a way that eliminates any danger of chipping, discoloration, cracks, fractures, and excessive bending stresses. Support panels in storage on firm blocking to protect the panel connection devices and the exposed exterior finish.

(6) Manufacture all units within the following tolerances:

(a) For panel dimensions, ensure that the position of panel connection devices is within 25 mm, and that all other dimensions are within 5 mm.

(b) For panel squareness, as determined by the difference between the two diagonals, do not exceed 13 mm.

(c) For panel surface finish, do not permit surface defects on smooth formed surfaces 1.5 m or more in length to exceed 3 mm. Do not permit surface defects on textured-finished surfaces 1.5 m or more in length to exceed 8 mm.

Concrete face panels with any or all of the following defects will be rejected:

- Defects that indicate imperfect molding.
- Defects that indicate honeycombed or open texture concrete.
- Cracked or severely chipped panels.
- Color variation on front face of panel due to excess form oil or for other reasons.

(b) Wire Facing. Fabricate wire facing from MW40 × MW15 welded wire fabric conforming to AASHTO M 55M, but ensure that the average shear value is not be less than 450 kPa. After fabrication, galvanize the wire mesh in accordance with AASHTO M 111.

(c) Backing Mat. Fabricate backing mat from MW10 × MW10 (minimum) welded wire fabric conforming to AASHTO M 55M. After fabrication, galvanize the backing mat in accordance with AASHTO M 111.

(d) Clevis Connector. Fabricate clevis connectors from cold-drawn steel wire conforming to AASHTO M 32 and welded in accordance with AASHTO M 55M. After fabrication, galvanize clevis connectors in accordance with AASHTO M 111.

(e) Connector Bars. Fabricate the connector bars from cold-drawn steel wire conforming to AASHTO M 32. Galvanize the bars in accordance with AASHTO M 111.

(f) Fasteners. Furnish 13-mm-diameter heavy hexhead bolts, nuts, and washers conforming to AASHTO M 164M. Galvanize the fasteners in accordance with AASHTO M 232.

(g) Hardware Cloth. Fabricate hardware cloth with maximum 7-mm square mesh openings from woven steel wire fabric conforming to ASTM A 740. After fabrication, galvanize the cloth in accordance with AASHTO M 111.

(h) Reinforcing Mesh. Fabricate the reinforcing mesh from cold-drawn steel wire conforming to AASHTO M 32. Weld the mesh into the finished mesh fabric in accordance with AASHTO M 55M. After fabrication, galvanize the reinforcing mesh in accordance with AASHTO M 111. Repair any damage to the galvanized coating before installation.

(i) **Reinforcing Strips.** Fabricate reinforcing strips from HSLA structural steel conforming to AASHTO M 223M, grade 65 (450), type 3. After fabrication, galvanize in accordance with AASHTO M 111.

(j) **Tie Strip.** Fabricate tie strips from hot rolled steel conforming to ASTM A 570, grade 50. Galvanize in accordance with AASHTO M 111.

720.02 Gabion Material

(a) **Basket Mesh.** Twist or weld the mesh from galvanized steel wire, class 3, soft temper, conforming to ASTM A 641M class 3, or from aluminized steel wire, soft temper, conforming to ASTM A 809. Use wire with a minimum tensile strength of 415 MPa when tested in accordance with ASTM A 370. The zinc or aluminum coating shall be applied after the mesh fabrication has been welded.

Fabricate baskets from either twisted wire mesh or welded wire mesh. Make the mesh openings with a maximum dimension of less than 120 mm and an area of less than 7,000 mm^2. Furnish baskets in the dimensions required with a dimension tolerance of ± 5 percent.

Where the length of the basket exceeds one and one-half times its width, equally divide the basket into cells less than or equal to the basket width using diaphragms of the same type and size mesh as the basket panels. Prefabricate each basket with the necessary panels and diaphragms secured so they rotate into place.

(1) Gabion Baskets 0.3 m or Greater in the Vertical Dimension. Fabricate the mesh for galvanized or aluminized coated gabions from wire with a diameter of 3.0 mm or greater in nominal size, and fabricate the mesh for epoxy or PVC-coated gabions from wire with a diameter of 2.7 mm or greater in nominal size.

(a) Twisted Wire Mesh. Form the mesh in a uniform hexagonal pattern with nonraveling double twists. Tie the perimeter edges of the mesh for each panel to a selvedge wire with a diameter of 3.9 mm or greater, or a selvedge wire with a diameter of 3.4 mm or greater for epoxy- or PVC-coated gabions, so that the selvedge is at least the same strength as the body of the mesh. Furnish selvedge wire from the same kind and type of material used for the wire mesh.

(b) Welded Wire Mesh. For mesh from wire with a diameter of 3.0 mm or greater in nominal size, weld each connection to obtain a minimum average weld shear strength of 2,600 N, with no value less than 2,000 N. For mesh for epoxy or PVC-coated gabions from wire with a diameter of 2.7 mm in nominal size, weld each connection to obtain a minimum average weld shear strength of 2,100 N, with no value less than 1,600 N.

(2) Gabion Mattresses. Fabricate the mesh from wire with a diameter of 2.2 mm or greater in nominal size.

(a) Twisted Wire Mesh. Form the mesh in a uniform hexagonal pattern with nonraveling double twists. Tie the perimeter edges of the mesh for each panel to a selvedge wire with a diameter of 2.7 mm or greater so that the selvedge is at least the same strength as the body of the mesh. Furnish selvedge wire from the same kind and type of material used for the wire mesh.

(b) Welded Wire Mesh. Weld each connection to obtain a minimum average weld shear strength of 1,300 N, with no value less than 1,000 N.

(3) PVC-Coated Gabions. Use either the fusion bonding or extrusion coating process to coat the galvanized or aluminized mesh.

Make the coating at least 0.38 mm in thickness. For PVC coating, make the color black or gray and conform to the following:

(a) Specific gravity, ASTM D 792 .... 1.20 to 1.40

(b) Tensile strength, ASTM D 638 .... 15.7 MPa min.

(c) Modulus of elasticity, ASTM D 638 .... 13.7 MPa min. at 100% strain

(d) Hardness—shore "A," ASTM D 2240 .... 75 min.

(e) Brittleness temperature, ASTM D 746 .... – 9 °C max.

(f) Abrasion resistance, ASTM D 1242,
method B, at 200 cycles,
CSI–A abrader tape, 80 grit .... 12% max. weight loss

(g) Salt spray (ASTM B 117) and .... No visual effect
ultraviolet light exposure — *(1)* $\Delta < 6\%$
(ASTM D 1499 and G 23 using — *(2)* $\Delta < 25\%$
apparatus type E and 63 °C — *(3)* $\Delta < 25\%$
for 3,000 h) — *(6)* $\Delta < 10\%$

(h) Mandrel bend, 360° bend at –18 °C around a
mandrel 10 times the wire diameter .... No breaks or cracks in coating

(b) Fasteners. For lacing wire, use wire with a diameter of 2.2 mm in nominal size and that is of the same type, strength, and coating as the basket mesh.

For welded wire mesh panels, form the spiral binders with wire that has at least the same thickness, strength, and coating as the basket mesh.

Furnish alternate fasteners that are acceptable to the gabion manufacturer and that remain closed when subjected to a 2,600-N tensile force when confining the maximum number of wires to be confined. Submit installation procedures and fastener test results.

(c) **Internal Connecting Wire.** Use lacing wire as described in Subsection 720.02(b) to reinforce side panels. Alternate stiffeners that are acceptable to the gabion manufacturer may also be used.

720.03 Metal Bin Type Crib Walls

Fabricate members of the type and kind of material SHOWN ON THE DRAWINGS. Conform to the following:

(a) Galvanized steel sheets .... AASHTO M 218

(b) Aluminum sheets .... AASHTO M 197M

(c) Fiber-bonded steel sheets 707.09

(d) Aluminum-coated steel sheets .... AASHTO M 274

(e) Bolts and nuts .... ASTM A 307, grade A

Furnish heavy hexagon heads and nuts without washers, or hexagon heads and nuts with two plate washers. Fabricate washers from 3.3-mm-thick round steel plate, including coating with holes not more than 1.6 mm larger than the bolt diameter. Galvanize the bolts, nuts, and washers in accordance with AASHTO M 232.

Section 722—Anchor Material

722.01 Ground Anchors

Furnish material for ground anchors in accordance with the specifications shown below.

(a) Tendons. Furnish ground anchor tendons for either single or multiple elements that conform to one of the following:

(1) Steel strand uncoated seven-wire stress relieved for prestressed concrete .... AASHTO M 203M

(2) Uncoated high-strength steel bar for prestressed concrete .... AASHTO M 275M

(3) Steel strand uncoated seven-wire compacted stress relieved[1] for prestressing concrete .... ASTM A 779M

[1] Ensure that elements also conform to the minimum requirements of AASHTO M 203M.

(b) Couplers. Furnish couplers for tendon sections that are capable of developing 95 percent of the minimum specified ultimate tensile strength of the tendon.

(c) Sheathing. Furnish sheathing of the tendon in accordance with one of the following:

(1) Unbonded Length. For unbonded length, meet the following requirements:

(a) Polyethylene Tube. Furnish polyethylene of type II, III, or IV, as defined by ASTM D 1248. Furnish tubing with a minimum wall thickness of 1.5 mm.

(b) Hot-Melt Extruded Polypropylene Tube. Furnish polypropylene with cell classification PP 210 B55542 11, as defined by ASTM D 4101. Furnish tubing with a minimum wall thickness of 1.5 mm.

(c) Hot-Melt Extruded Polyethylene Tube. Furnish polyethylene of high-density type III, as defined by ASTM D 3350 and D 1248. Furnish tubing with a minimum wall thickness of 1.5 mm.

(d) Steel Tubing. Furnish tubing that conforms to ASTM A 500 and has a minimum wall thickness of 5 mm.

(e) Steel Pipe. Furnish pipe that conforms to ASTM A 53, schedule 40, minimum.

(f) Plastic Pipe. Furnish pipe that conforms to ASTM D 1785, schedule 40, minimum.

(2) **Bonded Length.** For bonded length, meet the following requirements:

(a) High-Density Corrugated Polyethylene Tubing. Furnish tubing that conforms to AASHTO M 252 and has a minimum wall thickness of 0.75 mm.

(b) Corrugated PVC Tubes. Furnish PVC compounds that conform to ASTM D 1784, class 13464–B.

(c) Fusion-Bonded Epoxy. Furnish epoxy that conforms to AASHTO M 284M and has a minimum film thickness of 0.4 mm.

(d) Grease. Furnish grease that is compounded to provide corrosion-inhibiting and lubricating properties, and that conforms to the PTI "Post Tensioning Manual," table 3.2.1.

(e) Grout. Furnish grout that consists of a pumpable mixture of Portland cement, sand, water, and admixtures mixed in accordance with Subsection 701.03. Use type I, II, or III Portland cement in accordance with Subsection 701.01.

Chemical additives that can control bleed or retard set may be used, provided that the additives conform to Subsection 711.03 and are mixed in accordance with the manufacturer's recommendations.

Furnish grout that is capable of reaching a cube strength (AASHTO T 106) of 25 MPa in 7 days. Make grout cubes for testing from random batches of grout, as directed. Normally, strength testing will not be required, because system performance will be measured by proof-testing each anchor. Grout cube testing will be required if admixtures are used or irregularities occur in anchor testing.

(f) Centralizers. Centralizers and spacers may be fabricated from any type of material except wood that is not deleterious to the prestressing steel.

(g) Anchorage Devices. Furnish anchorage devices that conform to the PTI's "Post Tensioning Manual," section 3.2.3. Furnish anchorage devices for strand tendons that are designed to permit lift-off testing without the jack engaging the strand. Furnish bearing plate for anchorage devices that is steel plate conforming to AASHTO M 183M or M 222M.

Ensure that a pipe or trumpet extends from the anchor plate far enough to encapsulate the protective sheath. Furnish anchorage devices that are capable of developing 95 percent of the minimum specified ultimate tensile strength of the anchor tendon.

Section 723—Dust Palliative Materials

723.01 Magnesium or Calcium Chloride Brine

Furnish chloride brines consisting of water and magnesium and/or calcium chloride with the following chemical composition (percent by weight brine):

(a) Chloride concentration (sum of magnesium and calcium chloride):[1]

(1) Magnesium chloride products .... 28% min.
Calcium chloride products .... 36% min.

(2) Sulfate .... 4.3% max.

(3) Nitrate .... 5.0% max.

[1] Use test method R1–412/C1 (available on request from USDA Forest Service, Regional Materials Engineering Center, P.O. Box 7669, Missoula, Montana 59807).

Ensure that the pH is between 4.5 and 10.0, and that the temperature of the material is 5 °C or above when applied. Provide certification and sampling in accordance with Subsection 723.04.

723.02 Calcium Chloride Flake

Furnish calcium chloride flake in accordance with the following:

(a) Chemical composition (percent by weight):

(1) Calcium chloride ($CaCl_2$)[1] .... 77% min.

(2) Total alkali chlorides (as NaCl) ASTM E 449 ..... 3% max.

(3) Calcium hydroxide ($Ca(OH)_2$) ASTM E 449 .... 0.3% max.

[1] Use test method R1–412/C1 (available on request from USDA Forest Service, Regional Materials Engineering Center, P.O. Box 7669, Missoula, Montana 59807).

(b) Particle size (percent passing screen) by AASHTO T 27:

(1) 9.5-mm screen 100%

(2) 4.75-mm screen .... 80 to 100%

(3) 600-μm screen .... 0 to 5%

(c) Certification and sampling 723.04

723.03 Lignin Sulfonate

Furnish lignin sulfonate from the residue produced by the acid-sulfite pulping of wood. Ensure that its base cation is ammonium, calcium, or sodium, and supply it as a uniform mixture that is miscible with an equal weight of water and meets the following requirements:

(a) Undiluted lignin sulfonate:

(1) pH, AASHTO T 200 .... 4.5 min.

(2) Viscosity at 25 °C, AASHTO T 202 .... 20.5 poise max.

(3) Total lignin solids concentration[1] .... 48% min.

(b) Solids:

(1) Lignin sulfonate .... 50% min.

(2) Reducing sugars .... 25% max.

(c) Temperature during application .... 5 °C to 60 °C

(d) Certification and sampling 723.04

[1] Use test method R1–412/LS for total lignin solids concentration (available on request from USDA Forest Service, Regional Materials Engineering Center, P.O. Box 7669, Missoula, Montana 59807).

723.04 Certification & Sampling

(a) Certification With Shipments. When each load of dust palliative is delivered, furnish the CO with one copy of the Bill of Lading and a fully executed Certificate of Compliance containing the applicable information shown in figure 723-1. A separate Certificate of Compliance will not be required if the standard Bill of Lading contains the applicable information required on the certificate.

(b) Sampling. Sampling of dust palliative prior to any mixing with water may be required to validate certifications furnished by the Contractor. When sampling is directed by the Government, obtain the actual samples, and give the CO the opportunity to witness sampling. Construct all liquid delivery equipment to permit sampling in conformance with AASHTO T 40 sampling procedure.

CERTIFICATE OF COMPLIANCE

Consignee ____________________

Transportation ID (Truck No., etc.) ____________________

Product Concentration by Weight ____________________

Destination ____________________

Date ____________________

Magnesium Chloride: ____________________ %

Calcium Chloride: ____________________ %

Lignin Sulfonate: ____________________ %

Net Weight Total Shipment ____________________

Net Liters at 15 °C ____________________

Product Specific Gravity at 15 °C ____________________

This shipment of ______________ identified above and covered by this Certificate of Compliance complies with Forest Service Specifications applicable to Contract # ________ .

MSDS Identification Code: __________

Producer: ____________________

Signed By: ____________________
(Producer's Representative)

Figure 723-1.—Sample Certificate of Compliance.

Section 725—Miscellaneous Material

725.01 Water

Do not use water from streams, lakes, ponds, or similar sources without prior approval. Use water that conforms to the specifications shown below.

(a) Water for Mixing or Curing Cement Concrete, Mortar, or Grout. Furnish water that conforms to AASHTO M 157. Potable water of known quality may be used without testing in accordance with AASHTO T 26. Potable water is defined as water that is safe for human consumption, as described by the public health authority with jurisdiction.

(b) Water for Planting or Care of Vegetation. Furnish water that is free of substances injurious to plant life, such as oils, acids, alkalies, and salts.

(c) Water for Earthwork, Pavement Courses, Dust Control, & Incidental Construction. Furnish water that is free of substances detrimental to the work.

725.02 Calcium Chloride & Sodium Chloride

(a) Calcium Chloride for Concrete. Furnish material that conforms to ASTM D 98, type L, for the concentration specified.

(b) Sodium Chloride. Furnish material that conforms to ASTM D 632, type II, grade 1.

725.03 Hydrated Lime

(a) Lime for Masonry. Furnish hydrated lime that conforms to ASTM C 207, type N.

(b) Lime for Soil Stabilization & Paving. Furnish hydrated lime or quicklime that conforms to AASHTO M 216.

725.04 Fly Ash

Furnish fly ash and raw or calcined pozzolans that conform to AASHTO M 295.

725.05 Mineral Filler

Furnish mineral filler that conforms to AASHTO M 17.

725.06 Precast Concrete Curbing

Furnish units conforming to the lengths, shapes, and details SHOWN ON THE DRAWINGS, and to the following:

(a) Concrete 602

(b) Reinforcing steel 709.01

725.07 Clay or Shale Brick

Furnish clay or shale brick that conforms to one of the following:

(a) Sewer brick .... ASTM C 32, grade SM

(b) Building brick .... ASTM C 62, grade SW

725.08 Concrete Brick

Furnish concrete brick that conforms to ASTM C 55, grade N–I.

725.09 Concrete Masonry Blocks

Furnish rectangular or segmented concrete masonry blocks. When required, form the block ends to provide an interlock at vertical joints. Furnish blocks that conform to the following:

(a) Solid load-bearing blocks .... ASTM C 90

(b) Hollow load-bearing blocks .... ASTM C 90

(c) Nonload-bearing blocks .... ASTM C 129

725.10 Cellular Concrete Blocks

Furnish cellular concrete blocks that conform to ASTM C 90, normal weight; but use concrete that conforms to Section 602.

725.11 Precast Concrete Units

Cast the units in substantial permanent steel forms. Provide additional reinforcement as necessary to provide for handling the units. Use concrete that conforms to the following:

(a) 28-day strength, AASHTO T 22 .... 25 MPa min.

(b) Air content by volume, AASHTO T 152 .... 5% min.

Cure the units in accordance with AASHTO M 170M.

Cast a sufficient number of concrete cylinders from each unit to permit compression tests at 7, 14, and 28 days. Make at least three cylinders for each test. If the strength requirement is met at 7 or 14 days, the units will be certified for use 14 days from date of casting.

Do not use precast concrete units when:

- Representative cylinders do not meet the strength requirement by 28 days.
- Air content tests do not meet 5 percent minimum.
- Cracks or honeycombed or patched areas are larger than 0.02 m^2.

Furnish precast reinforced concrete manhole risers and tops conforming to AASHTO M 199M.

725.12 Frames, Grates, Covers, & Ladder Rungs

Fabricate metal grates and covers to evenly bear on the frames. Correct bearing inaccuracies by machining. Assemble all units before shipment. Mark all pieces to facilitate reassembly at the installation site. Uniformly coat all castings with asphalt varnish or a commercial preservative in accordance with the manufacture's standard practice. Conform to the following:

(a) Gray iron castings .... AASHTO M 105

(b) Carbon steel castings .... AASHTO M 103M

(c) Structural steel .... AASHTO M 183M

(d) Galvanizing .... AASHTO M 111

(e) Malleable iron castings .... ASTM A 47M

(f) Aluminum alloy ladder rung material .... ASTM B 221M, alloy 6061–T6

(g) Aluminum castings .... ASTM B 26M, alloy 356.0–T6

(h) Asphalt varnish .... FSS TT–V–51

725.13 Corrugated Metal Units

Furnish material that conforms to one of the following:

(a) Steel corrugated units .... AASHTO M 36M

(b) Aluminum corrugated units .... AASHTO M 196M

(c) Bituminous-coated corrugated units .... AASHTO M 190, type A

(d) Polymer-precoated corrugated units .... AASHTO M 245M, grade 250/250

(e) Fiber-bonded units 707.09

725.14 Protective Coatings for Concrete

Furnish protective coatings for bridge decks, curbs, sidewalks, and concrete portions of bridge railings. Provide one of the following coatings:

(a) Boiled Linseed Oil. Furnish boiled linseed oil in accordance with ASTM D 260, type I or II.

(b) Petroleum Spirits (Mineral Spirits). Furnish petroleum spirits (mineral spirits) in accordance with ASTM D 235.

725.15 PVC Pipe for Water Distribution Systems

Furnish material that conforms to the following for the designated sizes and strength schedules:

(a) PVC pipe .... ASTM D 1785

(b) Solvent cement for pipe and fittings .... ASTM D 2564

725.16 Polyethylene Pipe for Water Distribution Systems

Furnish material that conforms to ASTM D 2447 for the designated sizes and strength schedules.

725.17 Cast Iron Soil Pipe & Fittings

Furnish material that conforms to ASTM A 74, class SV, for the designated sizes.

725.18 Seamless Copper Water Tube & Fittings

Furnish material that conforms to ASTM B 88M, type L, for the designated sizes.

725.19 Plastic Lining

Furnish a film with a thickness of 175 ± 25 µm that conforms to one of the following:

(a) PVC plastic film .... ASTM D 1593, type II

(b) Polyethylene plastic film .... ASTM D 2103, type 02000

725.20 Bentonite

Furnish bentonite as sodium montmorillonite (sodium bentonite) in the form of a powder that meets the following requirements:

(a) Ensure that colloid content by AASHTO T88 or ASTM D422 is 60 percent minimum.

(b) Ensure that a sieve analysis in accordance with AASHTO T27 on a dry, unwashed, unpulverized sample yields the following:

(1) A minimum of 95 percent passing the 4.75-mm sieve.

(2) A minimum of 15 percent passing the 75-µm sieve.

725.21 Epoxy Resin Adhesives

Ensure that epoxy resin adhesives conform to AASHTO M 235.

725.22 Spray Finish

Furnish a commercial product that is specifically designed for color spraying concrete, and that consists of a pliolite resin base, fiberglass, perlite, mica additives, and durable tinting pigments capable of making a light gray color similar to the color of concrete containing 230 g of carbon black per bag of cement.

725.23 Color Coating

Furnish a semiopaque colored toner containing methyl methacrylateethyl acrylate copolymer resins or equivalent resins, solvents, and color toning pigments suspended in solution by a chemical suspension agent. Ensure that the color toning pigments consist of laminar silicates, titanium dioxide, and inorganic oxides. Furnish material that conforms to the following:

(a) Mass per liter, ASTM D 1475 .... 1 kg min.

(b) Solids by weight, ASTM D 2369 .... 30% min.

(c) Solids by volume .... 21% min.

(d) Drying time, ASTM D 1640 .... 30 min at 21 °C and 50% max. humidity

(e) Color change, ASTM D 822, 1,000 h .... No appreciable change

(f) Resistance to acids, alkalies, gasoline, and mineral spirits, ASTM D 543 .... Excellent

(g) Water vapor transmission from interior concrete, ASTM D 1653 .... Transmittable

(h) Exterior moisture absorption into the concrete surface pores, FSS TT–C–555 .... Reduces rate

(i) Oxidation over time .... None

725.24 Explosives & Blasting Accessories

For the transportation, handling, and storage of explosives, conform to 29 CFR, part 1926, subpart U.

Explosives and initiating devices include, but are not necessarily limited to, dynamite and other high explosives, slurries, water gels, emulsions, blasting agents, initiating explosives, detonators, and detonating cord.

725.25 Mineral Slurry (Driller's Mud)

Furnish commercially available sodium bentonite or attapulgite in a potable water. Use a mineral grain size that will remain in suspension with sufficient viscosity and gel characteristics so the mixture is capable of transporting excavated material to a suitable screening system.

725.26 Form Liner

Furnish a high-quality product that attaches easily to the forming system. Install the form liner so it does not compress more than 6 mm at a concrete pour rate of 3,650 kg/m^2.

725.27 Aluminum-Impregnated Caulking Compound

Conform to FSS TT–C–598, grade 1.

Appendices

Appendix A

ENGINEERING CONVERSION FACTORS*			
Quantity	**From Imperial Units**	**Multiply By**	**To Metric Units****
Mass	lb kip (1000 lb) ton (short)	0.453 592 0.453 592 0.907 184	kg metric ton (1000 kg) metric ton
Mass/unit length	plf	1.488 16	kg/m
Mass/unit area	psf lb/sy	4.882 43 0.542 492	kg/m^2 kg/m^2
Mass density	pcf	16.018 5	kg/m^3
Force	lb kip	4.448 22 4.448 22	N kN
Force/unit length	plf klf	14.593 9 14.593 9	N/m kN/m
Pressure, stress, modulus of elasticity	psf ksf psi ksi	47.880 3 47.880 3 6.894 76 6.894 76	Pa kPa kPa MPa
Bending moment, torque	ft-lb ft-kip	1.355 82 1.355 82	N•m kN•m
Volume/unit area	gal/sy	4.527 19	L/m^2

*Approximate conversions to SI units.

**To convert from metric units to imperial units, multiply by the reciprocal of the conversion factor.

Appendix B

SI* (MODERN METRIC) CONVERSION FACTORS

APPROXIMATE CONVERSIONS TO SI UNITS

Symbol	When You Know	Multiply By	To Find	Symbol
		LENGTH		
in	inches	25.4	millimeters	mm
ft	feet	0.305	meters	m
yd	yards	0.914	meters	m
mi	miles	1.61	kilometers	km
		AREA		
in^2	square inches	645.2	square millimeters	mm^2
ft^2	square feet	0.093	square meters	m^2
yd^2	square yards	0.836	square meters	m^2
ac	acres	0.405	hectares	ha
mi^2	square miles	2.59	square kilometers	km^2
		VOLUME		
fl oz	fluid ounces	29.57	milliliters	mL
gal	gallon	3.785	liters	L
ft^3	cubic feet	0.028	cubic meters	m^3
yd^3	cubic yards	0.765	cubic meters	m^3
NOTE: Volumes greater than 1000 L shall be shown in m^3.				
		MASS		
oz	ounces	28.35	grams	g
lb	pounds	0.454	kilograms	kg
T	short tons (2000 lb)	0.907	megagrams (or "metric ton")	Mg (or "t")
		TEMPERATURE (exact)		
°F	Fahrenheit temperature	5 (F – 32)/9 or (F – 32)/1.8	Celcius temperature	°C
		ILLUMINATON		
fc	foot-candles	10.76	lux	lx
fl	foot-Lamberts	3.426	candela/m^2	cd/m^2
		FORCE and PRESSURE or STRESS		
lbf	poundforce	4.45	newtons	N
lbf/in^2	poundforce per square inch	6.89	kilopascals	kPa

*SI is the symbol for the International System of Units. Appropriate rounding should be made to comply with section 4 of ASTM E 380.

Appendix B (cont.)

SI* (MODERN METRIC) CONVERSION FACTORS

APPROXIMATE CONVERSIONS FROM SI UNITS

Symbol	When You Know	Multiply By	To Find	Symbol
		LENGTH		
mm	millimeters	0.039	inches	in
m	meters	3.28	feet	ft
m	meters	1.09	yards	yd
km	kilometers	0.621	miles	mi
		AREA		
mm^2	square millimeters	0.0016	square inches	in^2
m^2	square meters	10.764	square feet	ft^2
m^2	square meters	1.195	square yards	yd^2
ha	hectares	2.47	acres	ac
km^2	square kilometers	0.386	square miles	mi^2
		VOLUME		
mL	milliliters	0.034	fluid ounces	fl oz
L	liters	0.264	gallon	gal
m^3	cubic meters	35.71	cubic feet	ft^3
m^3	cubic meters	1.307	cubic yards	yd^3
		MASS		
g	grams	0.035	ounces	oz
kg	kilograms	2.202	pounds	lb
Mg (or "t")	megagrams (or "metric ton")	1.103	short tons (2000 lb)	T
		TEMPERATURE (exact)		
°C	Celcius	1.8C + 32	Fahrenheit temperature	°F
		ILLUMINATON		
lx	lux	0.0929	foot-candles	fc
cd/m^2	$candela/m^2$	0.2919	foot-Lamberts	fl
		FORCE and PRESSURE or STRESS		
N	newtons	0.225	poundforce	lbf
kPa	kilopascals	0.145	poundforce per square inch	lbf/in^2

*SI is the symbol for the International System of Units. Appropriate rounding should be made to comply with section 4 of ASTM E 380.

INDEX

— A —

— B —

— C —

— D —

— E —

— F —

— G —

— H —

— I —

— J —

— L —

— M —

— O —

— P —

— Q —

— R —

— S —

— T —

— U —

— V —

— W —

☆ U.S. GOVERNMENT PRINTING OFFICE:1996-415-284

www.ingramcontent.com/pod-product-compliance
Lightning Source LLC
Chambersburg PA
CBHW060820220326
41599CB00017B/2240
9781907521904